Nonlinear Systems and Complexity

Series Editor
Albert C.J. Luo
Southern Illinois University
Edwardsville, IL, USA

For further volumes:
http://www.springer.com/series/11433

Jesús Cuevas-Maraver • Panayotis G. Kevrekidis •
Floyd Williams
Editors

The sine-Gordon Model and its Applications

From Pendula and Josephson Junctions to Gravity and High-Energy Physics

Springer

Editors
Jesús Cuevas-Maraver
Departamento de Físíca Aplicada I
University of Sevilla
Sevilla
Spain

Panayotis G. Kevrekidis
Mathematics and Statistics
University of Massachusetts
Amherst
Massachusetts
USA

Floyd Williams
Mathematics and Statistics
University of Massachusetts
Amherst
Massachusetts
USA

ISSN 2195-9994 ISSN 2196-0003 (electronic)
ISBN 978-3-319-06721-6 ISBN 978-3-319-06722-3 (eBook)
DOI 10.1007/978-3-319-06722-3
Springer Cham Heidelberg New York Dordrecht London

Library of Congress Control Number: 2014944067

Printed on acid-free paper

Springer is part of Springer Science+Business Media (www.springer.com)

Jesús Cuevas-Maraver dedicates this volume to Conchi Caballero for her love and patience during the birth of the book, and to Conchi Cuevas for the joy she gives to his life.

Panayotis G. Kevrekidis dedicates this volume to Maria Kevrekidis, for all her sacrifices that made this book possible, to Despina and Athena Kevrekidis for all the smiles they bring to his life and to George and Despina Kevrekidis for expecting this book, before it was even conceived.

Floyd Williams dedicates this volume to his mother, Mrs. Lee Edna Rollins, who at age 92 remains a source of Godly wisdom to him in matters great and small.

Preface

The sine-Gordon equation is, arguably, one of the most popular nonlinear wave models. This feature is due to a number of appealing traits of this system. One such characteristic is that the sine-Gordon equation arises in a very diverse range of applications. These started as early as the 1860s when it was discovered in the course of the study of surfaces of constant negative curvature through the so-called Gauss–Codazzi conditions for surfaces of curvature -1 in 3-space. It acquired renewed interest due to the classical study of Frenkel and Kontorova in the context of the discrete analog of the model in the 1930s in the theory of crystal dislocations. Subsequently, the work of A.C. Scott produced a mechanical analog of the system, through the realization of an array of coupled torsion pendula that proved extremely useful both in its visualization and in the experimental observation of its solutions. The realm of relevant applications kept expanding through the emergence of Josephson junction arrays and their fluxons, as well as breathers that were intensely studied in the 1990s and early 2000s. At that time there also surfaced a renewed interest in the discrete form of the model and its remarkable characteristics. Relevant applications of the model have continued to expand with recent proposals involving among others the orbits of a string of stars near the inner Lindblad resonance within a galaxy or the evolution of the electromagnetic field on neuronal microtubules.

Another major source of appeal of this equation consists of its intrinsic mathematical beauty and structure. The complete integrability of the model analyzed in detail in the 1970s, coupled with the ability to produce not only kink but especially (a very uncommon feature for PDE models) exact breather solutions, made the sine-Gordon quite unique. Using Bäcklund transformations one is able to generalize such solutions to multi-kink, multi-breather, and kink-breather type mixtures with remarkable robustness and elastic collision properties, again connected to its integrable underlying structure, Lax pair, infinite conservation laws, etc. Not only were these features identified in the classical realm but also semi-classical quantized versions of the model were proposed, the exact quantum scattering matrix was discovered and intriguing dualities to other well-known models such as the Thirring model were revealed. Such connections have continued to expand even in recent years, e.g., among others with a remarkable set of transformations connecting the sine-Gordon with the short pulse equation (SPE), a recent, intriguing model reduction of the Maxwell equations for the description of few cycle optical pulses. Such connections have been recently utilized to demonstrate the integrability of the SPE, the identification of its solutions, to propose integrable discretizations thereof and so on and so forth. It seems rather extraordinary to us that even 150 years since the initial inception of this deceivingly simple-looking nonlinear variant of the wave equation, the sine-Gordon model still continues to provide mathematical surprises, while in parallel continuing to emerge in novel and ever-expanding physical applications.

In light of all of the above, we decided that it would be a good time to attempt to summarize some of the important recent developments in this field, while also capturing some of the historical perspective of the studies of this PDE and the diverse and broad appeal of its applications. It was with that goal in mind that we pursued the creation of this volume, targeting a number of solicited mini-reviews in different areas of mathematical and physical interest where the sine-Gordon has played a key role. Undoubtedly the resulting chapters have a substantial personal flavor and reflect the particular view and interests of the contributors; nevertheless, we hope and do believe that they give a sense of the excitement, interest and stir that this model equation has brought to different sub-disciplines of mathematics and physics.

The first chapter "The sine-Gordon Model: General Background, Physical Motivations, Inverse Scattering, and Solitons" by Malomed presents an overview of the background of the equation, its motivation, and its principal solutions. It also presents a number of particular mathematical topics, such as perturbations to the model and the development of a perturbation theory for its solutions, as well as discrete and higher dimensional variants of the model.

The second chapter "sine-Gordon Equation: From Discrete to Continuum" by Chirilus-Brückner et al. focuses more particularly on the discrete form of the equation and the transition from that to the well-known continuum limit thereof. This transition is analyzed from the perspective of the fundamental kink and breather solutions, while at the end some very recent variant has been added in the form of $\mathcal{PT}$-symmetric forms of the sine-Gordon model.

The third chapter "Soliton Collisions" by Dmitriev and Kevrekidis presents an overview of the richly studied theme of solitonic collisions in the sine-Gordon and related models. The role of internal modes, of various forms of integrability-breaking perturbations, and of other mechanisms such as the radiationless energy exchange is analyzed in this context.

The discrete model is once again the focus of the fourth chapter "Effects of Radiation on sine-Gordon Coherent Structures in the Continuous and Discrete Cases" by Cisneros-Ake and Minzoni. In that case, the prototypical problem of traveling of kinks in a discrete sine-Gordon medium and their radiation analysis and potential pinning is explored and also two-dimensional standing and traveling structures are proposed as per recent work of the authors.

The fifth chapter "Experimental Results for the sine-Gordon Equation in Arrays of Coupled Torsion Pendula" by English gives an overview of experimental developments on the sine-Gordon model as captured by its prototypical realization of coupled torsion pendula. Features such as the modulational instability, the famous Peierls–Nabarro barrier, and the realization (and potential decay due to damping) of solitary waves are presented in different configurations such as lines and rings.

The sixth chapter "Soliton Ratchets in sine-Gordon-Like Equations" by Quintero focuses on a recent theme, namely the realization of ratchet effects in the sine-Gordon context. The role of symmetries in the ratcheting behavior is elucidated and the asymmetric transport of solitary waves in different driven (and damped) variants of the model and relevant applications is explored.

The seventh chapter "The sine-Gordon Equation in Josephson-Junction Arrays" by Mazo and Ustinov presents another major experimental playground associated with the sine-Gordon model, namely the realm of Josephson junction arrays. The dynamics of fluxons, quantum solitons, and other such states is examined in different types of experimentally relevant configurations.

Finally, the chapters "Some Selected Thoughts Old and New on Soliton-Black Hole Connections in 2d Dilaton Gravity," "Dressing with Control: Using Integrability to Generate Desired Solutions to Einstein's Equations," and "A Planar Skyrme-Like Model" present a different perspective motivated by the sine-Gordon equation by emphasizing its potential relevance to gravity and high energy physics. In the chapter "Some Selected Thoughts Old and New on Soliton-Black Hole Connections in 2d Dilaton Gravity," Williams connects sine-Gordon solitons (kinks, oscillating antikinks, breathers, etc.) and static bright solitons to black hole solutions and black hole vacua in dilaton gravity. In the chapter "Dressing with Control: Using Integrability to Generate Desired Solutions to Einstein's Equations," Beheshti and Tahvildar-Zadeh present a variant of the dressing method which they apply to the sine-Gordon equation while subsequently extending it to Einstein equations and Kerr spacetimes. Then in the chapter "A Planar Skyrme-Like Model," Cova focuses on a planar Skyrme-like model and presents its solitons and their interactions. Although the chapters "Dressing with Control: Using Integrability to Generate Desired Solutions to Einstein's Equations" and "A Planar Skyrme-Like Model" involve the study of nonlinear sigma models, these models are referred to as harmonic maps in the chapter "Dressing with Control: Using Integrability to Generate Desired Solutions to Einstein's Equations."

We hope that this broad and diverse (although by no means comprehensive) exposition may benefit both young researchers, including graduate students and post-doctoral fellows, and more seasoned researchers in the field. We are certain that the sine-Gordon equation holds many more surprises in store for the future and would be delighted if this book, intended as a rather partial (in both senses of the word) account of its life so far, allows to springboard further developments along this vein.

Sevilla, Spain — Jesús Cuevas-Maraver
Amherst, MA — Panayotis G. Kevrekidis
Amherst, MA — Floyd Williams
February, 2014

Contents

Acronyms

BT	Bäcklund Transformation
DNLS	Discrete Nonlinear Schrödinger
FK	Frenkel–Kontorova
ILM	Intrinsic Localized Mode
IST	Inverse Scattering transform
JJ	Josephson junction
J-T	Jackiw–Teitelboim
KG	Klein–Gordon
NLS	Nonlinear Schrödinger
PDE	Partial Differential Equation
PN	Peierls–Nabarro
sG	sine-Gordon
VA	Variational Approximation

The sine-Gordon Model: General Background, Physical Motivations, Inverse Scattering, and Solitons

Boris A. Malomed

Abstract This chapter offers an overview of the vast research area developed around the sine-Gordon (sG) equation, including solution methods and various nonlinear modes generated by this equation, viz., topological and dynamical solitons (kinks and breathers), cnoidal waves (chains of kinks), and others. Also included is a survey of physical applications of the sG equation, and an outline of the perturbation theory used for the analysis of physical models based on the sG equation but differing from the ideal integrable form of this equation. Topics presented in the chapter, with some details or in a brief form, are the inverse scattering, transform and the perturbation theory based on it, the Bäcklund transform, energy- and momentum-balance methods for the analysis of the soliton dynamics in perturbed versions of the sG equation, the double sG equation, quantum sG systems, multidimensional sG equations, systems of coupled sG equations, and others.

Keywords Bäcklund transform • Breathers • Charge-density waves • Frenkel–Kontorova model • Inverse scattering transform • Josephson junctions • Kinks • Magnetic solitons • Perturbation theory • Self-induced transparency

1 Exact and Perturbed sine-Gordon Equations: Fundamental Models of the Contemporary Mathematical and Theoretical Physics

One of commonly known classical equations of mathematical physics is the one-dimensional (1D) Klein–Gordon (KG) equation, i.e., the simplest (and, in that sense, universal) linear model for the propagation of dispersive waves:

B.A. Malomed (✉)
Department of Physical Electronics, School of Electrical Engineering, Faculty of Engineering, Tel Aviv University, Tel Aviv 69978, Israel
e-mail: malomed@post.tau.ac.il

J. Cuevas-Maraver et al. (eds.), *The sine-Gordon Model and its Applications*, Nonlinear Systems and Complexity 10, DOI 10.1007/978-3-319-06722-3_1,

$$\phi_{tt} - \phi_{xx} + \phi = 0, \tag{1}$$

where $\phi(x,t)$ is a real wave field, with t and x standing, as usual, for time and the spatial coordinate. The substitution of $\phi(x,t) = \phi_0 \exp(ikx - i\omega t)$ into Eq. (1) (the formal use of the complex solution of this real equation simplifies the analysis) yields the dispersion equation between the frequency, ω, and wavenumber, k:

$$\omega^2 = 1 + k^2, \tag{2}$$

which is characterized by the presence of the *gap* in frequency spectrum: $\omega^2(k) \geq \omega_{\text{gap}}^2 \equiv 1$. In the multidimensional geometry, the spatial derivative in Eq. (1) is replaced by the Laplacian, $\partial^2/\partial x^2 + \partial^2/\partial y^2 + \partial^2/\partial z^2$.

The 1D sine-Gordon (sG) equation is a nonlinear counterpart of Eq. (1):

$$\phi_{tt} - \phi_{xx} + \sin\phi = 0. \tag{3}$$

Another known form of the 1D sG equation is written in the *light-cone coordinates*, $\xi \equiv (x+t)/2$, $\tau \equiv (x-t)/2$, instead of the laboratory coordinates, x and t:

$$\phi_{\xi\tau} = \sin\phi. \tag{4}$$

In the latter form, the sG equation was first discovered by French mathematician Jacques Edmond Émile Bour in 1862, as the Gauss-Codazzi equation describing two-dimensional (2D) surfaces with constant negative curvature (*pseudospheres*) embedded into the 3D Euclidean space [1]. The integrability of the 1D sG equation in the form of Eq. (4), i.e., practically speaking, a possibility to find vast families of exact analytical solutions, was discovered in the nineteenth century too, in the form of *Bäcklund transformations*, by Swedish mathematician Albert Victor Bäcklund, in the same context of the differential geometry of pseudospheres [2]. The most remarkable solutions to the 1D sG equation are solitons (i.e., solitary waves, which are the main subject of the present chapter). The Bäcklund transformation generates $(n+1)$-soliton solutions of the sG equation from the n-soliton ones, thus making it possible to generate an infinite hierarchy of solutions with increasing complexity, starting from the trivial solution, $\phi \equiv 0$, which plays the role of the zero-soliton state.

The second advent of the sG equation had occurred in 1939, in the form of the continuous limit of the Frenkel–Kontorova (FK) model, i.e., a celebrated model of dislocations in solid state [3]. It was derived as a discrete sG equation for lattice wave field $\phi_n(t)$ (n is the discrete coordinate):

$$\frac{d^2\phi_n}{dt^2} - C(\phi_{n+1} + \phi_{n-1} - 2\phi_n) + \sin\phi_n = 0, \tag{5}$$

where C is a constant of the intersite coupling. Then, a similar model was developed for a chain of adsorbed atoms on a metallic surface [4]. Later, many other physical applications of the FK models have been elaborated [5]. The continuous limit of Eq. (5) implies that discrete coordinate n may be considered as a set of discrete values of a continuous one. Then, the substitution of the truncated Taylor expansion:

$$\phi\,(n \pm 1) \approx \phi(n) \pm d\phi/dn + (1/2)d^2\phi/dn^2 \tag{6}$$

into Eq. (5) yields Eq. (3) with continuous coordinate $x \equiv n/\sqrt{C}$.

The sG equation had received great popularity in the 1970s as a result of the discovery of its integrability by means of the inverse scattering transform (IST) method [6, 7] and development of the understanding of the significance of this equation as a model of many important physical systems [8]. Around that time, the name "sine-Gordon" was coined, as a pun based on the pattern of KG. Probably, the first paper in the title of which "sine-Gordon equation" had appeared was published by J. Rubinstein in 1970 [9] (see also a historical review [10] by A. Scott).

Eventually, it has been understood that the 1D sG equation is one of the most fundamental models of modern mathematical physics. This commonly accepted opinion is based on two basic facts. First, the sG equation is one of few relatively simple nonlinear equations solvable by means of the IST. In fact, together with the Korteweg-de Vries and nonlinear Schrödinger (NLS) equations, it belongs to the small set of most fundamental integrable equations [11]. Second, the sG equation is a universal model for media combining the wave dispersion (the same as in the KG equation) and the nonlinearity which is a periodic function of the field variable, i.e., this variable has the meaning of a phase in the respective physical setting (an overview of physical applications of the sG equation is given below).

Of course, this universality may be violated if the nonlinear term contains additional harmonics of the phase variable. The simplest form of the accordingly *perturbed* version of the sG equation is the *double sG equation*, which is usually taken in the following form:

$$\phi_{tt} - \phi_{xx} + \sin\phi = \epsilon \sin(\phi/2) \tag{7}$$

(in this case, $\sin(\phi/2)$ actually represents the fundamental harmonic, while $\sin\phi$ is the second harmonic, but it is assumed that coefficient ϵ in front of the fundamental harmonic is a small parameter). The double sG equation was identified as a separate (nonintegrable) model no later than in 1976 [12].

The dispersion relation for the linearized version of Eq. (7) features the frequency gap like the KG equation, cf. Eq. (2): $\omega^2 = \omega_{\text{gap}}^2 + k^2$, with $\omega_{\text{gap}}^2 = 1 - \epsilon/2$. Another property which the sG equations share with their linear KG counterpart is the invariance with respect to the Lorentz boost,

$$x' = (x - ct)/\sqrt{1-c^2},\; t' = (t - cx)/\sqrt{1-c^2}, \tag{8}$$

i.e., the sG equation and its generalizations are relativistically invariant equations (hence they may be used as simple classical-field-theory models).

A commonly known principle states that a physically relevant equation should be derived from the underlying Lagrangian, and should conserve the corresponding Hamiltonian. Here, for the sake of generality, the Lagrangian and Hamiltonian are written for the double sG equation (7):

$$L = \int_{-\infty}^{+\infty} \left\{ \frac{1}{2} \left(\phi_x^2 - \phi_t^2 \right) + (1 - \cos\phi) - 2\epsilon \left[1 - \cos\left(\phi/2 \right) \right] \right\} dx, \tag{9}$$

$$H = \int_{-\infty}^{+\infty} \left\{ \frac{1}{2} \left(\phi_x^2 + \phi_t^2 \right) + (1 - \cos\phi) - 2\epsilon \left[1 - \cos\left(\phi/2 \right) \right] \right\} dx. \tag{10}$$

In addition to the Hamiltonian (energy), the sG equation, including its perturbed version (7), conserves the momentum, whose form is universal (it does not depend on the particular shape of the nonlinearity in the sG equation, nor even on the fact that the nonlinearity is a periodic function of the field variable, ϕ):

$$P = -\int_{-\infty}^{+\infty} \phi_x \phi_t dx. \tag{11}$$

The nonintegrable equation (7) has exactly two dynamical invariants, H and P. However, the integrable equation (3) has an infinite set of conserved quantities, which is a specific feature of integrable equations. Nevertheless, only the conservation of H and P is essential for physical applications, as higher-order invariants do not admit a clear physical interpretation, unlike the energy and momentum.

2 Mathematical Techniques: The Inverse Scattering Transform, Bäcklund Transform, and Others

The integrability of the sG equation in the form of Eq. (3) is based on the possibility to represent it as the compatibility condition of the system of two linear matrix equations for the two-component complex *Jost functions*, $\Psi(x,t) = \left\{ \Psi^{(1)}(x,t), \Psi^{(2)}(x,t) \right\}$:

$$\Psi_x = \frac{i}{2}\lambda \begin{pmatrix} 1 & 0 \\ 0 & -1 \end{pmatrix} \Psi + \frac{i}{4}(\phi_x - \phi_t) \begin{pmatrix} 0 & 1 \\ 1 & 0 \end{pmatrix} \Psi - \frac{i}{8\lambda} \begin{pmatrix} \cos\phi & -i\sin\phi \\ i\sin\phi & -\cos\phi \end{pmatrix} \Psi, \tag{12}$$

$$\Psi_t = -\frac{i}{2}\lambda \begin{pmatrix} 1 & 0 \\ 0 & -1 \end{pmatrix} \Psi - \frac{i}{4}(\phi_x - \phi_t) \begin{pmatrix} 0 & 1 \\ 1 & 0 \end{pmatrix} \Psi - \frac{i}{8\lambda} \begin{pmatrix} \cos\phi & -i\sin\phi \\ i\sin\phi & -\cos\phi \end{pmatrix} \Psi, \tag{13}$$

where λ is a (generally, complex) *spectral parameter*. The IST technique is based on the analysis of the scattering problem for Eq. (12), where coefficient functions ϕ_x, ϕ_t and $\sin\phi$ are assumed to be localized, i.e., they all decay fast enough at $|x| \to \infty$. The sG equation with periodic boundary conditions, rather than subject to the localization, is integrable too, on the basis of the Lax-pair representation, but the mathematical technique used in that case is completely different, and actually more complex, being based on methods of algebraic geometry for hyperelliptic Riemann surfaces [13].

The localization of coefficient functions ϕ_x, ϕ_t and $\sin\phi$ makes it possible to define solutions to Eq. (12) for the Jost functions by their asymptotic form at $|x| \to \infty$, where ϕ_x, ϕ_t and $\sin\phi$ vanish, while $\cos\phi$ becomes equal to $+1$ or -1, hence coefficients of Eq. (12) are asymptotically constant. Accordingly, solutions to the linear ordinary differential equation with constant coefficients reduce to plane waves. In particular, at $x \to -\infty$ they can be defined as

$$\Psi^{(1)} = 0,\ \Psi^{(2)} = \exp\left(-\frac{i}{2}\left(\lambda - \frac{1}{4\lambda}\right)x + \frac{i}{2}\left(\lambda + \frac{1}{4\lambda}\right)t\right). \tag{14}$$

The most general asymptotic form of the same solution at $x \to +\infty$ may be represented as a linear combination of two independent plane-wave solutions of the respective linear equation with constant coefficients:

$$\Psi^{(1)} = B(\lambda, t)\exp\left(\frac{i}{2}\left(\lambda - \frac{1}{4\lambda}\right)x + \frac{i}{2}\left(\lambda + \frac{1}{4\lambda}\right)t\right), \tag{15}$$

$$\Psi^{(2)} = A(\lambda, t)\exp\left(-\frac{i}{2}\left(\lambda - \frac{1}{4\lambda}\right)x + \frac{i}{2}\left(\lambda + \frac{1}{4\lambda}\right)t\right). \tag{16}$$

In this way, one can define the *scattering data* for Eq. (12), i.e., the reflection and transmission coefficients, $B(\lambda)$ and $A(\lambda)$, introduced in Eqs. (16) and (15), which connect the asymptotic forms of the same solution at $x \to \pm\infty$. While these coefficients are originally defined for real values of λ, a fundamental property of the scattering problem based on Eq. (12) is the possibility to analytically extend the scattering data and Jost functions into the upper complex half-plane of the spectral parameter, i.e., to $\mathrm{Im}(\lambda) > 0$. Obviously, the scattering data are determined by the localized physical fields, which induce mixing between the solutions defined by their plane-wave asymptotic forms at $x \to \pm\infty$—similar to how a localized potential determines the scattering matrix in the usual one-dimensional linear Schrödinger equation. Thus, the localized physical fields, viz., ϕ_t, ϕ_x and $\sin\phi\ (x)$, are effectively mapped into the scattering data in the upper half-plane of λ, determined by these fields. This mapping, together with the solution of the respective *inverse problem*. i.e., finding the physical field corresponding to a given set of the scattering data, are the core ingredients of the IST technique (in particular, the latter *inverse problem* had lent the IST technique its name).

In the general case, the inverse problem is formulated in the form of a rather complex system of linear integral equations for the Jost functions, which are usually called Gelfand-Levitan-Marchenko equations [14]. Explicit exact solutions, both for the Jost functions and the underlying physical field, $\phi\ (x)$, are available for the most important particular case, when $B(\lambda)$ vanishes at real values of λ, while $A(\lambda)$ is fully determined by its zeros in the complex plane, each zero corresponding to a soliton, see below.

The fundamental asset of the IST is the fact that the temporal evolution of the scattering data, which is determined by Eq. (13) for the evolution of the Jost functions, is actually trivial, contrary to the complex evolution of the physical field in the x-space. As well as the asymptotic forms of the solution to Eq. (12), which are given by simple expressions (14) and (16), (15), the evolution of the scattering data is determined by the consideration of the asymptotic limits of Eq. (13) at $x \to \pm\infty$, where coefficients of the equation become effectively constant, due to the underlying condition of the localization of the physical fields. Substituting the asymptotic expression (14) for the Jost functions into the accordingly simplified asymptotic form of Eq. (13) at $x \to -\infty$ demonstrates that this expression satisfies the equation. On the other hand, substituting the asymptotic expressions (15) and (16) for the same Jost-function solution at $x \to +\infty$, where the scattering coefficients are assumed to be functions of time, i.e., $A = A(\lambda; t)$ and $B = B(\lambda; t)$, into the asymptotic limit form of Eq. (13) with constant coefficients, which is valid at $x \to \infty$, readily gives rise to the following linear evolution equations:

$$\frac{dA(\lambda)}{dt} = 0, \ \frac{dB(\lambda)}{dt} = -i\left(\lambda + \frac{1}{\lambda}\right) B(\lambda). \tag{17}$$

Solutions to Eq. (17) are evident:

$$A(\lambda; t) = A(\lambda; t = 0);\ B(\lambda; t) = B(\lambda; t = 0) \exp\left(-i\left(\lambda + \frac{1}{\lambda}\right)t\right). \tag{18}$$

These solutions are valid both for real values of λ, as well as in the upper complex half-plane of λ, into which the scattering data may be analytically extended, as stated above.

Equations (17) and (18) demonstrate that, as a matter of fact, the IST resembles the classical Fourier-transform technique, which reduces the relatively complex dynamics of wave fields in the coordinate space, governed by linear dispersive equations with constant coefficients [for instance, Eq. (1)], to the trivial evolution of the Fourier transform in the space of wave vectors, quite similar to that described by solution (18). The difference is that the Fourier transform applies to any linear equation with constant coefficients, while the IST method solves only exceptional nonlinear partial differential equations, such as the sG equation.

The IST applies as well to the sG equation written in terms of the light-cone coordinates, see Eq. (4). In fact, the two versions of the IST technique, for Eqs. (3) and (4), can be obtained from each other by a simple transformation.

As mentioned above, the most important component of the scattering data is a set of eigenvalues of λ at which the transmission coefficient $A(\lambda)$, defined as per Eq. (16), vanishes. It is easy to demonstrate that the Jost functions corresponding to these eigenvalues exponentially vanish at $|x| \to \infty$. Due to the symmetry imposed by the fact that physical field $\phi(x, t)$ is real, it is also possible to demonstrate that these eigenvalues are located, in the upper half-plane of λ, symmetrically with respect to the imaginary axis. Accordingly, they may be isolated eigenvalues corresponding to purely imaginary values of the spectral parameter, or symmetric pairs:

$$\lambda_0 = i\ \mathrm{Im}(\lambda);\ \lambda_\pm = i\ \mathrm{Im}(\lambda) \pm \mathrm{Re}(\lambda), \tag{19}$$

with $\mathrm{Im}(\lambda) > 0$. Note that the eigenvalues remain constant in time, according to Eq. (18). Basic exact solutions of the sG equation and their relation to the eigenvalues of the spectral parameter, as well as to other components of the scattering data, are outlined in the next section.

The quantum version of the sG equation for the operator field can be defined too. If introduced properly, this quantum model is also integrable (i.e., it admits finding exact spectra of eigenstates of the corresponding quantum Hamiltonian). Without going into details of the technically involved analysis, it is relevant to mention that the exact solution of the quantum sG equation is provided by a quantum version of the IST technique [15].

In the connection to the use of the sG equation as a quantum-field model, it is relevant to mention its equivalence to another 1D quantum-field theory, namely, the *massive Thirring model*, which is based on the following Lagrangian [16]:

$$L = \sum_{\mu=0,1} \int_{-\infty}^{+\infty} \left[\bar{\psi} \left(\gamma^\mu \partial_\mu - m \right) \psi - \frac{g}{2} \bar{\psi} \gamma^\mu \psi \bar{\psi} \gamma_\mu \psi \right] dx, \tag{20}$$

where $\psi(x)$ is an operator spinor field describing 1D fermions, $\partial_0 \equiv \partial_t$, $\partial_1 \equiv \partial_x$, m is the mass, $g > 0$ is a constant of the attractive interaction, and $\gamma^0 \equiv \sigma_1$, $\gamma^1 \equiv -i\sigma_2$, where σ_1 and σ_2 are two of the Pauli matrices. The map between the Thirring and the sG models (alias the *bosonization of the fermions* [17]) is performed according to the following relation between the respective operator fields [18]:

$$: \exp\left(\pm i\phi(x)\right) : \ = \frac{m}{2} \bar{\psi}(x) \left(1 \mp \sigma_3\right) \psi(x), \tag{21}$$

with :: standing for the normally-ordered operator product.

As mentioned above, about 100 years prior to the discovery of the integrability of the sG equation by means of the IST method, the solvability of this equation with the help of the Bäcklund transform was already known [2]. This transform, which remains an important tool of the analysis nowadays [19], can be represented in a simple form for the sG equation written in the light-cone coordinates (4). Namely, if $\phi(\xi, \tau)$ is any solution to Eq. (4), then the following system of equations for function $\chi(\xi, \tau)$ is solvable and yields a new solution of the sG equation:

$$\chi_\xi = \phi_\xi + 2a \sin\left(\frac{\chi + \phi}{2}\right),\ \chi_\tau = -\phi_\tau + \frac{2}{a} \sin\left(\frac{\chi - \phi}{2}\right), \tag{22}$$

where a is an arbitrary constant. This transform (which is called *auto-Bäcklund*, in the case when it produces a new solution of the same equation) can be derived from the IST technique. Other transformations which allow one to generate more complex exact solutions from simpler ones, such as Darboux and Hirota methods (see, e.g., [20] and [21]), can also be derived from the IST technique.

3 Soliton Modes in One Dimension: Kinks, Breathers, and Cnoidal Waves

The simplest exact solution of the sG equation can be found by substituting the elementary ansatz for traveling-wave states, $\phi(x,t) = \phi(x-ct)$, into Eq. (3). The result is a family of solutions which are usually called *kinks*:

$$\phi(x,t) = 4\arctan\left(\exp\left(\sigma\frac{x-ct-x_0}{\sqrt{1-c^2}}\right)\right), \tag{23}$$

where $\sigma = \pm 1$ is the *polarity* of the kink ($\sigma = -1$ defines an *antikink*), and x_0 an arbitrary shift of its center. In accordance with the above-mentioned fact that the sG equation is a relativistically invariant one, this solution features a typical relativistic form, and exists for velocities $-1 < c < +1$, i.e., $c = 1$ plays the (formal) role of the speed of light. Obviously, the moving kink can be generated from the quiescent one, with $c = 0$, by means of the Lorentz transform, as per Eq. (8). The kinks may be considered as *topological solitons*, with topological charge defined by the difference of values of $\phi(x)$ at $x \to \pm\infty$: $[\phi(x=+\infty) - \phi(x=-\infty)]/(2\pi) \equiv \sigma$, i.e., the topological charge is identical to the polarity of the kink. The topological charge is related to the above-mentioned fact that, in terms of the corresponding quantum field theory, transformation (21) maps the kinks into fermionic solitons of the Thirring model.

Less obvious, but nevertheless quite simple too, is a family of exact time-periodic spatially localized solutions to the sG equation (3) in the form of *breathers* (nontopological solitons):

$$\phi(x,t) = 4\arctan\left\{(\tan\mu)\frac{\sin(\Psi\cos\mu)}{\cosh(Z\sin\mu)}\right\}, \tag{24}$$

$$\Psi \equiv \frac{t-cx}{\sqrt{1-c^2}} + \Psi_0,\; Z \equiv \frac{x-ct-x_0}{\sqrt{1-c^2}} \tag{25}$$

where c is the same velocity as in the kink solution (23), x_0 and Ψ_0 are arbitrary shifts, and μ, with values from the interval of $0 < \mu < \pi/2$, is the breather's amplitude, which simultaneously determines its width, $\sim 1/\sin\mu$, and intrinsic frequency, $\omega_{br} = \cos\mu$. Note that the frequency takes values in interval $|\omega_{br}| < 1$, which is exactly the gap in the dispersion relation (2) for the linear waves. The latter

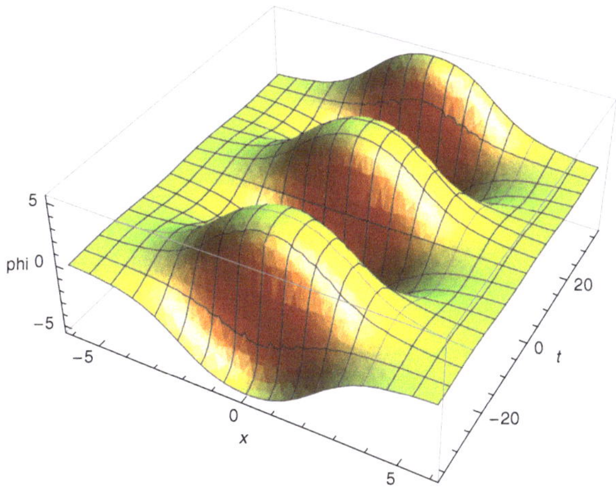

Fig. 1 An example of the time-periodic evolution of field $\phi(x,t)$ in a breather with zero velocity, $c = 0$, and $\mu = 0.43\pi$. Reproduced with permission from A. Fischer and O. Lechtenfeld, "The noncommutative sine-Gordon breather", J. Math. Phys. **50**, 095201 (2009). Copyright 2009, AIP Publishing LLC

fact implies that the breather does not decay into radiation because its frequency does not fall into resonance with any linear wave.

In particular, $\mu \ll 1$ corresponds to a small-amplitude breather,

$$\phi(x,t) \approx 4\mu \sin\left(\left(1 - \frac{\mu^2}{2}\right)\Psi\right) \operatorname{sech}(\mu Z), \qquad (26)$$

and in the opposite limit case, with $\zeta \equiv \pi/2 - \mu \ll \pi/2$, solution (24) takes the form of a low-frequency breather,

$$\phi(x,t) \approx 4 \arctan\left\{\zeta^{-1} \sin(\zeta\Psi) \operatorname{sech} Z\right\}. \qquad (27)$$

In expressions (26) and (27), Ψ and Z are defined as per Eq. (25). A generic example of the breather solution is displayed in Fig. 1.

Breathers may be considered as kink-antikink bound states, which perform oscillations at frequency $\cos\mu$. Accordingly, the limit form of solution (27) corresponding to $\zeta \to 0$ represents an exact solution in the form of the pair of a kink and antikink, which begin the collision with zero relative velocity and infinitely large separation at $t \to -\infty$, then collide, passing through each other at $t = 0$, and again separate at $t \to \infty$:

$$\phi(x,t) = 4 \arctan(t \operatorname{sech} x). \qquad (28)$$

The separation between the kink and antikink corresponding to this exact solution grows at $|t| \to \infty$ as $L \approx \ln(|t|)$.

The energy and momentum of the kink and breather, defined as per Eqs. (10) and (11) are, respectively,

$$E_{\text{kink}} = 8/\sqrt{1-c^2}\ , \quad P_{\text{kink}} = 8c/\sqrt{1-c^2}, \tag{29}$$

$$E_{\text{br}} = 16\,(\sin\mu)\,/\sqrt{1-c^2}\ , \quad P_{\text{br}} = 16c\,(\sin\mu)\,/\sqrt{1-c^2}. \tag{30}$$

From here, it is obvious that the kink and breather may be considered as relativistic particles with effective masses

$$m_{\text{kink}} = 8, \ m_{\text{br}} = 16\sin\mu, \tag{31}$$

respectively. Further, comparing the energies of the breather and far separated kink and antikink, one concludes that the *binding energy* of the kink-antikink pair, which form the breather as the bound state, is

$$E_{\text{bind}} \equiv 2E_{\text{kink}} - E_{\text{br}} = 16\,(1-\sin\mu) \tag{32}$$

(here $c = 0$ is set, for the sake of clarity).

Both the kink and breather are readily produced by the IST method as solutions corresponding, respectively, to the single and double eigenvalues (19), with

$$\{\text{Im}(\lambda)\}_{\text{kink}} = \frac{1}{2}\sqrt{\frac{1+c}{1-c}}; \ \lambda_{\text{br}} = \pm\frac{1}{2}\sqrt{\frac{1+c}{1-c}}\exp\left(\pm i\mu\right). \tag{33}$$

Furthermore, scattering-data sets containing several eigenvalues make it possible to construct exact solutions which describe collisions between kinks and breathers. These solutions corroborate the celebrated principle that collisions between solitons in integrable models are completely elastic—the solitons emerge from the collisions with precisely the same shapes, amplitudes, and velocities that they had prior to the collision, the only effect being the shift of the soliton as a whole. In particular, the collision-induced shifts of two kinks moving with velocities $\pm c$ are

$$\Delta x = \pm\sqrt{1-c^2}\ln\left(1/c^2\right), \tag{34}$$

so that each kink is shifted by the collision ahead in the direction of its motion, and the shifts do not depend on polarities of the kinks. Nevertheless, the actual character of the collisions is different for the kink-kink and kink-antikink pairs: as seen in Fig. 2, unipolar kinks *repel each other*, so that they actually bounce back, while the kinks with opposite polarities *interact attractively*, passing through each other and restoring their shapes afterwards.

The IST method makes it also possible to construct exact localized solutions of the sG equation in the form of more sophisticated complexes. In particular, the scattering data including a single eigenvalue λ_0 and one twin pair $\lambda_\pm$ with $|\lambda_0| = |\lambda_\pm|$, i.e., with equal velocities corresponding to these eigenvalues, see Eqs. (19) and (33), give rise to a solution in the form of a "wobbling kink", i.e., a nonseparating superposition of a kink and breather [22]. However, such complexes are unstable, as an arbitrary infinitesimal perturbation of the initial conditions will

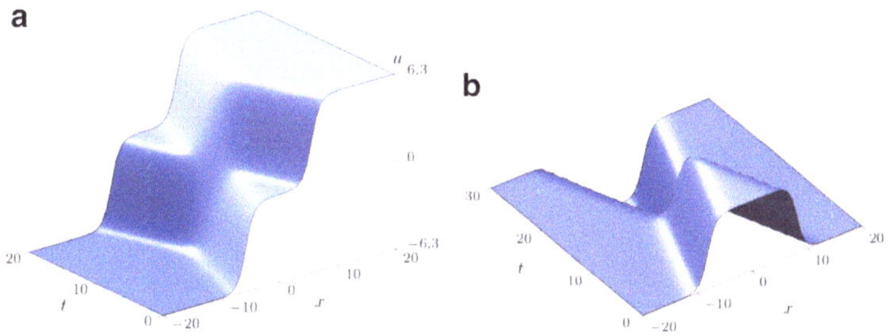

Fig. 2 Typical examples of the kink-kink (**a**) and kink-antikink (**b**) collisions as produced by exact solutions of the sG equation. The figures from the website of Prof. L.N. Trefethen, people.maths.ox.ac.uk/trefethen/ (Oxford University Mathematical Institute)

give rise to an infinitesimal difference between $|\lambda_0|$ and $|\lambda_\pm|$, hence the kink and breather will acquire a small relative velocity, causing the splitting of the wobbling kink.

In addition to the solitons (kinks, antikinks, and breathers), the general solution of the sG equation contains a radiation component, i.e., small-amplitude (quasi-linear) spreading dispersive waves. In terms of the IST, the solitons correspond, as said above, to discrete eigenvalues, i.e., to the discrete component of the scattering data, while the radiation is represented by the continuous component, which is represented by reflection coefficient $B(\lambda)/A(\lambda)$, taken at real values of λ [see Eqs. (15) and (16)]. In the limit of $t \to \infty$, the local amplitude of the radiation component decays as $t^{-1/2}$, hence only the solitons remain conspicuous features in the asymptotic configuration of the wave field at large values of t.

In addition to the solitons, spatially periodic solutions to the sG solutions are interesting too (as mentioned above, the sG equation subject to the spatially periodic boundary conditions, is also integrable, although the respective mathematical technique is quite complex [13]). Such solutions, in their simplest form, which is most relevant to physical applications, may be built as periodic chains of alternating kinks and antikinks, or as arrays of unipolar kinks. The kink-antikink chains are always unstable, due to the above-mentioned attractive interactions between kinks with opposite polarities. However, unipolar arrays are stable, therefore they are meaningful solutions. The respective exact solution can be expressed in terms of the Jacobi's amplitude function with elliptic modulus k,

$$\phi(x) = \pi - 2\,\mathrm{am}\left((x-\xi)/k;k\right), \tag{35}$$

where ξ is an arbitrary constant shift, and the period of the solution (which may be understood as the spacing between kinks in the array) is

$$L = 2kK(k), \tag{36}$$

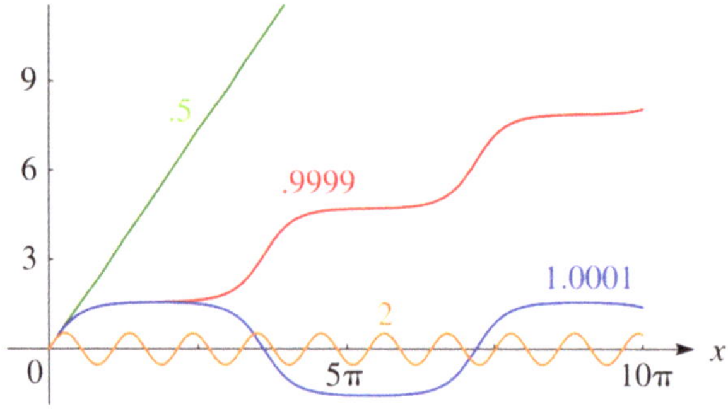

Fig. 3 Typical shapes of cnoidal waves (35) are illustrated by plots of the Jacobi's amplitude function, am $(x; k)$ for four values of the elliptic modulus, $k = 0.5, 0.9999, 1.0001$, and 2. The two former values correspond to stable densely and rarely-packed kink chains, while the two latter values of k correspond to unstable kink-antikink chains. The plot is taken from the Digital Library of Mathematical Functions, dlmf.nist.gov/22.19.F1.mag

with $K(k)$ standing for the complete elliptic integral of the first kind. This solution, which is written with zero velocity, can be boosted by means of the Lorentz transformation (8). Solutions of this type are often called "cnoidal waves", due to their relation to the Jacobi's elliptic functions, cn and sn. In the limit of $k \to 1$ the array turns into a chain of kinks with large distance between them,

$$L \approx 2\ln\left(4/\sqrt{1-k^2}\right). \tag{37}$$

In the opposite limit of $k \to 0$, the spatial period is small, $L \approx \pi k$, and solution (35) for the densely packed kink array is close to a linear function of x with a large slope:

$$\phi(x) \approx 2x/k - (k/2)^2 \sin(2x/k). \tag{38}$$

Basic shapes of the cnoidal waves are adequately illustrated by a set of plots of function am $(x; k)$, which are displayed in Fig. 3. In particular, the nearly linear graph pertaining to $k = 0.5$ represents the densely-packed chain (38), while the staircase-shaped plot, corresponding to $k = 0.9999$, shows the rarefied chain with distance L (37) between them. The plots for $k = 1.0001$ and $k = 2$ represent unstable arrays built of alternating kinks and antikinks.

As mentioned above, the sG equation with periodic boundary conditions gives rise to very complex exact solutions which can be constructed in terms of algebraic geometry [13]. Those solutions involve a great deal of challenging mathematics, but are not closely related to straightforward physical applications.

As mentioned above, the quantum version of the sG equation is an integrable model too, therefore exact quantum bound states can be found in it, in the form of breathers with a discrete spectrum of binding energies [15], in contrast to the continuous spectrum in the classical breather solution, see Eq. (32). Before finding these quantum states in the exact form, the discrete spectrum of the breathers was predicted with the help of the semi-classical Bohr-Sommerfeld quantization rule [23], which is a natural approximation for the quantization of time-periodic states:

$$\int_{-T/2}^{+T/2} dt \int_{-\infty}^{+\infty} dx\phi_t^2 (x,t) = 2\pi n\varepsilon, \tag{39}$$

where $\int_{-T/2}^{+T/2}$ implies the integration over the temporal period, n is a (large) integer quantum number, ε is a (small) coupling constant, and the Planck's constant is scaled to be 1. A noteworthy finding is that the Bohr-Sommerfeld spectrum turns out to be *identical* to the exact one in this model, unlike, for instance, the commonly known result for the harmonic oscillator, where the semi-classical quantization fails to predict the term $\hbar\omega/2$ in the energy of the ground state.

4 Physical Realizations: Josephson Junctions, Self-induced Transparency, Spin Waves in Magnetics, Charge-Density Waves, the Frenkel–Kontorova Model, etc.

4.1 *Long Josephson Junctions*

As mentioned above, the sG equation is a profoundly important model not only due to its integrability and the remarkable ability to produce a vast variety of exact solutions, but also because it emerges as a natural model in several fundamentally important physical settings. Among these physical applications, perhaps the most significant one is the sG model of long Josephson junctions (JJs). The specific dynamics of tunneling was predicted by Josephson [24, 25] in the junction built as two bulk superconductors separated by a thin dielectric layer, across which Cooper pairs of superconducting electrons may tunnel keeping their coherence, i.e., the thickness of the dielectric barrier is much smaller than the correlation length of the superconducting state. Soon afterwards, the Josephson effect was observed experimentally [26]. A typical scheme of the long JJ is displayed in Fig. 4a.

The sG equation as the model of the long JJ was derived, from the fundamental Ginzburg-Landau equations for the bulk superconductors, by Josephson himself [24]. The extended form of the sG model, written in scaled units, is

$$\phi_{tt} - \phi_{xx} + [1 + \epsilon f(x)] \sin\phi = -\alpha\phi_t - \gamma(t), \tag{40}$$

where $\phi(x,t)$ determines two physical variables, which are proportional to each other: the jump of the phase of the wave function of superconducting electrons across the JJ (absolute values of the wave functions are equal), and the magnetic flux trapped in the junction (the local strength of the magnetic field, $H \sim \phi_x$, is directed

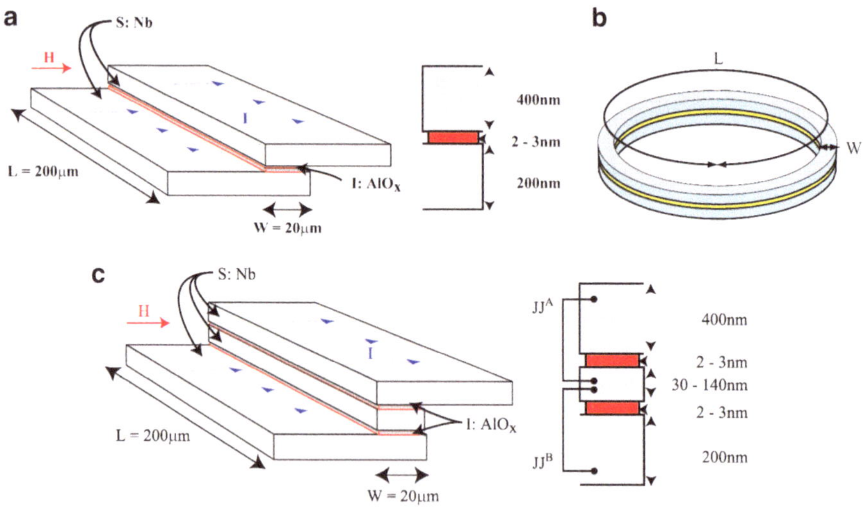

Fig. 4 (**a**) Schematic drawings of a typical long Josephson junction (JJ) and its cross section. In this case, the JJ is made of bulk niobium superconductors (S) and aluminum oxide used as the dielectric barrier. Typical values of the junction's length (L), width (W) and thickness are indicated in the figure. The bias current (I) driving the JJ, and the magnetic field (H) applied at its edges, are designated too. (**b**) A long circular JJ. (**c**) The same as in (**a**), but for a stack of two parallel magnetically coupled junctions. The figures from: A. Wallraff, *Fluxon dynamics and radiation emission in twofold long Josephson junction stacks*, Diplomarbeit in Physik, Institut für Schicht und Ionentechnik des Forschungszentrums Jülich (KFA), 1997

perpendicularly to the junction in the plane of the overlapping bulk superconductors, see Fig. 4a). In physical units, the largest velocity admitted by the sG equation, which is $c_{\max} = 1$ in the scaled units, see Eqs. (23) and (24), corresponds to the *Swihart's velocity*, i.e., the largest group velocity of the propagation of the linear electromagnetic waves (alias *plasma waves*) in the long JJ, which is usually ~ 0.01 of the speed of light in vacuum [25].

The terms on the left-hand side of Eq. (40) represent, respectively, the capacitance of the dielectric barrier (the local voltage across the barrier is $\sim \phi_t$), the electromagnetic inductance of the junction, and the density of the superconducting current tunneling across the barrier. The part $\epsilon f(x)$ of the coefficient in front of the latter term takes into regard possible inhomogeneity of the long JJ [27]. Further, the first term on the right-hand side of Eq. (40), which plays the role of dissipation, represents the normal component of the tunnel current. According to the Ohm's law, it is proportional to the local voltage, ϕ_t. Further, $\gamma(t)$ is the density of the external (*bias*) current flowing across the junction, as shown by arrows in Fig. 4a. The bias current, which is necessary to compensate the Ohmic loss, may be constant (dc) or variable (ac), or a combination of both.

The sG kink represents a physical state in the long JJ in the form of a current loop, formed by two spatially separated superconducting tunneling currents. Directions of these current are opposite, corresponding to opposite signs of $\sin\phi(x)$ in different portions of the kink solution (23). These two currents are connected by two mutually opposite superconducting currents flowing parallel to junction's direction in the bulk superconductors, thus forming a loop. The current loop induces magnetic flux, with the local magnetic field oriented, as said above, perpendicular to the direction of the junction, see Fig. 4a. In fact, each kink carries a quantum of magnetic flux, $\Phi_0 = h/(2e)$, where h is the Planck's constant, therefore it is usually called *fluxon* (and the antikink is called *antifluxon)*, in terms of the JJ model.

The profound significance of fluxons for the studies of the dynamics of long JJs was first highlighted by Fulton and Dynes [28]. The first direct experimental observations of isolated fluxons in long JJs were reported in 1982 and 1983 [29,30]. These publications were followed by a flurry of experimental and theoretical work in this area [31, 32]. In particular, the first experiments with fluxons in circular long JJs (see Fig. 4b), where the dynamics is not affected by edges, were reported in 1985 [33]. A convenient tool for the work with the circular long junction of radius R is offered by magnetic field, H, oriented in the plane of the circle, which induces an additional free term $H\cos(x/R)$ in Eq. (40) (moreover, rotating magnetic field induces a traveling term, $H\cos((x-ct)/R)$, which can drag a fluxon along the circular junction [34]). Quantum properties of the fluxons (their tunneling across an effective potential barrier) in the circular system have been demonstrated experimentally too [35].

An important related physical setting is a system of parallel magnetically coupled long JJs, see Fig. 4c. The system is described by coupled sG equations [36, 37]:

$$\phi_{tt} - \phi_{xx} + \sin\phi = \epsilon\psi_{xx}, \tag{41}$$

$$\psi_{tt} - \psi_{xx} + \sin\psi = \epsilon\phi_{xx}. \tag{42}$$

This is an interesting dynamical model, which naturally gives rise to *bi-fluxons* [31, 38]. Much attention has also been drawn to studies of multicore stacked long JJs [39–41] with the magnetic coupling between adjacent cores. In particular, the study of the stacked configurations is suggested by a similar structure of high-temperature superconductors [42].

4.2 Self-induced Transparency

Another "clean" and therefore popular physical application of the sG equation models the self-induced transparency (SIP) in nonlinear optics (the "cleanness" means that, like in the case of JJs, the sG equation provides a very accurate model for this physical setting). The SIP effect, as predicted theoretically and demonstrated experimentally by McCall and Hahn [43], implies that a nonlinear pulse of the electromagnetic field can propagate, rather than being absorbed, in the resonant two-level medium. This medium is described by the equation for the unidirectional propagation of the electric field (E), coupled to the Bloch equations for the microscopic polarization (P) and population inversion of the medium (N):

$$E_X + E_T = \beta P,\ P_T = EN,\ N_T = -EP, \tag{43}$$

where β is the scaled atomic density of the medium. Boundary conditions relevant for the pulse propagation are $E(|X| = \infty) = 0, P(|X| = \infty) = 0$, $N(|X| = \infty) = -1$. Equations (43) can be exactly reduced to the single sG equation (3) by substitution $E = \phi_T,\ P = -\sin\phi,\ N = -\cos\phi,\ x = \sqrt{\beta}(T - 2X),\ t = \sqrt{\beta}T$ [44]. Then, the kink plays the role of the stably propagating "2π pulse" (2π refers to pulse's area, $A = \int_{-\infty}^{+\infty} E(X,T)\,dT \equiv \pm 2\pi$, if E is substituted by $\partial\phi/\partial T$ here), whose energy is perfectly absorbed by the resonant medium, and then perfectly released in the propagation direction.

4.3 Solid-State Realizations: Magnetics and Charge-Density-Wave Conductors

In solid-state physics, the sG equation and its extended versions find various realizations as models of spin waves in magnetic media. Rather straightforward are the applications of the sG equation to the description of easy-plane ferromagnets [45–47] and antiferromagnets [48], where $\phi(x,t)$ is an angle of the local orientation of the magnetization, and kink solutions correspond to domain-wall states. In a similar way, the sG equation serves as a dynamical model of ferroelectrics [49]. An additional term in the sG model of the antiferromagnet appears, in the form of $h(t)\sin(\phi/2)$ [cf. the double sG equation (7)], if an external magnetic field, $h(t)$ (it may be time-dependent), which is directed perpendicular to axis x and aligned with the magnetization at $x = \pm\infty$, is taken into account [48]. On the other hand, another additional term, $\epsilon\sin(2\phi)$, accounts for weak anisotropy in the model of the easy-plane ferromagnet [50].

Another possible realization of the sG equation in solid-state physics dates back to 1974, when it was demonstrated that Hamiltonian (10) naturally appears in the model of charge-density waves (CDWs) in "one-dimensional metals", provided that the wavenumber accounting for Peierls' instability is commensurate with the inverse spacing of the underlying ionic lattice [51]. In this case, dynamical variable $\phi(x,t)$ is the local residual misfit between the electronic CDW and the lattice. Using this model, a possibility of the existence of CDW phase solitons, corresponding to the sG kinks, was predicted in 1976 [52]. However, clear experimental evidences of the existence of such solitons are missing.

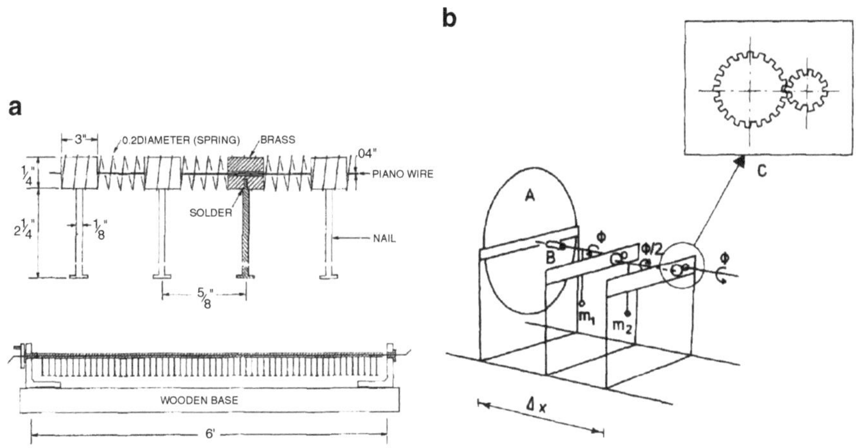

Fig. 5 (**a**) The experimental realization of the sG system in the form of the chain of pendula attached to an elastic support, as per the original paper by A. C. Scott [53] (Reproduced with permission from “A Nonlinear Klein–Gordon Equation”, by A. C. Scott, American Journal of Physics **37**, 52 (1969); doi: 10.1119/1.1975404. Copyright 1969, American Association of Physics Teachers). (**b**) A segment of the mechanical realization of the double sG equation in the form of a chain of pendula developed by M. Salerno [54]. The alternation of pendula rotating by angles ϕ and $\phi/2$, which is the characteristic feature defining the double sG system, is provided by mechanism C with gear ratio 1 : 2

4.4 The Frenkel–Kontorova Model and Other Realizations

As mentioned above, the sG equation was, as matter of fact, first introduced in physics as the continuous limit of the FK model [3, 4]. This model was experimentally implemented in a simple form as a chain of pendula suspended on a single elastic rod, built by A. C. Scott and reported in 1969 [53] (Fig. 5a). Historically, this work had played an important role in drawing attention of many researchers to the sG equations and its physical applications. A similar but more sophisticated mechanical setup, implementing the double sG equation (7), was later built by M. Salerno [54], see Fig. 5b. In this context, it is relevant to mention that the “residue” of the discreteness in the continuous limit of models of the FK type is represented by an extra term in the sG equation, $\sim \phi_{xxxx}$, i.e., the higher-order dispersion, which is naturally produced by the extension of expansion (6) [55, 56]. As well as other perturbation terms, e.g., $\sin(\phi/2)$ in Eq. (7), this term breaks the integrability of the ideal sG equation.

Other physical realizations of the sG equation are known too. These include, in particular, liquid crystals [57], spin waves in liquid helium [58, 59], and hydrodynamics, where the sG equations appear as models of a rapidly rotating baroclinic fluid [60] and spatially forced thermal convection [61].

5 The Perturbation Theory for sine-Gordon Solitons

5.1 The Adiabatic Approximation for Kinks and Breathers

As mentioned above, in various physical models small terms (*perturbations*) should be added to the ideal sG equation, to take into regard various weak but indispensable effects, see, e.g., Eqs. (7) and (40). Such small additional terms break the integrability of the sG equation, which naturally suggests the possibility to develop a perturbation theory, employing exact sG kink and breather solutions as the zero-order approximation. First-order effects induced by the perturbations make it possible to predict various results for the solitons which are impossible in ideal integrable models [31, 62–65]. Thus, the perturbation theory is relevant not only because it offers the way to explore realistic models of physical systems, but also as a tool to reveal dynamical effects for solitons which are never produced by strictly integrable equations. In this section basic results produced by the perturbation theory for sG solitons are briefly surveyed, chiefly following review [31].

Perturbation effects that do not or do include radiation waves, in addition to solitons, may be naturally categorized as adiabatic and radiative ones, respectively. Further, it is natural to distinguish between single- and multi-solitons configurations to which the perturbation theory is applied.

The simplest but physically important problem is the motion of a single fluxon under the action of the dc bias current, $\gamma = \text{const}$, in the presence of the Ohmic loss ($\alpha > 0$), in the uniform long JJ ($\epsilon = 0$), see Eq. (40). By calculating the time derivative of momentum (11) and substituting the unperturbed wave form (23) for the fluxon, the effective driving and friction forces acting upon it are obtained:

$$\frac{dP}{dt} = 2\pi\sigma\gamma - \frac{8\alpha c}{\sqrt{1-c^2}}. \tag{44}$$

The equilibrium velocity (which determines the voltage across the junction produced by the motion of the fluxon) corresponds to $dP/dt = 0$ in Eq. (44), which leads to the *McLaughlin-Scott formula* [65],

$$c = \frac{\pi\sigma\gamma}{\sqrt{16\alpha^2 + (\pi\gamma)^2}}. \tag{45}$$

Because the fluxon's velocity is proportional to the dc voltage across the long JJ, Eq. (45) predicts the current-voltage characteristic corresponding to the motion of the fluxon in the long junction (recall that γ is the density of the bias current applied to the junction) [25]. This dependence between the voltage and current was very accurately corroborated by experiments [32].

Another problem of major physical interest is to consider the fluxon dynamics in the JJ model (40) with the extra term corresponding to $f(x) = \delta(x)$. This term with $\epsilon > 0$ or $\epsilon < 0$ represents a localized defect in the long JJ, which is called, respectively, microshort or microresistor (i.e., a short segment where the tunneling of the superconducting current is enhanced or suppressed). Adding the respective term to the sG Hamiltonian (10), it is straightforward to derive the effective potential of the interaction of the fluxon with the local defect:

$$U(\xi) = 2\epsilon \operatorname{sech}^2 \xi, \tag{46}$$

where ξ is the coordinate of the fluxon's center (for solution (23), $\xi = ct + x_0$; note that the potential does not depend on the fluxon's polarity). In the case of the microshort ($\epsilon > 0$), the fluxon dragged by the bias current may hit the potential barrier and get trapped in the full potential, which is a sum of the repulsive potential (46) and the linear potential, $-2\pi\sigma f \xi$, corresponding to the dragging force in Eq. (44). The critical condition for the trapping is determined by equating the kinetic energy of the "nonrelativistic" fluxon moving with velocity $|c| \ll 1$, $E_{\text{kin}} \approx 4c^2$ [recall the fluxon's mass is 8, see Eq. (31)], to the height of the potential barrier (46). The result is the prediction for the smallest (threshold) value of the bias-current density which will drag the fluxon across the microshort, preventing its capture [65]:

$$f_{\min}^2 = (8/\pi)\alpha^2\epsilon, \tag{47}$$

which is relevant under condition $\epsilon \gg \alpha^2$. Such thresholds were indeed observed experimentally in long JJs [32].

In the case of $\epsilon < 0$ (the microresistor), expression (46) represents a potential well. In this case, the capture is possible too, but the calculation is more elaborate. It aims to calculate the loss of the kinetic energy by the fluxon under the action of the friction force [see Eq. (44) and cf. Eq. (50) written below in a similar context] and equate it to the kinetic energy of the incident fluxon, the latter being determined by Eq. (45). The final result is [66] $f_{\min}^2 = \left(16\sqrt{2}/\pi\right)\sqrt{|\epsilon|}\alpha^3$, which is much smaller than its counterpart for the microresistor, given by Eq. (47), as condition $|\epsilon| \gg \alpha^2$ is adopted here.

An important setting is also a periodic lattice of microshorts of microresistors with spacing a [67]. In this case, the effective periodic potential for the fluxon can be expressed in terms of the Jacobi's elliptic sine function sn [68]:

$$2\epsilon \sum_{n=-\infty}^{+\infty} \operatorname{sech}^2(\xi - an) = \frac{8\epsilon}{a^2}\left\{K(k)\left[1 - k^2 \operatorname{sn}^2\left(\frac{2K(k)}{a}\xi\right)\right] - K(k)E(k) + \frac{a}{2}\right\}, \tag{48}$$

where K and E are the complete elliptic integrals, and the elliptic modulus k is determined by equation $\pi/a = K\left(\sqrt{1-k^2}\right)/K(k)$.

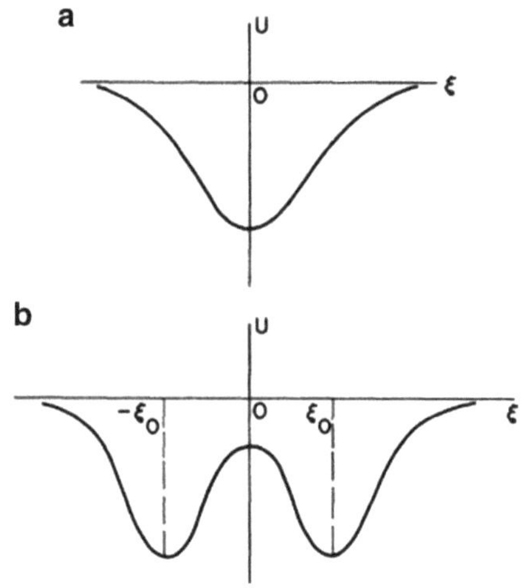

Fig. 6 The time-average potential of the interaction of breather (24) with the local defect described by $f(x) = \delta(x)$ in Eq. (40) with $\epsilon < 0$ (the microresistor in the long Josephson junction), as per [69]. The breather's amplitude is $\tan\mu < \sqrt{2}$ in (**a**), and $\tan\mu > \sqrt{2}$ in (**b**)

The interaction of breather (24) with the above-mentioned local defect, corresponding to $f(x) = \delta(x)$ in Eq. (40), gives rise to the following effective potential averaged with respect to intrinsic oscillations of the breather:

$$U(\xi) = \frac{8\epsilon\,(\cot\mu)\cosh(\xi\sin\mu)}{\left[1 + (\cot^2\mu)\cosh^2(\xi\sin\mu)\right]^{3/2}}, \tag{49}$$

where ξ is the coordinate of the breather's center, cf. the interaction potential (46) for the fluxon. The shape of this potential, displayed in Fig. 6 for $\epsilon < 0$ (the microresistor type of the defect), shows its noteworthy property: while the broad breather with a relatively small amplitude, $\tan\mu < \sqrt{2}$, may have a single stable stationary position, $\xi = 0$, in the case of $\epsilon < 0$, the breather with a larger amplitude, $\tan\mu > \sqrt{2}$, has a stable position at the center, $\xi = 0$, at $\epsilon > 0$ (microshort) and, on the other hand, it has two asymmetric stable positions at $\xi_0 = \pm(\sin\mu)^{-1}\ln\left(\left(\tan\mu + \sqrt{\tan^2\mu - 2}\right)/\sqrt{2}\right)$, for $\epsilon < 0$ [69].

Another fundamental perturbation-induced effect is the merger of colliding fluxon and antifluxon into a breather, in the framework of the model of the long JJ based on Eq. (40). Indeed, according to Eq. (45), the fluxon and antifluxon move with opposite velocities. The analytical prediction for the merger threshold is based on the calculation of the total Ohmic-term-induced energy loss for the exact solution (28) which separates exact solutions for breathers and freely colliding kink-antikink pairs:

$$\Delta E = \alpha\int_{-\infty}^{+\infty} dt \int_{-\infty}^{+\infty} dx\phi_t^2 = 8\pi^2\alpha. \tag{50}$$

Equating this expression to the total kinetic energy of the nonrelativistic fluxon-antifluxon pair, $E_{\mathrm{kin}}^{(\mathrm{pair})} \approx 8c^2 \approx 2(\pi\gamma/\alpha)^2$, which is given by Eq. (45), yields the following prediction: the "fluxon-antifluxon plasma" in the long JJ exists if the bias current density exceeds the critical value [70],

$$\gamma_{\mathrm{cr}} \approx 4\alpha^{3/2}. \tag{51}$$

The concept of the effective interaction potential is also relevant for two far separated fluxons (in that case, the interaction force is exponentially small) [71, 72], and for fluxons belonging to the different junctions in the coupled system described by system (41), (42). The additional term in the system's Hamiltonian accounting for the linear coupling is $H_{\mathrm{int}} = \epsilon \int_{-\infty}^{+\infty} dt \int_{-\infty}^{+\infty} dx \phi_x \psi_x$. The substitution of the unperturbed fluxon solutions (23) in each junction, with central coordinates $\xi_{1,2}$ and polarities $\sigma_{1,2}$, gives rise to the potential of the interaction between them, in the lowest approximation:

$$U_{\mathrm{int}} = 8\epsilon\sigma_1\sigma_2 (\xi_1 - \xi_2) / \sinh(\xi_1 - \xi_2). \tag{52}$$

The use of this potential makes it possible to predict a number of effects for the interaction of the two fluxons, such as their coalescence into a bound state in the form of a bi-fluxon [38]. The dynamics of a transverse fluxon chain in the multilayer stack of coupled long JJs was also analyzed by means of the perturbation theory [73].

In the case when the fluxon is present only in one junction, it creates its "shadow" in the coupled one. In particular, the "shadow" of the fluxon residing in the second junction, with polarity σ and velocity c, can be found in an approximate analytical form in the first junction, assuming that the coupling constant, ϵ, is small [38]:

$$\phi_{\mathrm{shadow}}(x,t) = \frac{2\sigma\epsilon}{1-c^2}\left[(\sinh(|z|))\ln(2\cosh z) - |z|\cosh z\right]. \tag{53}$$

where $z \equiv (x - ct)/\sqrt{1-c^2}$. Note that expression (53), although it may seem divergent, actually vanishes at $|z| \to \infty$.

In the case of the double sG equation (7), the perturbation term, $\sin(\phi/2)$, prevents the existence of 2π kinks, such as those given by Eq. (23), but it gives rise to stable bound states of two 2π kinks in the form of a 4π kink. It may be found in an exact form, in spite of the nonintegrability of the double sG equation. For practical purposes, the approximate form of the solution, valid for small ϵ, is sufficient [74, 75]:

$$\phi(x,t) \approx 4\arctan\left(\sqrt{\epsilon/2}\sinh\left(\frac{x-ct}{\sqrt{1-c^2}}\right)\right). \tag{54}$$

This bound state may exist in an excited state, with the two constituent 2π kinks featuring relative oscillations of their positions with frequency which was predicted by means of the perturbation theory too, $\omega = \sqrt{\epsilon}$ [74] (such an excited state is sometimes called a "wobbler" [75]).

5.2 Radiative Effects

As an example of the perturbation-induced radiative effect, it is relevant to mention the prediction of the largest (threshold) velocity of the freely moving kink, c_{thr}, that will be trapped due to its collision with the microresistor, i.e., the perturbation represented by $f(x) = \delta(x)$ with $\epsilon < 0$ in Eq. (40), while the bias current and loss are absent, $\alpha = \gamma = 0$. The threshold is determined by equating the total radiation loss of the energy of the kink interacting with the microresistor to the kinetic energy of the incident kink. The total loss is exponentially small with respect to ϵ, which is a consequence of the existence of the gap in the spectrum (2) of the linear waves. The analytical result is $c_{\mathrm{thr}}^2 = 2^{21/4}\sqrt{\pi}|\epsilon|^{3/4}\exp\left(-2\sqrt{2/|\epsilon|}\right)$ [70].

Another basic perturbation-induced radiative effect is the merger of colliding kink and antikink into the breather in the absence of the dissipative term and driving field. The merger is caused by radiative losses accompanying the kink-antikink collision if the integrability of the sG equation is broken by an additional Hamiltonian term, such as the above-mentioned one, $\epsilon \sin(2\phi)$ (unlike the term $\epsilon \sin(\phi/2)$, this one does not preclude the existence of 2π kinks). Numerically, the possibility of the kink-antikink annihilation into breathers in the sG equation with Hamiltonian perturbations was discovered by Peyrard and Campbell [76]. The analytical calculation for the perturbation term $\epsilon \sin(2\phi)$ was reported in [70], yielding the threshold value of velocities $\pm c$ of the kink-antikink pair below which the merger due to radiative loss is expected to occur: $c_{\mathrm{cr}} \approx 1.112\epsilon$. An essential difference of this result from a similar one for the kink-antikink merger under the action of dissipative loss, see Eqs. (50) and (51), is that c_{cr} is of the first order in ϵ (as the radiative energy loss scales as ϵ^2), while the dissipation coefficient α gives rise to the energy loss $\sim \alpha$, hence the respective threshold velocity is $\sim \sqrt{\alpha}$, i.e., of order 1/2 with respect to the underlying small perturbation parameter.

Numerical simulations of kink-antikink collisions in strongly perturbed versions of the sG equation [76, 77] have revealed another remarkable effect, viz., the existence of *multiple thresholds*, $c_0 < c_1 < c_2 < c_3 < \ldots$, so that the collisions give rise to the annihilation of the kink-antikink pair in intervals $c < c_0$, $c_1 < c < c_2, \ldots$, while in the intermediate "windows" the collisions are quasi-elastic, with the kink and antikink passing through each other without the merger. An explanation to this intriguing dynamical feature is based on an approximation which treats the colliding kink and antikink as a dynamical system with a few degrees of freedom [56, 77].

5.3 The Multidimensional sine-Gordon Equation

Multidimensional equations, with ϕ_{xx} replaced by the Laplacian, may be reduced to the radial sG equation, assuming the axial or spherical symmetry, in the 2D and 3D cases, respectively:

$$\phi_{tt} - \left(\phi_{rr} + \frac{D-1}{r}\phi_r\right) + \sin\phi = 0, \tag{55}$$

where r is the radial coordinate, and $D = 2$ or 3 is the spatial dimension. One can start the analysis of this equation by adopting a configuration in the form of a radial kink of a large radius, R_0, in which case the additional term, $(D-1)\,r^{-1}\phi_r$, may be treated as a small perturbation [78]. Then, assuming that emission of radiation and a change of the kink's shape against the unperturbed 1D profile (23) may be neglected, the law of the inward motion of the radial kink, shrinking under the action of the "surface tension", was found for the 2D and 3D cases:

$$R_{2D}(t) = R_0 \cos\left(t/R_0\right), \; R_{3D}(t) = R_0 \mathrm{cn}\left(\sqrt{2}t/R_0\right), \tag{56}$$

where cn is the Jacobi's elliptic cosine function with modulus $k = 1/\sqrt{2}$ [78]. Further, numerical simulations of the multidimensional sG equations [79–82] demonstrate that the collapsing 2D radial kink is quickly destroyed after a few reflections from the center, while the 3D kink reflects in an almost unscathed form (and without polarity inversion). Thus, the formal periodic regime of radial oscillations predicted by Eq. (56) is irrelevant in 2D, but makes sense as an approximation in 3D.

Separately from the simulations of the evolution of the radial kinks, an intriguing numerical result was reported by Geicke [83]: in a certain interval of energies, the 2D sG equation gives rise to an axially symmetric "pulson", which performs extremely robust periodic oscillations, losing virtually no energy in the form of radiation after thousands of oscillation periods. This observation challenges the apparent nonintegrability of the 2D sG equation. On the other hand, the 3D spherically symmetric pulson gradually decays into radiation, as might be naturally expected [83].

5.4 The Frenkel–Kontorova Model

The perturbation theory can also be applied to dynamics of discrete solitons in the FK model. In particular, its variety including the friction and ac (time-periodic) driving force is of obvious significance to physical applications:

$$\frac{d^2\phi_n}{dt^2} - C\left(\phi_{n+1} + \phi_{n-1} - 2\phi_n\right) + \sin\phi_n = -\alpha\frac{d\phi_n}{dt} - \Gamma\cos\left(\omega t\right), \tag{57}$$

cf. the continuous counterpart (40). In this model, the interplay of the underlying discreteness suggests that the ac drive gives rise to resonant velocities, $c = (m/M)(\omega/2\pi)$, at *super-* and *subharmonic* resonance orders, $m = \pm 1, \pm 2, \ldots$ and $M = \pm 1, \pm 2, \ldots$, at which the progressive motion of a discrete kink or antikink (in other words, dislocation in the FK lattice [5]) may be expected [84]. The resonance actually takes place between the periodic passage of lattice sites by the moving discrete kink and the periodic action of the ac driving force, which makes it possible to balance the friction force, even though the drive has no dc component. The stability of such ac-driven regimes was confirmed by direct simulations [85]. This regime was experimentally demonstrated for fluxons in a long circular ac-driven JJ with a periodic lattice of local defects [86].

Another physically relevant effect in the FK model is the resonance between a moving discrete kink and linear lattice waves ("phonons"), which are taken as $\phi_n \sim \exp(ipn - i\chi t)$. This consideration is most relevant for the circular lattice of integer length L, hence the phonon wavenumbers are selected by relation $pL = 2\pi m$, with arbitrary integer m. Then, setting $\alpha = \Gamma = 0$ in Eq. (57) gives rise to the dispersion relation for the small-amplitude phonon waves, $\chi^2 = 1 + 4C\sin^2(\pi m/L)$, with the respective phase velocity,

$$v \equiv \frac{\omega}{p} = \frac{L}{2\pi m}\sqrt{1 + 4C\sin^2\left(\frac{\pi m}{L}\right)}. \tag{58}$$

The resonance takes place when the velocity of a discrete fluxon coincides with one of values (58).

This qualitative prediction was implemented in detailed simulations of Eq. (57) with the ac driving term replaced by the dc bias-current density, γ, as shown in Fig. 7 [87]. Because, as mentioned above, ϕ_t is proportional to the voltage across the junction, the fluxon's velocity determines the voltage, hence the corresponding experiment can be performed (as usual [25, 32]) by measuring the current-voltage characteristic of the system, i.e., the dependence between the current density, γ, and the voltage. A remarkable manifestation of the resonance, clearly seen in Fig. 7, is a system of hysteresis loops in the current-voltage dependence. To produce this picture, γ was swept many times up and down. These numerical results were confirmed by direct experimental measurements [87].

6 *Supersolitons*: Collective Excitations in Trapped Chains of Kinks, and the *Jacobi-sine-Gordon* Equation

Long arrays of kinks are solutions which are naturally generated by the sG equation, in its ideal and perturbed forms alike [see (35) for the solution which represents the high-density chain]. If such an array is pinned by an underlying periodic lattice, it represents an effective medium, similar to chains of interacting particles (such as

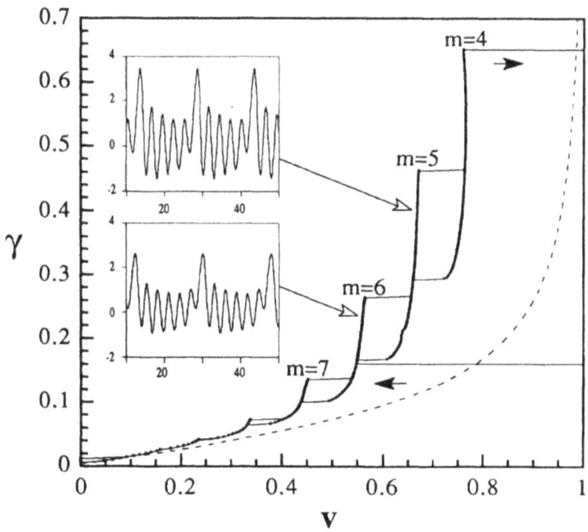

Fig. 7 The numerically simulated current-voltage (I-V) characteristic for a circular array built of 10 coupled Josephson junctions, which corresponds to Eq. (57) with $C = 1$, $\alpha = 0.1$, and dc bias-current density γ [which replaces the ac-driving term in Eq. (57)], as per [87]. The dashed curve shows the I-V characteristic for the continuous counterpart of this system (long Josephson junction). Numbers m correspond to the resonance order in Eq. (58), insets showing the time evolution of the local voltage in particular dynamical regimes. Solid arrows indicate directions of switching in the hysteretic loop

the Newton's cradle), which may support spatially periodic and localized collective excitations propagating along the pinned array. This situation suggests a possibility to study "supersolitons", in the form of broad soliton-like excited states propagating in the underlying array of "primary" solitons. In particular, the collective excitations traveling in an array of fluxons (kinks), i.e., topological solitons, may carry their own topological charge, which makes it possible to identify these excitations as "superfluxons". Such collective modes were observed experimentally in a long JJ with a chain of fluxons pinned to a periodic lattice of local defects [67], and investigated in detail theoretically in the sG model of this a system [88].

The basic sG model of the long JJ corresponding to the setting outlined above is based on the following equation:

$$\phi_{tt} - \phi_{xx} + \sin\phi = -\epsilon \sum_{n=-\infty}^{+\infty} \delta\left(x - an\right)\sin\phi, \tag{59}$$

where the bias current and friction terms are neglected, to focus on the most interesting effects introduced by the periodic lattice. Recall that this lattice gives rise to the periodic potential (48) for the single kink. The unperturbed solution for the array of kinks (corresponding to $\epsilon = 0$) may be taken in the form of Eq. (35). The pinning of the array to the lattice implies commensurability between the array's period (36) and lattice spacing a, which is taken as $a = pL$, with integer p. A deformable array is introduced by assuming that ξ in Eq. (35) is not a constant, but rather a slowly varying function of x and t. After that, using the Hamiltonian of Eq. (59), one can derive an effective Hamiltonian, averaged with respect to fast spatial oscillations of the wave form (35), and, eventually, the evolution equation for $\xi(x,t)$ [88]:

$$\frac{\partial^2 \Xi}{\partial T^2} - \frac{\partial^2 \Xi}{\partial \Xi^2} - 2\mathrm{sgn}(\epsilon)\mathrm{sn}\left(\frac{\Xi}{2}\right)\mathrm{cn}\left(\frac{\Xi}{2}\right)\mathrm{dn}\left(\frac{\Xi}{2}\right), \tag{60}$$

where $\Xi \equiv 2\xi/k$, $T \equiv \sqrt{\epsilon K(k)/aE(k)}k^{-2}x$, $X \equiv \sqrt{\epsilon K(k)/aE(k)}k^{-2}x$, while sn, cn, and dn are the Jacobi's elliptic functions with the same elliptic modulus k as in the underlying solution (35).

Equation (60) may be called the *Jacobi-sine-Gordon equation* [88]. In fact, there are two different equations of this type, corresponding to $\epsilon > 0$ and $\epsilon < 0$. Most plausibly, these are nonintegrable equations. Indeed, the expansion of the equation with $\mathrm{sgn}(\epsilon) = -1$ for small k yields the sG equation with a nonintegrable perturbation which was discussed above, $\Xi_{TT} - \Xi_{XX} + \sin \Xi = -(5/8)k^2 \sin(2\Xi) + \mathcal{O}(k^4)$. Nevertheless, *exact* kink-like solutions of Eq. (60) can be found, in terms of the incomplete elliptic integral of the first kind, F. In comparison with the underlying fluxon array, these solutions may be considered as *superfluxons*. In terms of the original variables, $\xi(x,t)$, the superfluxons take the following form:

$$\xi(x,t) = kF\left(\arcsin\left\{\frac{1}{\sqrt{(1-k^2)\cosh^2\left(\sqrt{\frac{|\epsilon|K(k)}{aE(k)}}\frac{x}{k^2}\right)+k^2}}\right\}\right), \quad \text{at} \quad -\infty < x < 0,$$

$$\xi(x,t) = 2kK(k) - \xi(-x), \quad \text{at} \quad 0 < x < +\infty, \tag{61}$$

for $\epsilon < 0$; and, for $\epsilon > 0$,

$$\xi(x,t) = -kF\left(\arccos\left\{\sqrt{\frac{1-k^2}{\cosh^2\left(\sqrt{(1-k^2)\frac{\epsilon K(k)}{aE(k)}}\frac{x}{k^2}\right)-k^2}}\right\}\right), \quad \text{at} \quad -\infty < x < 0,$$

$$\xi(x,t) = -\xi(-x), \quad \text{at} \quad 0 < x < +\infty. \tag{62}$$

The topological charge of these solutions is determined by the fact that both of them obey the following boundary condition:

$$\xi(x = +\infty) - \xi(x = -\infty) = 2kK(k) \equiv L, \tag{63}$$

cf. Eq. (36). This implies that each superfluxon, (61) and (62), may be realized as a "hole" (missing element) in the underlying chain of fluxons. The superfluxons of the opposite polarities, which are given by the same solutions with opposite signs, correspond to dislocations (surplus fluxons) in the chain.

Because Eq. (60) is Lorentz invariant, as well as the underlying unperturbed sG equation, one can generate moving holes and dislocations from the quiescent ones (61) and (62) by the application of the Lorentz boost (8). The availability of the moving superfluxons suggests a possibility to numerically simulate collisions between them. Such simulations, performed in the framework of Eq. (60) with periodic boundary conditions [67], had demonstrated that, in spite of the apparent nonintegrability the equation, the collisions are virtually completely elastic: the colliding pair would survive 200 collisions without any conspicuous energy loss.

Lastly, if the usual bias-current and friction terms [see Eq. (40)] are added to Eq. (59), it is possible to derive the respective counterparts of the McLaughlin-Scott formula (45) for the equilibrium velocity of the dc-driven motion of the superfluxons. The analysis makes it possible to explain experimental measurements performed for the long chain of fluxons in the long JJ with the built-in periodic lattice of local defects, as reported in [89].

The propagation of the collective excitations in the chain of fluxons with repulsive interactions between adjacent ones resembles the transmission of excitations in the *Newton's cradle*. This observation stimulated search for realizations of Newton's cradles as soliton chains in other models. Thus, the cradles and the creation of supersolitons in them were predicted in a chain of alternating solitons described by two repulsively interacting NLS/Gross-Pitaevskii equations [90], and, most recently, in the NLS equation including a third-order-dispersion (TOD) term [91]. In the latter case, a nonuniform cradle built of solitons with different amplitudes is formed as a result of the TOD-induced fission of a higher-order N-soliton input.

7 Conclusion

This chapter gives a brief overview of the research activity that has been performed for the sG equation, its various modifications, physical models based on such equations, and nonlinear states (chiefly, solitons) produced by these models. While the sG equation per se is exactly integrable, physical models originating from it usually include additional terms which break the integrability, and make it necessary to use the perturbation theory (and direct simulations, of course).

A majority of basic results for solitons in the perturbed versions of the sG equation were obtained relatively long time ago, which makes the overview presented in this chapter somewhat "historical". Nevertheless, those results remain fully relevant for the ongoing theoretical and experimental research work in diverse areas. A number of problems still wait to be solved. Just as an example, it is pertinent to mention the aforementioned surprising stability of oscillating axially symmetric pulsons in the 2D version of the sG equation [83]. It remains to understand what mechanism supports this stability (the absence of any tangible radiation loss) in the apparently nonintegrable equation.

References

1. E. Bour, J. Ecole Imperiale Polytechnique **19**, 1 (1862)
2. A.V. Bäcklund, Math. Ann. **9**, 297 (1876)
3. J. Frenkel, J. Phys. (USSR) **1**, 137 (1939)
4. F.C. Frank, J.H. van der Merwe, Proc. R. Soc. London **198**, 205 (1949)
5. O.M. Braun, Yu.S. Kivshar, *The Frenkel-Kontorova Model: Concepts, Methods, and Applications* (Springer, Berlin, 2004)
6. M.J. Ablowitz, D.J. Kaup, A.C. Newell, H. Segur, Phys. Rev. Lett. **31**, 125 (1973)
7. V.E. Zakharov, L.A. Takhtajan, L.D. Faddeev, Doklady AN SSSR **219**, 1334 (1973)
8. R.K. Dodd, J.C. Eilbeck, J.D. Gibbon, H.C. Morris, *Solitons and Nonlinear Wave Equations* (Academic, London, 1982)
9. J. Rubinstein, J. Math. Phys. **11**, 258 (1970)
10. A.C. Scott, Riv. Nuovo Cim. **27**, 1 (2004)
11. M.J. Ablowitz, H. Segur, *Solitons and the Inverse Scattering Transform* (SIAM, Philadelphia, 1981)
12. S. Duckworth, R.K. Bullough, P.J. Caudrey, J.D. Gibbon, Phys. Lett. A **57**, 19 (1976)
13. P.G. Grinevich, S.P. Novikov, Commun. Pure Appl. Math. **56**, 956 (2003)
14. V.E. Zakharov, S.V. Manakov, S.P. Novikov, L.P. Pitaevskii, *Theory of Solitons: The Inverse Scattering Method* (Nauka, Moscow, 1980) [English translation: Consultants Bureau, New York, 1984]
15. L.D. Faddeev, V.E. Korepin, Phys. Rep. **42**, 1 (1978)
16. W. Thirring, Ann. Phys. **3**, 91 (1958)
17. A.O. Gogolin, A.A. Nersesyan, A.M. Tsvelik, *Bosonization and Strongly Correlated Systems* (Cambridge University Press, Cambridge, 2004)
18. S. Coleman, Phys. Rev. D **11**, 2088 (1975)
19. C. Rogers, W.K. Schief, *Bäcklund and Darboux Transformations: Geometry and Modern Applications in Soliton Theory* (Cambridge University Press, New York, 2002)
20. K. Imai, J. Phys. Soc. Jpn. **68**, 355 (1999)
21. K.W. Chow, Wave Motion **35**, 71 (2002)
22. H. Segur, J. Math. Phys. **24**, 1439 (1983)
23. R. Rajaraman, *Solitons and Instantons* (North Holland, Amsterdam, 1982)
24. B.D. Josephson, Phys. Lett. **1**, 251 (1962)
25. A. Barone, G. Paternò, *Physics and Applications of the Josephson Effect* (Wiley, New York, 1982)
26. P.W. Anderson, J.M. Rowell, Phys. Rev. Lett. **10**, 230 (1963)
27. G.S. Mkrtchyan, V.V. Shmidt, Solid State Commun. **30**, 791 (1979)
28. T.A. Fulton, R.C. Dynes, Solid State Commun. **12**, 57 (1973)
29. A. Matsuda, S. Uehara, Appl. Phys. Lett. **41**, 770 (1982)
30. M. Scheuermann, J.R. Lhota, P.K. Kuo, J.T. Chen, Phys. Rev. Lett. **50**, 74 (1983)
31. Yu.S. Kivshar, B.A. Malomed, Rev. Mod. Phys. **71**, 763 (1989)
32. A.V. Ustinov, Physica D **123**, 315 (1998)
33. A. Davidson, B. Dueholm, B. Krygger, N.F. Pedersen, Phys. Rev. Lett. **55**, 2059 (1985)
34. N. Grønbech-Jensen, B.A. Malomed, M.R. Samuelsen, Phys. Rev. B **46**, 294 (1992)
35. A. Wallraff, A. Lukashenko, J. Lisenfeld, A. Kemp, M.V. Fistul, Y. Koval, A.V. Ustinov, Nature **425**, 155 (2003)
36. M.V. Mineev, G.S. Mkrtchyan, V.V. Shmidt, J. Low Temp. Phys. **45**, 497 (1981)
37. A.F. Volkov, Pis'ma Zhurn. Eksp. Teor. Fiz. **45**, 299 (1987) [JETP Lett. **45**, 376 (1987)]
38. Yu.S. Kivshar, B.A. Malomed, Phys. Rev. B **37**, 9325 (1988)
39. A.V. Ustinov, H. Kohlstedt, M. Cirillo, N.F. Pedersen, G. Hallimanns, G. Heident, Phys. Rev. B **48**, 10614 (1993)
40. S. Sakai, A.V. Ustinov, H. Kohlstedt, A. Petraglia, N.F. Pedersen, Phys. Rev. B **50**, 12905 (1994)

41. S. Savel'ev, V.A. Yampol'skii, A.L. Rakhmanov, F. Nori, Rep. Prog. Phys. **73**, 026501 (2010)
42. R. Kleiner, P. Müller, Phys. Rev. B **49**, 1327 (1994)
43. S.L. McCall, E.L. Hahn, Phys. Rev. **183**, 517 (1969)
44. G.L. Lamb Jr., Rev. Mod. Phys. **43**, 99 (1971)
45. V. Enz, Helv. Phys. Acta **37**, 245 (1964)
46. J.K. Kjems, M. Steiner, Phys. Rev. Lett. **41**, 1137 (1978)
47. H.I. Mikeska, J. Phys. C **11**, L29 (1978)
48. V.G. Bar'yakhtar, B.A. Ivanov, M.V. Chetkin, Usp. Fiz. Nauk **146**, 417 (1985) [Sov. Phys. Uspekhi **28**, 564 (1985)]
49. J. Pouget, G.A. Maugin, Physica D **14**, 88 (1984)
50. V.M. Eleonskii, N.N. Kirova, N.E. Kalugin, Zhurn. Eksp. Teor. Fiz. **75**, 2210 (1978) [Sov. Phys. JETP **48**, 1113 (1978)]
51. P.A. Lee, T.M. Rice, P.W. Anderson, Solid State Commun. **14**, 703 (1974)
52. M.J. Rice, A.R. Bishop, J.A. Krumhansl, S.E. Trullinger, Phys. Rev. Lett. **36**, 432 (1976)
53. A.C. Scott, Am. J. Phys. **37**, 52 (1969)
54. M. Salerno, Physica D **17**, 227 (1985)
55. Y. Ishimori, T. Munakata, J. Phys. Soc. Jpn. **51**, 3367 (1982)
56. M. Peyrard, M.D. Kruskal, Physica D **14**, 88 (1984)
57. L. Lei, C. Shu, J. Shen, P.M. Lam, H. Yun, Phys. Rev. Lett. **49**, 1335 (1982)
58. K. Maki, Phys. Rev. B **11**, 4264 (1975)
59. R.K. Bullough, P.J. Caudrey, in *Nonlinear Evolution Equations Solvable by the Inverse Scattering Transform*, ed. by F. Calogero (Pitman, London, 1978), p. 180
60. J.D. Gibbon, I.N. James, I.M. Moroz, Phys. Scr. **20**, 402 (1979)
61. P. Coullet, P. Huerre, Physica D **23**, 27 (1986)
62. D.J. Kaup, SIAM J. Appl. Math. **31**, 121 (1976)
63. V.I. Karpman, E.M. Maslov, Zhurn. Eksp. Teor. Fiz. **73**, 537 (1977) [Sov. Phys. JETP **46**, 281 (1977)]
64. D.J. Kaup, A.C. Newell, Proc. R. Soc. London A **361**, 413 (1978)
65. D.W. McLaughlin, A.C. Scott, Phys. Rev. A **18**, 1652 (1978)
66. Yu.S. Kivshar, B.A. Malomed, Phys. Lett. A **129**, 443 (1988)
67. B.A. Malomed, V.A. Oboznov, A.V. Ustinov, Zh. Eksp. Teor. Fiz. **97**, 924 (1990) [Sov. Phys. JETP **70**, 518 (1990)]
68. B.A. Malomed, Phys. Rev. B **38**, 9242 (1988)
69. B.A. Malomed, Physica D **27**, 113 (1987)
70. B.A. Malomed, Physica D **15**, 385 (1985)
71. V.I. Karpman, N.A. Ryabova, V.V. Solov'ev, Phys. Lett. A **85**, 251 (1981)
72. V.I. Karpman, E.M. Maslov, V.V. Solov'ev, Zhurn. Eksp. Teor. Fiz. **84**, 289 (1983) [Sov. Phys. JETP **57**, 167 (1983)]
73. I.B. Khalfin, B.A. Malomed, B.Ya. Shapiro, Phys. Rev. B **50**, 9351 (1994)
74. A.C. Newell, J. Math. Phys. **18**, 922 (1977)
75. R.K. Bullough, P.J. Caudrey, H.M. Gibbs, in *Solitons*, ed. by R.K. Bullough, P.J. Caudrey (Springer, Berlin, 1980), p. 107
76. M. Peyrard, D.K. Campbell, Physica D **9**, 32 (1983)
77. D.K. Campbell, M. Peyrard, P. Sodano, Physica D **19**, 165 (1986)
78. O.H. Olsen, M.R. Samuelsen, Phys. Lett. A **77**, 95 (1980)
79. I.L. Bogolyubskii, V.G. Makhan'kov, Pis'ma Zh. Eksp. Teor. Fiz. **25**, 120 (1977) [JETP Lett. **25**, 107 (1977)]
80. P.L. Christiansen, P.S. Lomdahl, Physica D **2**, 482 (1981)
81. J. Geicke, Phys. Lett. A **98**, 147 (1983)
82. J.G. Caputo, M.P. Soerensen, Phys. Rev. E **88**, 022915 (2013)
83. J. Geicke, Phys. Scr. **29**, 431 (1984)
84. L.L. Bonilla, B.A. Malomed, Phys. Rev. B **43**, 11539 (1991)
85. G. Filatrella, B.A. Malomed, J. Phys. Cond. Matt. **11**, 7103 (1999)
86. A.V. Ustinov, B.A. Malomed, Phys. Rev. B **64**, 020302(R) (2001)

87. A.V. Ustinov, M. Cirillo, B.A. Malomed, Phys. Rev. B **47**, 8357 (1993)
88. B.A. Malomed, Phys. Rev. B **41**, 2616 (1990)
89. B.A. Malomed, A.V. Ustinov, J. Appl. Phys. **67**, 3791 (1990)
90. D. Novoa, B.A. Malomed, H. Michinel, V.M. Pérez-García, Phys. Rev. Lett. **101**, 144101 (2008)
91. R. Driben, B.A. Malomed, A.V. Yulin, D.V. Skryabin, Phys. Rev. A **87**, 063808 (2013)

sine-Gordon Equation: From Discrete to Continuum

M. Chirilus-Bruckner, C. Chong, J. Cuevas-Maraver, and P.G. Kevrekidis

Abstract In the present chapter, we consider two prototypical Klein–Gordon models: the integrable sine-Gordon equation and the non-integrable ϕ^4 model. We focus, in particular, on two of their principal solutions, namely the kink-like heteroclinic connections and the time-periodic, exponentially localized in space breather waveforms. Two limits of the discrete variants of these models are contrasted: on the one side, the analytically tractable original continuum limit, and on the opposite end, the highly discrete, so-called anti-continuum limit of vanishing coupling. Numerical computations are used to bridge these two limits, as regards the existence, stability and dynamical properties of the waves. Finally, a recent variant of this theme is presented in the form of $\mathcal{PT}$-symmetric Klein–Gordon field theories and a number of relevant results are touched upon.

Keywords ϕ^4 model • Anti-continuum limit • Breathers • Continuum and discrete models • Kinks • Klein–Gordon lattices • Klein–Gordon PDEs • Nanopteron • PT-symmetry

---

M. Chirilus-Bruckner (✉)
School of Mathematics & Statistics F07, University of Sydney, Sydney, NSW 2006, Australia
e-mail: martinac@maths.usyd.edu.au

C. Chong • P.G. Kevrekidis
Department of Mathematics and Statistics, University of Massachusetts, Amherst, MA 01003-9305, USA
e-mail: chong@math.umass.edu; kevrekid@math.umass.edu

J. Cuevas-Maraver
Nonlinear Physics Group, Departamento de Física Aplicada I, Universidad de Sevilla. Escuela Politécnica Superior. C/ Virgen de África, 7, 41011, Sevilla, Spain

Instituto de Matemáticas de la Universidad de Sevilla (IMUS), Edificio Celestino Mutis. Avda. Reina Mercedes s/n, 41012, Sevilla, Spain
e-mail: jcuevas@us.es

J. Cuevas-Maraver et al. (eds.), *The sine-Gordon Model and its Applications*, Nonlinear Systems and Complexity 10, DOI 10.1007/978-3-319-06722-3_2,

1 Introduction

The sine-Gordon (sG) equation of the form

$$u_{tt} = u_{xx} - \sin(u)\,, \tag{1}$$

with $x, t, u = u(x,t) \in \mathbb{R}$, is a prototypical mathematical model with a wide range of applications [1, 2]; here, the subscripts t, x denote partial derivatives with respect to time and space, respectively.

One such example is the dynamics of magnetic flux propagation in Josephson junction (JJ) transmission lines which are a promising way of transmitting, storing and processing information. The behavior of such elements of magnetic flux (often referred to as fluxons) can be accurately modeled by the propagation of kink-like, heteroclinic solutions of the sine-Gordon equation. In fact, even mechanical analogs based on systems of pendula of such fluxon propagation on a long Josephson junction, including adjustable torques on the pendula (from air jets emulating the current of typical JJ experiments) also exist [3]. The connection of the sG equation and the JJ setting merits its own overview which will be given in another chapter of this volume.

Another intriguing application of such a model lies in the problem of charge density waves (CDWs) [4, 5]. Such CDWs originate from the fact that in several quasi-1d conducting materials, it is favorable under a certain temperature to undergo a phase transition to a state in which the electron density develops small periodic distortions which are followed by a modulation of the ion equilibrium positions. As a result, a CDW condensate is formed which can be subsequently pinned by impurities or interchain coupling. In this setting too, the phase of the fluctuations of the collective wavefunction of the condensate may encounter a periodic potential and kink type solutions may thus be suitable to model the behavior of a CDW condensate.

These are only some among the many examples where the sine-Gordon continuum model has been used as a prototypical system. On the other hand, a discrete variant of the model (DsG)

$$\ddot{u}_n = \epsilon \Delta_2(u_{n-1}, u_n, u_{n+1}) - \sin(u_n)\,, \tag{2}$$

with $u_n = u_n(t) \in \mathbb{R}, n \in \mathbb{Z}$, and where Δ_2 is the central difference operator, is of particular interest in its own right. Here ϵ is the coupling between adjacent sites and the subscript indexes the lattice sites. In that light, the DsG has been originally proposed as a model for the dynamics of dislocations in crystal-lattices, under the form of the celebrated Frenkel–Kontorova model [6]; see also the comprehensive book of [7] and its discussion of some of the historic origins of the model. The prototypical realization of such a discrete system via an array of coupled torsion pendula was originally proposed in the contributions of [8]; see also [9]. This insightful connection has provided a simple playground for a wide variety of research studies that remains very active to this day exploring e.g. the role of external driving and damping (including in the stability of different types of breathing solutions of the model) [10], or that of longer-range interactions and how they modify the nearest neighbor ones [11].

The sine-Gordon equation constitutes one of the integrable nonlinear partial differential equations through the inverse scattering transform [12], which makes it a model of particular interest in mathematical physics. However, it is also relevant to compare/contrast such a model with variants that are *not* integrable, i.e., their exact analytical solution cannot be prescribed on the basis of suitable initial data in the general case. Perhaps one of the most well-known Klein–Gordon examples of this form is the so-called ϕ^4 model [13] of the form:

$$u_{tt} = u_{xx} + 2(u - u^3)\,. \tag{3}$$

This model has been physically argued as being of relevance in describing domain walls in cosmological settings [14], but also structural phase transitions, uniaxial ferroelectrics, or even simple polymeric chains; see, e.g., [15] and references therein. At this continuum limit, one of the particularly intriguing features that were discovered early on was the existence of a fractal structure [14] in the collisions between the fundamental nonlinear waves, once again, a kink and an antikink in this model. Notice here the fundamental contrast of such a fractal structure with the completely inoccuous collisions arising in the sine-Gordon integrable model, whereby the integrability and presence of an infinite number of conservation laws dictates a completely elastic collision outcome (and a mere phase-shifting of the kink or even breather solitons). This topic of the complex collisional outcomes in the context of the ϕ^4 model was initiated by the numerical investigations of [15, 16] and was later studied in [13, 14] and is still under active investigation see, e.g., the careful recent mathematical analysis of the relevant mechanism provided in [17, 18].

It should be added here that a discrete variant of the ϕ^4 model

$$\ddot{u}_n = \epsilon \Delta_2(u_{n-1}, u_n, u_{n+1}) + 2(u_n - u_n^3)\,, \tag{4}$$

is also a model of both mathematical and physical interest. For instance, discrete double well models arise in various physical settings such as electronic excitations in conducting polymers [19], structural phase transitions in ferroic materials [20] i.e., on crystal lattices in ferroelectrics, ferromagnets, and ferroelastics.

Since this chapter considers a variety of equations, we summarize them in Table 1.

Our aim in the present chapter is to present a view of the principal solutions—kinks, breathers and solutions akin to them—of two different Klein–Gordon (KG) equations: the one-dimensional sine-Gordon equation, $f(u) = \sin(u)$, and its non-integrable counterpart, namely the ϕ^4 model, $f(u) = \alpha u + \beta u^3$, from the complementary perspectives of the continuum and the highly-discrete model, the so-called anti-continuum (AC) limit. As alluded to, our goal is to outline how the analytically more tractable continuum version and (AC) limit allow insight into the

Table 1 The (KG) equation in its continuum, discrete and anti-continuum form

(KG)	$u_{tt} = u_{xx} - f(u)\,,\ x, t, u = u(x,t) \in \mathbb{R}$
Discrete (KG)	$\ddot{u}_n = \epsilon \Delta_2(u_{n-1}, u_n, u_{n+1}) - f(u_n)\,,\ u_n = u(t) \in \mathbb{R}, n \in \mathbb{Z}$
(AC) limit	$\ddot{u}_n = -f(u_n)$

We consider the sG variant where $f(u) = \sin(u)$ and the ϕ^4 variant where $f(u) = -2(u - u^3)$

respective discrete version and how the different scenarios compare. In that light, in Section 2, we explore the continuum and discrete kinks, while in Section 3, we focus on the continuum and discrete breathers, in both cases displaying a collection of numerical and analytical results concerning existence and (spectral) stability. Finally, in Section 4, we present a fairly timely alternative variant of the models in the form of $\mathcal{PT}$-symmetric Klein–Gordon field theories. These are models, which in the spirit of the original proposal of C. Bender and his collaborators for Schrödinger models [21, 22] (see also the review [23]) are invariant under the joint action of the symmetries of parity ($\mathcal{P}$) and time reversal ($\mathcal{T}$), potentially bearing in this way a real spectrum even for an "open" (i.e., bearing gain and loss) system.

2 The Kink Case

2.1 *The Continuum Limit and Its Spectral Properties*

We start our considerations from the original continuum limit Eq. (1). This limit possesses standing wave solutions that satisfy the ODE of the form $u_{xx} - \sin(u) = 0$. This ODE has homogeneous steady states which are $u = 0 \bmod(2\pi)$ and unstable ones which are $u = \pi \bmod(2\pi)$. The heteroclinic connections that connect 0 with 2π are typically referred to as kinks and their explicit functional form is given by

$$u(x) = 4\arctan(e^{x - x_0}). \tag{5}$$

The presence of an undetermined constant x_0 reveals the effective *translational invariance* of the model, according to which the kink can be centered equivalently at any point x_0 along the one-dimensional line. In accordance with the well established theory of Noether, this invariance (with respect to translations) is tantamount to the existence of a conservation law for the field theory (which in this case is the conservation of the linear momentum $P = -\int u_t u_x dx$). Another important conservation law among the infinitely many of the sG equation is the conservation of energy (which corresponds to invariance under time shifts). In this context, the energy or Hamiltonian

$$H = \int \frac{1}{2}u_t^2 + \frac{1}{2}u_x^2 + (1 - \cos(u))\, dx, \tag{6}$$

is a constant of motion. The D'Alembertian structure of the underlying linear differential operator leads to invariance under the Lorentz transformations of the form:

$$x' = \gamma(x - vt), \qquad t' = \gamma(t - vx), \tag{7}$$

where $\gamma = 1/\sqrt{1-v^2}$. The equivalence of the wave operator in the original variables (x, t) vs. the new variables (x', t') establishes that the transformation of any of the above standing kinks centered at x_0 enables the formation of a *traveling variant* of the same solution

$$u(x,t) = 4\arctan(e^{\gamma(x-x_0-vt)}), \tag{8}$$

which propagates through the one dimensional line with speed v. Given the equivalence of this traveling kink of the continuum problem [upon the transformation (7)] with the static kink, we will focus predominantly on the latter in what follows.

One can linearize around static solutions of the sG to determine the fate of small perturbations to such solutions. This can be achieved using the ansatz $u(x,t) = u_0(x) + \epsilon w(x,t)$, whereby w satisfies the linear PDE $w_{tt} = w_{xx} - \cos(u_0(x))w$, which can be solved via separation of variables $w(x,t) = e^{\lambda t} v(x)$, which leads to the eigenvalue problem

$$\lambda^2 v = \left[\partial_x^2 - \cos(u_0)\right] v. \tag{9}$$

This linearized equation yields the above professed stability of the $u_0 = 0$ and instability of the $u_0 = \pi$ homogeneous states, as the former leads to $\lambda = \pm i\sqrt{1+k^2}$ where k is the wavenumber of a plane wave $v(x) = e^{ikx}$, while the latter leads to $\lambda = \pm\sqrt{1-k^2}$. It is clear that the former state will bear a continuous spectrum along the imaginary axis of the spectral plane (λ_r, λ_i) of the eigenvalues $\lambda = \lambda_r + i\lambda_i$, while the latter will feature a band of unstable eigenvalues arising for wavenumbers $|k| < 1$.

Interestingly, the spectrum is explicitly available even in the case of the kink, as it leads to a Schrödinger problem with a (special case of a) Rosen-Morse type potential which is well-known from quantum mechanics. The spectrum features a pair of eigenmodes at $\lambda = 0$, which are directly associated with the translational invariance, since the corresponding eigenfunction is $v = u_{0,x} = (1/\sqrt{2})\text{sech}(x - x_0)$ and arises directly from the differentiation of the steady state equation $u_{0,xx} - \sin(u_0) = 0$ (the derivative operator being the generator of the translation group). This mode is the only localized mode of the point spectrum around the kink. The remainder of the spectrum consists of purely continuous spectrum of non-decaying eigenfunctions whose explicit form is given by:

$$v^{(k)} = e^{ik(x-x_0)} \frac{k + i\tanh(x - x_0)}{\sqrt{2\pi}(k+i)}. \tag{10}$$

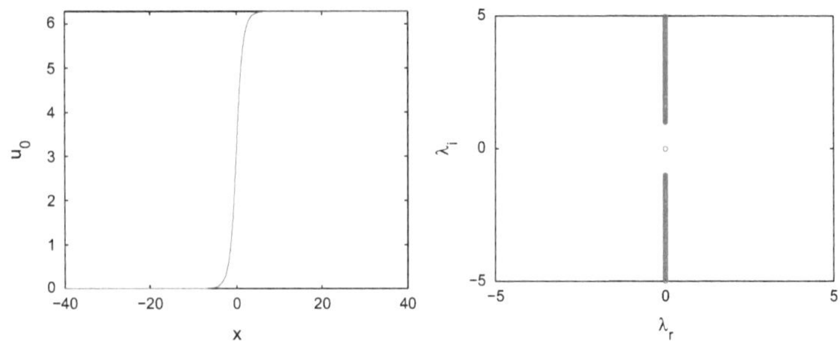

Fig. 1 The *left panel* shows the profile u_0 of the kink of the sG model as a function of x. The *right panel* shows the spectral plane (λ_r, λ_i) of the linearization eigenvalues $\lambda = \lambda_r + i\lambda_i$; see the detailed discussion in the text

The continuous spectrum of the linearization around the kink shares exactly the same eigenvalues $\lambda = \pm i\sqrt{1+k^2}$ as the ones of the spectrum around 0 or 2π i.e., the two homogeneous states that the kink heteroclinic orbit connects. A typical example of the kink of the continuum sG equation and of its corresponding spectrum is shown in Fig. 1.

In a sense, one may think of the spectrum of the kinks as consisting of two separate "ingredients". On the one hand, there is the point spectrum (which consists purely of the neutral mode at $\lambda = 0$) which pertains to the coherent structure itself. On the other hand, there is the continuous spectrum which corresponds to the background on which the waveform "lives". The background for the kink is the 0 state on the left and the 2π on the right. For an anti-kink, where the opposite sign exists in the exponent of the right hand side of Eq. (5) (and which connects 2π with 0), this is reversed. We note that the parsing of the spectrum into localized and extended segments is of particular relevance when considering e.g. the effect of collisions between kinks and anti-kinks and how these reflect the integrability of the dynamics through the well-known notion of elasticity of soliton collisions [24].

Results for kinks and their spectrum in the case of the ϕ^4 model are similar to the sG case. In particular, in the case of Eq. (3), the kink assumes the well-known form of the heteroclinic solution to the Duffing oscillator

$$u_0(x) = \tanh(x - x_0). \tag{11}$$

Once again, the invariance with respect to translations, and associated conservation law are present (a general feature of Klein–Gordon field theories of the form $u_{tt} = u_{xx} - V'(u)$ for arbitrary field-dependent potentials $V(u)$). In this case as well, the linearized problem can be written in the form of an eigenvalue problem,

$$\lambda^2 v = \left[\partial_x^2 + 2(1 - 3u_0^2)\right] v. \tag{12}$$

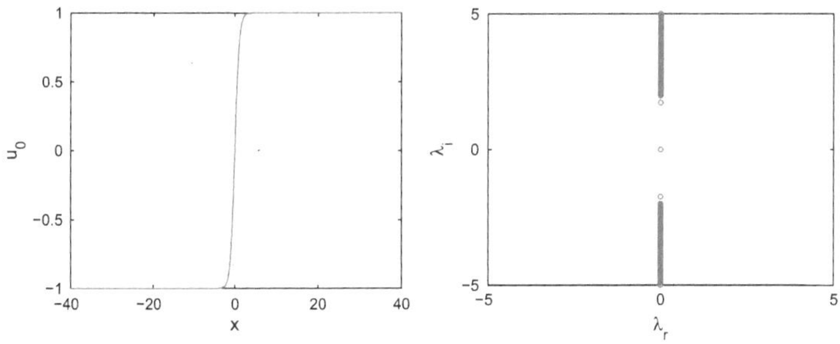

Fig. 2 Same as Fig. 1, but now for the kink of the ϕ^4 model. Notice on the *right panel*'s spectral plane the additional presence of the point spectrum eigenvalue at $\lambda = \pm i\sqrt{3}$

Here the homogeneous steady states are ± 1 and 0, with the first two being stable with $\lambda = \pm i\sqrt{4+k^2}$, while the latter is unstable with $\lambda = \pm\sqrt{2-k^2}$.

The spectrum of the kink is also known in this case, although it is more complex than that of the sG problem. In particular, the neutral mode with $\lambda = 0$ has an eigenfunction $v(x) = \sqrt{3/4}\text{sech}^2(x-x_0)$. The continuous spectrum pertains to the band $\lambda = \pm i\sqrt{4+k^2}$ (again, similarly to the underlying homogeneous problem) with the eigenfunctions

$$v^{(k)} = N_k e^{ikx}\left(3\tanh^2(x-x_0) - 1 - k^2 - 3ik\tanh(x-x_0)\right), \tag{13}$$

where $N_k^{-2} = 4\pi(2(k^2/2+1)^2 + k^2/2)$ [25]. However, the key difference of the spectrum of the ϕ^4 kink from that of the sG is the presence of a point spectrum mode, corresponding to a localized eigenfunction, in the *gap* between the origin of the spectral plane and the continuous spectrum (see Fig. 2). This internal (often referred to also as shape) mode of the kink has an eigenvalue $\lambda = \pm\sqrt{3}i$ and a corresponding eigenfunction,

$$v(x) = \sqrt{\frac{3}{2}}\tanh(x-x_0)\text{sech}(x-x_0). \tag{14}$$

It is this localized "shape" mode that has been argued to be typically responsible for the very rich phenomenology of collisions of ϕ^4 kinks and anti-kinks as explored e.g. in [14–18], among other works.

2.2 *Anti-Continuum Limit and Its Spectral Properties*

We now turn to the opposite end of the continuum limit to construct a complementary picture of the kink existence and stability properties, so that we can connect the two in the next subsection. In particular, we focus now on Eqs. (2) and (4), which correspond to the discrete models. Here, one way to interpret ϵ (e.g. in applications such as coupled torsion pendula) is that of the spring constant linearly connecting adjacent such pendula. A way of interpretation closer to numerical analysis is that of adjacent nodes of a lattice, separated by a distance Δx, such that $\epsilon = 1/\Delta x^2$. This represents then a finite difference scheme for the spatial discretization of the corresponding continuum models of Eqs. (1) and (3) and as such the continuum limit is approached when $\Delta x \to 0 \Rightarrow \epsilon \to \infty$.

However, what was pioneered by MacKay and Aubry about 20 years ago [26], was a technique that offered a completely alternative viewpoint to the one approaching the continuum as above. In particular, they advocated using the *opposite* limit, namely that of uncoupled oscillators at $\epsilon = 0$. This was dubbed the anti-continuum (AC) limit. There, our system consists of isolated oscillators of the form $\ddot{u}_n = -\sin(u_n)$ for the sG and $\ddot{u}_n = 2(u_n - u_n^3)$ for ϕ^4. The steady states now are *solely* 0 and π (mod(2π) as usual) and ± 1 and 0 for the two models, respectively. Hence one can think of ways to "initiate" a kink at this limit, which if ϵ starts becoming finite, upon continuation (using a bifurcation theory analogy), the kink will start gradually picking more sites progressively of more nontrivial ordinate. In this way, the kink will be progressively "fleshed out" (i.e., the spine of the heteroclinic connection will be populated) and will start looking more like its continuum analog, which will be eventually traced as $\epsilon \to \infty$. This kind of approach was used e.g. in [27] to construct a variety of kink and multi-kink solutions. Here, we will only explore the example of states leading (as ϵ is increased) to the single kink, and persisting from the AC to the continuum limit.

Upon some toying with the relevant background states, one realizes that there are two prototypical possibilities for creating such a kink. One of them concerns the so-called *intersite-centered* configuration, initiated at the AC limit through a sequence $(0, \ldots, 0, 0, 2\pi, 2\pi, \ldots, 2\pi)$ in the sG and similarly $(-1, \ldots, -1, -1, 1, 1, \ldots, 1)$ in the ϕ^4. On the other hand, there is the *onsite-centered* configuration, which incorporates a point of the unstable steady state in the middle, namely $(0, \ldots, 0, 0, \pi, 2\pi, 2\pi, \ldots, 2\pi)$ in the sG and similarly $(-1, \ldots, -1, -1, 0, 1, 1, \ldots, 1)$ in the ϕ^4.

If we now examine such configurations in the AC limit, it will be straightforward to infer their stability since the linearization operator under the discretization becomes an infinite dimensional matrix operator (or from a computational perspective it is restricted to a finite but large number of nodes domain). In this setting the linearization Jacobian matrix will be tri-diagonal with the following diagonal and super/sub-diagonal elements:

$$J_{i,i} = -2\epsilon - V'(u_i), \qquad J_{i,i+1} = J_{i,i-1} = \epsilon. \tag{15}$$

Here $V'(u_i) = -\cos((u_0)_i)$ for the sG and $V'(u_i) = 2(1 - 3(u_0)_i^2)$ for the ϕ^4. In the AC limit ($\epsilon = 0$) the matrix becomes diagonal and it is evident that the eigenvalues λ^2 will be negative if the kink involves only the stable steady states $(0, 2\pi$ or $\pm 1)$ and will have as many unstable eigenvalue pairs as many sites there are at the unstable fixed points π for the sG and 0 for the ϕ^4.

Now, the key question becomes what can we expect in terms of the kink stability as we depart from the AC limit, i.e., for finite ϵ. To determine this, we resort to a well-known theorem for matrix eigenvalue problems, namely Gerschgorin's theorem [28]. In particular, if the order of the Jacobian matrix is N, the eigenvalues λ^2 thereof belong to the sets

$$Z_i = \{z \in C, \quad |z - J_{i,i}| < r_i, \quad \text{with} \quad r_i = \sum_{j=1, j \neq i}^{N} |J_{i,j}|\}. \tag{16}$$

Applying this to our setting, we find that $|\lambda^2 + 2\epsilon + V'(u_i)| \leq 2\epsilon$, which in turn implies

$$- V'(u_i) - 4\epsilon \leq \lambda^2 \leq -V'(u_i). \tag{17}$$

In the case of the stable fixed points (at $0, 2\pi$ or at ± 1, respectively for the 2 models), this inequality properly outlines the edges of the continuous spectrum. Hence, this indicates stability for $N \to \infty$, at least for small ϵ of the inter-site centered kink.

However a critical point is that it also establishes the instability of the site-centered kink that *always* bears one site centered at π or 0 in the sG or, respectively, the ϕ^4 case. For this eigenvalue, we have $1 - 4\epsilon \leq \lambda^2 \leq 1$ for sG, while $2 - 4\epsilon \leq \lambda^2 \leq 2$. Hence, this approach *guarantees* the instability of the onsite-centered kink for $\epsilon \leq 1/4$ for the sG problem and for $\epsilon \leq 1/2$ in the ϕ^4 case.

We now turn to the merging of the two pictures in the subsection that follows.

2.3 Merging the Two Pictures

In the intermediate regime between the continuum limit of $\Delta x \to 0$ and $\epsilon \to \infty$ and the AC limit of $\Delta x \to \infty$ and $\epsilon \to 0$, we anticipate the following. For the site centered kink, we expect that the unstable eigenvalue, while bounded away from 0 along the real line, according to our above estimate, will start approaching the origin of the spectral plane as the continuum limit is approached. While it is not guaranteed from our considerations herein that the eigenvalue will have a monotonic behavior, we find that indeed it appears to, arriving at the origin as the asymptotic limit of the continuum regime is reached. The approach of this eigenvalue towards the spectral plane origin is shown in the left panel of Fig. 3.

On the other hand, in a similar manner, for the stable inter-site centered kink, the approach towards the continuum limit needs to involve an eigenvalue pair that approaches the origin as $\Delta x \to 0$ and the translational invariance is restored. Since in this case all eigenvalues initially start at $\pm i$ (at $\epsilon \to 0$) what occurs is that an eigenvalue pair bifurcates from the continuous spectrum and starts approaching the origin of the spectral plane. In this case too, although not necessarily so, the

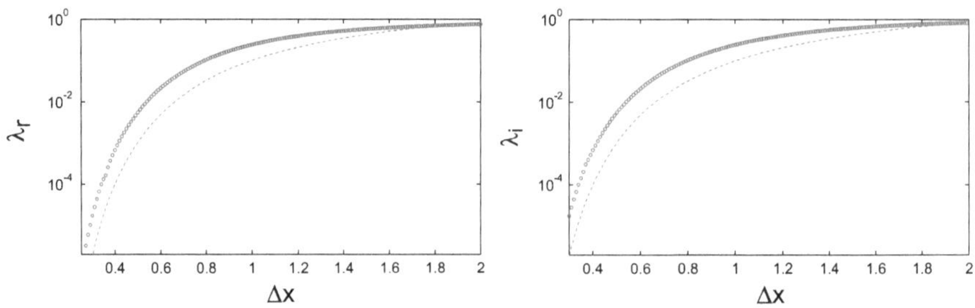

Fig. 3 The *left panel* shows the real eigenvalue pertaining to the breaking of TI for the onsite centered kink. The *blue circles* denote the numerical computation of this eigenvalue and the *red dashed line* provides the corresponding theoretical prediction of Eq. (19). The *right panel* shows the imaginary eigenvalue similarly associated to the breaking of TI for the stable intersite centered kink (the *circles/lines* have the same designation as for the *left panel*)

approach to the origin is monotonic (and asymptotic) as Δx is decreased. This case is shown in the right panel of Fig. 3.

The eigenvalue associated to instability (resp. stability) of the site (resp. intersite) centered kinks is related to the translational invariance (TI) in the continuum limit. It is the breaking of the translational invariance once the system becomes discrete that is "responsible" for the instability/stability of these two different kinks. The difference in the energies associated to these two kinks is defined as the celebrated Peierls–Nabarro (PN) barrier, which is discussed at considerable length in the physical literature [7] as the barrier needed to be overcome for a dislocation to move by a lattice site.

One of the typical issues that arise in the calculation of the PN barrier is that it is not straightforward to perturb off of the continuum limit in the direction of discreteness. This problem was bypassed in [29] by perturbing off of an exceptional (translationally invariant) discretization of the sG model originally proposed by Speight and Ward in [30]. This was in the form:

$$\ddot{u}_n = 2\epsilon \left(\sin(\frac{u_{n+1} - u_n}{2}) - \sin(\frac{u_n - u_{n-1}}{2}) \right) - \frac{1}{2} \left(\sin(\frac{u_{n+1} + u_n}{2}) + \sin(\frac{u_n + u_{n-1}}{2}) \right). \tag{18}$$

The fact that this model is TI implies that it possesses kinks which can be located arbitrarily, as is in fact evidenced by the exact solutions $u_n = 4\arctan(e^{a(n-\xi)})$ where $a = \log((2 + \Delta x)/(2 - \Delta x))$ and ξ is a free parameter. It also suggests that the relevant TI eigenvalue pair is at $\lambda^2 = 0$, yet because of its inherent discrete nature, the model is amenable to being perturbed to be reshaped in the form of Eq. (3). Following this path, [29] predicted that the relevant eigenvalue should be given by

$$\lambda_r \approx \pm\frac{13.96}{\Delta x^{1/2}}e^{-\frac{\pi^2}{\Delta x}}, \qquad \lambda_i \approx \pm i\frac{13.96}{\Delta x^{1/2}}e^{-\frac{\pi^2}{\Delta x}}. \tag{19}$$

The first of these expressions applies to the real eigenvalue of the onsite-centered kink, while the second to the imaginary one of the intersite-centered kink. Both are applicable when approaching the continuum limit. Yet, it was recognized in [29] that while the functional form of the dependence should be suitable, the prefactor may sustain additional contributions from higher order terms which should appear in the same order of eigenvalue dependence. It is for this reason that the dashed lines describing this theoretical dependence of Eq. (19) capture the appropriate functional dependence as the limit is approached, but clearly the prefactor is smaller than it should be.

Entirely similar features also arise for the ϕ^4 case, which is thus not examined further here. Instead, we now turn to the corresponding examination of the breather waveform.

3 The Breather Case

The other fundamental solution of the sG model is a state which is also exponentially localized in space (similarly to the kink) but periodically varying in time. This is the so-called breather state whose exact profile is given by

$$u(x,t) = 4\arctan\left(\frac{\sqrt{1-\omega^2}}{\omega}\sin(\omega(t-t_0))\mathrm{sech}(\sqrt{1-\omega^2}(x-x_0))\right). \tag{20}$$

Aside from translations with respect to space and time (discussed in the previous section), this solution has a free parameter ω, associated with the frequency of its "breathing" satisfying $0 < \omega < 1$. Interestingly, the breather seems like the result of the merger of a kink and an antikink. However, the energy of the kink and antikink is 8 such that the sum of their energies is $E = 16$, whereas the energy of their bound state, i.e. the breather, is always less than that i.e., $E = 16\sqrt{1-\omega^2}$.

3.1 Continuum Limit: Genuine Breathers and Modulating Pulses

For the sG equation, which is an infinite dimensional completely integrable Hamiltonian system, breather solutions can be obtained through its auto-Bäcklund transformation (BT). This procedure is, of course, limited to integrable systems and does not help in the quest for breathers in general (KG) equations

$$u_{tt} = u_{xx} - u + f(u), \tag{21}$$

with $x, t, u = u(x,t) \in \mathbb{R}$, and some nonlinearity $f(u)$. To this end, various techniques have been employed and developed during the past decades to elucidate the existence of breathers or solutions akin to them. Roughly speaking, one can distinguish between three approaches. The first approach is the analysis via series expansion techniques (see [31, 32]) and the study of perturbed sG equations (see [33–35]), whose findings indicated the non-persistence of breather families and the continuation of breathers as solutions which lack the spatial localization property. Out of these results emerged the more general second approach of constructing for (21) so-called *modulating pulses*, which are moving breathers that are not necessarily spatially localized, but feature oscillating tails; see e.g. [36, 37]. Yet a different, third approach exploits the wave packet structure of breathers. In their small-amplitude limit they can be thought of as wave packets whose envelope is described by a Nonlinear Schrödinger (NLS) equation (see [38] for a rigorous approximation result). While perturbation arguments in the first approach rely heavily on integrable systems techniques and/or the explicit solution formula known for breathers, the second approach follows the realm of infinite dimensional dynamical systems and inspires an extension of center manifold theory beyond the classical case. The methodology of the third approach belongs to the theory of reduction to amplitude/modulation equations and plays an important role far beyond the setting discussed here.

Before we review each approach in more detail, let us briefly summarize the sG-breather construction. As alluded to, the sG equation admits an auto-Bäcklund transformation, which is, loosely speaking, a means of obtaining solutions to the sG equation by a procedure that is in some sense easier than solving a PDE (i.e. it might involve solving ODEs or algebraic equations). To be more precise, the BT maps "solutions to solutions", i.e., it is a mapping B with

$$B(u_1; a) = u_2\,, \qquad u_1, u_2 \text{ solutions of (sG)}$$

where a is some parameter. One can, for instance, use the trivial solution as a seed to grow non-trivial solutions like kinks or anti-kinks

$$B(0,1) = 4\arctan(e^x)\,, \qquad B(0,-1) = 4\arctan(e^{-x})\,. \tag{22}$$

It is even possible to derive a surprisingly simple relation between four different solutions u, U, u_1, u_2 of the (sG) equation by using the commutativity theorem schematically depicted in Fig. 4 (cf., for instance, [39]).

Using this scheme, one can derive a formula for (a family of) standing breathers (see Fig. 5)

$$u_*(x,t;\omega) = 4\arctan\left(\frac{m}{\omega}\,\frac{\sin(\omega t)}{\cosh(mx)}\right)\,, \quad m^2 + \omega^2 = 1\,, \tag{23}$$

with $u = 0, u_{1/2} = \arctan\left(e^{mx \pm i\omega t}\right)$ and $U = u_*$. Note that choosing $\omega = i\tilde{\omega} \in i\mathbb{R}$ turns u_* into

$$\tan\left(\frac{u+U}{4}\right) = \left(\frac{a_1+a_2}{a_1-a_2}\right)\tan\left(\frac{u_1+u_2}{4}\right)$$

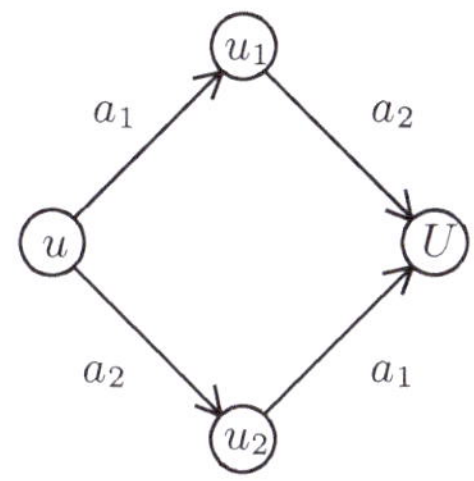

Fig. 4 Scheme of commutativity of the BT: $B(B(u, a_1), a_2) = B(B(u, a_2), a_1)$

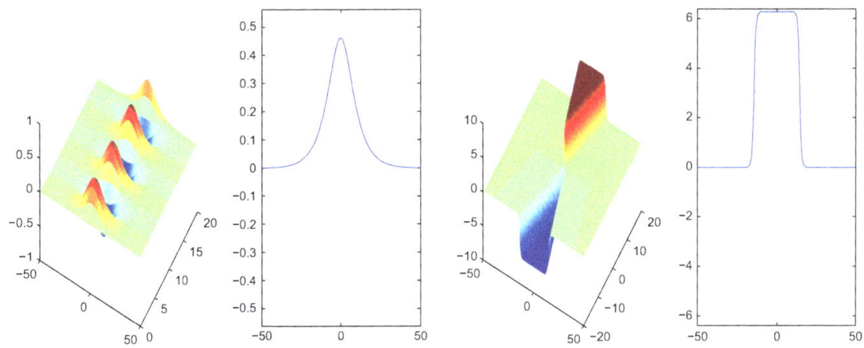

Fig. 5 *Top panel*: Standing breather with $\omega = 0.99$ in (23). *Bottom panel*: Kink/anti-kink interaction with $\omega = i0.99$ in (24)

$$u_*(x, t; i\tilde{\omega}) = 4 \arctan\left(\frac{m}{i\tilde{\omega}}\frac{\sin(i\tilde{\omega}t)}{\cosh(mx)}\right) = 4 \arctan\left(\frac{m}{\tilde{\omega}}\frac{\sinh(\tilde{\omega}t)}{\cosh(mx)}\right), \qquad (24)$$

which describes a kink/anti-kink solution (see Fig. 5). Iterating this procedure results in more complicated solutions given by complexes of kinks and/or breathers.

As alluded to, one can derive kink solutions by a completely different viewpoint, namely, by making a traveling wave ansatz $u(x, t) = y(x - ct)$, $0 \leq c < 1$, and examining the traveling wave ODE whose heteroclinics are given by $y(\xi) = 4\arctan(e^{\pm\xi})$, $\xi = \gamma(x - ct)$, $\gamma = (1 - c^2)^{-1/2}$. For $c = 0$ we recover (22), while the traveling kink can be viewed also as a consequence of the Lorentz transformation (7), as discussed above. Note, however, that breathers (or, more generally, any multikink solution) cannot be obtained from the traveling wave ODE since they are not stationary in any fixed co-moving frame. Hence, breathers can be viewed as genuinely infinite dimensional structures. While the explicit expressions (23) or (24) rely heavily on PDE tools for integrable systems, the general non-integrable case calls for different PDE techniques. Breather existence could, for instance, be explored via the construction of small-amplitude wave packets, which can be approximately performed for a large class of Klein–Gordon equations (21) via reduction to amplitude/modulation equations. Despite their widespread use, rigorous justifications of such procedures are more sparse, with an early example being [38], where it was demonstrated that there are solutions u to the (sG) equation for which

$$\sup_{t\in\left[0,\frac{T_0}{\varepsilon^2}\right]} ||u(\cdot,t) - u_{\mathrm{a}}(\cdot,t)||_{C_b^r} \leq C\varepsilon^{3/2},$$

with $0 < \varepsilon \ll 1$ being a small perturbation parameter and where u_{a} is a small-amplitude wave packet

$$u_a(x,t) = \varepsilon\, A(\varepsilon(x - \omega'(k_0)t), \varepsilon^2 t)\, e^{ik_0 x - i\omega(k_0)t} + c.c. + h.o.t., \tag{25}$$

with $\omega(k) = \sqrt{k^2+1}$; the envelope $A = A(X,T) \in \mathbb{C}$ is obtained by solving the Nonlinear Schrödinger (NLS) equation

$$2i\omega(k_0)\partial_T A = (\omega'(k_0)^2 - 1)\partial_X^2 A - 3|A|^2 A\,. \tag{26}$$

This result is also valid for general KG equations of the form given in (21) since small-amplitude solutions do not feel to leading order the exact structure of the nonlinearity. Note that, if we choose the NLS soliton as an envelope, the wave packet ansatz u_a resembles the traveling sG breather in the small-amplitude limit $0 < m = \varepsilon \ll 1$. Note that, although the approximation result treats a large class of solutions—namely, those whose envelopes are given by solutions of the NLS—of which the breather is just a special case, the limited time-validity of the construction (namely, $t \in \left[0, \frac{T_0}{\varepsilon^2}\right]$) does not give the optimal result (that is, global existence in time) in the breather case. Such an improvement of the perturbation procedure has so far not been reported in the literature and would be a particularly interesting direction for future work.

The analogy between breathers and general wave packets becomes even more remarkable when one considers interaction. Using the BT, one can construct, for instance, a 2-breather describing the elastic collision of two individual breathers. Such an interaction for small amplitude wave packets in the general KG setting (21) has striking similarities to the breather interaction: to leading order, the shape of the wave packets remains unchanged after collision, the only interaction effects being a shift of envelope and carrier waves (see [40] for KG with constant coefficients and [41] for KG with spatially periodic coefficients). In the case of large amplitudes, however, the situation is far more complex, as is attested also by the complexity of interactions of simpler structures such as kinks in non-integrable KG models [13–18]. Since a large class of KG equations support (approximate) small-amplitude wave packets which seem akin to the sG breather, it had been long speculated that exact breathers may exist for other KG equations as well. This belief has been proven wrong in numerous case examples from several different viewpoints.

The work of [31] contrasted the sG setting with the ϕ^4 model. By employing the Fourier representation

$$u(x,t) = \sum_{n\in\mathbb{Z}} \hat{u}_n(x)\, e^{in\omega_* t}, \tag{27}$$

for time-periodic solutions with some temporal frequency $\omega_*/2\pi$, they derive an infinite-dimensional system of equations for the Fourier coefficients $\hat{u}_n$ and conclude that only the sG equation allows spatially localized solutions $\hat{u}_n(x) \to 0, x \to \pm\infty$, which in turn give rise to breathers, while the ϕ^4 model only supports approximate breathers that slowly radiate their energy to $\pm\infty$.

The work of [32] considered a more general KG setting with a force term $f(u)$ in (21) and presented a detailed convergence analysis of different breather series expansions relating the corresponding expansion coefficients to the coefficients f_n in $f(u) = \sum_{n=1}^{\infty} f_n\, u^n$. The main result singles out the cases $f(u) = Cu, C\sinh(au), C\sin(au)$ which are the only choices yielding analyticity of the solution, while (again) only $f(u) = C\sin(au)$ gives breathers. In the early 1990s, the works of [33] and [34, 35] studied perturbed sG equations

$$u_{tt} - u_{xx} + \sin(u) = \varepsilon P(u;\varepsilon), \quad 0 < \varepsilon \ll 1, \tag{28}$$

and concluded that breathers can persist (in families) if and only if the perturbation is equivalent to a rescaling of the sG equation. Consequently, the non-persistence results indicated that breathers perturb into solutions that are similar in nature, but not localized. This intuition was confirmed in [36] (and later also in [37]) by using spatial dynamics and a blend of invariant manifold, partial normal form and bifurcation theory for KG equations (21). They study so-called modulating pulse solutions, that is,

$$u(x,t) = v(x - (1/c_{\text{ph}})t; x - \left(c_{\text{ph}} + \gamma_1\varepsilon^2\right)t) = v(\xi, y),$$

where v is periodic in y with period $2\pi/k_0$ for some $k_0 > 0$. Note that the sG-breathers fall into this class with the additional property of being localized in ξ, a feature that seems rather exceptional in the following way. The infinite dimensional system for (v, v_x) on the phase space $\chi = \{(v,w) \in H^{s+1}_{\text{per}} \times H^s_{\text{per}}\}, s \geq 0$, (where now x is the dynamic variable, hence, the name spatial dynamics and H^s_{per} are L^2-based Sobolev spaces of periodic functions) has a Hamiltonian structure whose linearization features infinitely many eigenvalues on the imaginary axis and exactly two real eigenvalues $\lambda_{1,2} = \pm\varepsilon\rho$. The leading order analysis of the full nonlinear system restricted to the subspace associated with $\lambda_{1,2}$ reveals a homoclinic which is, however, unlikely to persist for the full system. So, in general, one expects that resonances related to the infinitely many eigenvalues on the imaginary axis generate oscillatory tails, hence, preventing the existence of truly localized breathers for the full PDE. The special structure of the sG-nonlinearity seems to remedy this spectral picture, allowing the homoclinic to persist. Nevertheless, other mechanisms to avoid such resonances are possible in KG equations with e.g. periodic coefficients [42].

3.2 Spectral Properties of the Breather: From the Anti-Continuum to Finite Coupling

In a chain of N oscillators described by e.g. Eq. (2) or (4), discrete breathers in the AC limit consist of p (excited) oscillators describing periodic orbits of frequency $\omega \neq \omega_0$ and $N-p$ oscillators at rest. Here, ω_0 is the frequency of the oscillations at the linear limit around the equilibrium position $\tilde{u}$, with $\tilde{u}=0$ in the sine-Gordon and $\tilde{u}=\pm 1$ in the ϕ^4 case. Breathers with $p>1$ are usually dubbed *multibreathers*; in the special case when $p=N$, they constitute an anharmonic phonon and are also known as *phonobreathers*. The MacKay–Aubry theorem [26] establishes that if $n\omega \neq \omega_0$ with n being an integer, a time-reversible breather or multibreather can be continued to a nonzero coupling (i.e. $\epsilon \neq 0$). The time-reversible multibreather can be expressed as a Fourier series:

$$u_n(t) = \sum_{k=0}^{\infty} z_{k,n} \cos(k\omega t). \tag{29}$$

In the case of the sine-Gordon equation, due to the spatial symmetry of the potential, all the even coefficients are zero and the resonance condition mentioned above restricts to only odd multiples of ω.

Among all the localized multibreathers supported by the lattice, there are only two that can be continued up to $\epsilon \to \infty$: (1) the single-site breather, i.e. the breather with $p=1$ (also known as the site-centered breather or Sievers–Takeno mode [43]), and (2) the two-site breather, i.e. multibreather with $p=2$, whose excited sites are adjacent and oscillate in phase, which is also dubbed as the bond-centered breather or Page mode [44].

Very recently, it was proven that non-time-reversible multibreathers cannot exist in Hamiltonian lattices if the number of neighbors of each oscillator is equal to two and the boundary conditions are not periodic [45]. This proof does not forbid, however, the existence of non-time-reversible multibreathers in two-dimensional lattices where discrete vortices or percolating clusters [46–49] have been found in the past. In addition, non-time-reversible multibreathers have also been demonstrated in one-dimensional chains with periodic boundary conditions [50] or long-range interactions [51].

In order to obtain periodic orbits of the lattice equations (2) or (4), a fixed-point method (such as the Newton-Raphson method) can be implemented either in real or in Fourier space. In the real space case (see e.g. [48]), orbits are found as roots of the map $(\{u_n(T)\}, \{\dot{u}_n(T)\}) - (\{u_n(0)\}, \{\dot{u}_n(0)\})$ with $T = 2\pi/\omega$. The Jacobian of this map is $\mathcal{M} - I$ where $\mathcal{M}$ is the monodromy matrix (see e.g. (33)) and I is the identity. Fourier space methods consist of introducing the Galerkin truncation to index k_m of the Fourier series expansion (29) into the lattice equations so that they can be cast as a set of algebraic equations of the form [50]:

$$-k^2 z_{k,n}^2 = \epsilon \Delta_2 z_{k,n} - \mathcal{F}_{k,n}, \tag{30}$$

with $\mathcal{F}_{k,n}$ being the discrete cosine Fourier transform,

$$\mathcal{F}_{k,n} = \frac{1}{N} \sum_{n=-k_m}^{k_m} V' \left(\sum_{p=-k_m}^{k_m} z_{p,n} \cos\left[\frac{2\pi pn}{N}\right] \right) \cos\left[\frac{2\pi kn}{N}\right], \tag{31}$$

with $N = 2k_m + 1$ and $V'(u) = \sin(u)$ for the sine-Gordon model and $V'(u) = u^3 - u$ for the ϕ^4 equation.[1] Fourier space methods have the advantage of providing with an analytical expression for the Jacobian, whereas it must be numerically generated by means of numerical integrators (as Runge-Kutta) in real space methods. On the contrary, the dimension of the Jacobian in Fourier space $[(k_m + 1)N, (k_m + 1)N]$ is larger than in the real space (N, N) which makes the inversion intractable for large lattices.

Spectral stability of periodic orbits can be determined by means of Floquet analysis. This consists firstly of introducing a perturbation ξ_n to a given solution $u_{n,0}$ of the lattice equations (2), (4). Then, the equation for the perturbation reads:

$$\ddot{\xi}_n = \epsilon \Delta_2 \xi_n - V''(u_{n,0})\xi_n, \tag{32}$$

with $V''(u_n) = \cos u_n$ and $V''(u_n) = 3u_n^2 - 1$ for, respectively, the sine-Gordon and ϕ^4 models. The stability properties are given by the spectrum of the Floquet operator $\mathcal{M}$ (whose matrix representation is called *monodromy*) defined as:

$$\begin{pmatrix} \{\xi_n(T)\} \\ \{\dot{\xi}_n(T)\} \end{pmatrix} = \mathcal{M} \begin{pmatrix} \{\xi_n(0)\} \\ \{\dot{\xi}_n(0)\} \end{pmatrix}. \tag{33}$$

The $2N$ monodromy eigenvalues $\Lambda = \exp(\mathrm{i}\theta)$ are dubbed as *Floquet multipliers* and θ are denoted as *Floquet exponents*. Due to the symplectic nature of the Floquet operator for our Hamiltonian systems of interest, if Λ is a multiplier, so is Λ^{-1}, and due to its real character if Λ is a multiplier, so is Λ^*. Consequently, Floquet multipliers always come in complex quadruplets $(\Lambda, 1/\Lambda, \Lambda^*, 1/\Lambda^*)$ if $|\Lambda| \neq 1$ and is not real, or in pairs $(\Lambda, 1/\Lambda)$ if Λ is real, or (Λ, Λ^*) if $|\Lambda| = 1$ and is not real. In addition, due to the time translation invariance of the system there is always a pair of multipliers at $+1$. Therefore, for a breather to be stable, all the multipliers must lie on the unit circle.

In the AC limit, Floquet multipliers lie in three bundles. The two conjugated ones, corresponding to the oscillators at rest, lie at $\theta = \pm 2\pi\omega_0/\omega$, while the third one lies at $+1$ and consists of p multiplier pairs corresponding to the excited oscillators. For $\epsilon \neq 0$, the bundle corresponding to the oscillators at rest split and their multipliers move to form the phonon band, while the multipliers corresponding to the excited

[1] Notice in order to use the same notation as in classical papers as [31] and [52], the factor 2 that was introduced in (4) has been removed. However, the results are equivalent by rescaling $x \to \sqrt{2}x$ and $t \to \sqrt{2}t$.

oscillators can move along, either in the unit circle (stability), or along the real axis (instability). However, as mentioned above, a pair of multipliers remains at $+1$ corresponding to the, so-called, phase and growth modes of the whole system. This implies that single-site breathers are stable for low coupling [53].

Whereas the stability at low coupling of single-site breathers is trivial, the situation for multibreathers is less clear. The theorems developed in [54, 55] provide insight regarding this question. One can derive an expression of the Floquet multipliers corresponding to the excited oscillators of multibreathers close to the AC limit. In the case of Page modes, the relevant Floquet exponent is given by:

$$\theta = \pm\frac{2\pi}{\omega}\sqrt{2\epsilon\frac{J}{\omega}\frac{\partial\omega}{\partial J}}, \tag{34}$$

with

$$J = \frac{1}{2\pi}\int_0^T [\dot{u}(t)]^2 \mathrm{d}t = \frac{\omega}{2}\sum_{k\geq 1} k^2 z_k^2, \tag{35}$$

being the action of an isolated oscillator. Applying a rotating wave approximation (RWA) $u(t) \approx z_0 + z_1\cos(\omega t)$ we obtain the following Fourier coefficients for the sG model [56]:

$$z_0 = 0, \qquad z_1 = 12 - 4\sqrt{3[3 - 4(1-\omega^2)]}, \tag{36}$$

whereas the following coefficients are found for the ϕ^4 model:

$$z_0 = 1 - \frac{3}{5}(2-\omega^2), \qquad z_1 = \frac{2}{5}(2-\omega^2). \tag{37}$$

Using the values from (34), the relevant Floquet exponent at low coupling for the sG model can now be approximated by:

$$\theta = \pm\frac{2\pi}{\omega}i\sqrt{2\epsilon\frac{\omega^2\sqrt{3(4\omega^2-1)} - (2\omega^2-1)(4\omega^2-1)}{16\omega^4 - 7\omega^2 + 1}}, \tag{38}$$

whereas for the ϕ^4 equation we have

$$\theta = \pm\frac{2\pi}{\omega}i\sqrt{2\epsilon\frac{\omega^2-2}{2-3\omega^2}}. \tag{39}$$

Figure 6 compares the Floquet exponents of the Page mode found by integrating the perturbation equation (33) with those of the analytical approximation (34) and the subsequent, fully analytical RWA (38)–(39) for low coupling. Good agreement is found in the case of small values of the coupling constant ϵ.

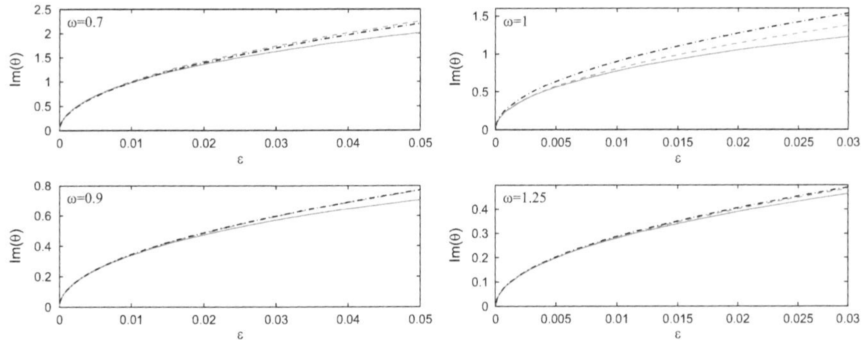

Fig. 6 Floquet exponent for the Page mode in the sine-Gordon (*left*) and ϕ^4 (*right*) models. Full line corresponds to the numerically exact result, whereas the stability theorem predictions are given by the *dashed line*. The explicit analytical predictions of the RWA approach for the stability exponents are depicted by the *dash-dotted line*

3.3 *Continuation in the Coupling: Going Towards the Continuum*

When the coupling constant is increased to finite values of $\epsilon > 0$, stable multibreathers that are close to the AC limit destabilize generally by means of Hamiltonian Hopf bifurcations (also referred to as Krein crunches); in addition, every multibreather ceases to exist for a finite value of the coupling constant by means of fold bifurcations. As discussed above, the two exceptions to these facts are given by the Sievers-Takeno (site-centered breathers) and Page modes (bond-centered breathers) which can be continued up to the continuum limit ($\epsilon \to \infty$). During this continuation, however, both modes experience important changes both in their shape and in their spectral properties (see Fig. 7).

The site-centered and bond-centered modes can be continued until resonance with the linear phonon band occurs. For periodic boundary conditions and assuming that the lattice has an *even* number of particles N, the phonon band is given by:

$$\omega_{\mathrm{ph}}^2(\epsilon) = \omega_0^2 + 4\epsilon \sin^2 \frac{q}{2}, \qquad \text{with} \qquad q = \left(1 - \frac{2n}{N}\right)\pi,\ n = 0, 1, 2, \ldots N/2 \tag{40}$$

where the modes are sorted by frequency and those from $n = 1$ to $n = N/2 - 1$ are doubly degenerate; $\omega_0^2 = 1$ ($\omega_0^2 = 2$) in the sG (ϕ^4) model. Therefore, the first resonance, upon variation of ϵ for fixed ω, must occur when $n = 0$, i.e. with the $q = \pi$ linear mode. In the case of ϕ^4 potential, resonances occur when the second harmonic of the breather frequency coincides with a phonon frequency ($m\omega = \omega_{\mathrm{ph}}$, $m = 2$), i.e. the first resonance occurs at $\epsilon_\pi = \omega^2 - 1/2$. However, in the sine-Gordon potential, the Fourier coefficients corresponding to even harmonics are

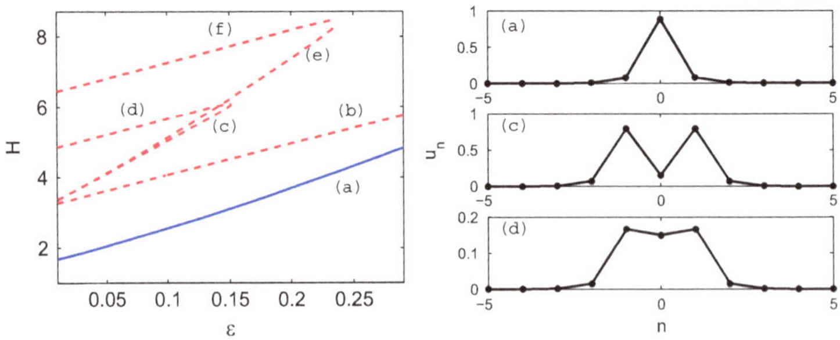

Fig. 7 *Left*: Hamiltonian energy (H) of breathers of the discrete sG equation as the coupling parameter ϵ is increased for $\omega = 0.7$. The labels correspond to breathers which have the following excited nodes in the AC limit: (**a**) $(\ldots, \bullet, \bullet, \bullet, \uparrow, \bullet, \bullet, \bullet, \ldots)$ (**b**) $(\ldots, \bullet, \bullet, \bullet, \uparrow, \uparrow, \bullet, \bullet, \ldots)$ (**c**) $(\ldots, \bullet, \bullet, \uparrow, \bullet, \uparrow, \bullet, \bullet, \ldots)$ (**d**) $(\ldots, \bullet, \bullet, \uparrow, \uparrow, \uparrow, \bullet, \bullet, \ldots)$ (**e**) $(\ldots, \bullet, \bullet, \uparrow, \bullet, \bullet, \uparrow, \bullet, \bullet, \ldots)$ and (**f**) $(\ldots, \bullet, \bullet, \uparrow, \uparrow, \uparrow, \uparrow, \bullet, \bullet, \ldots)$. Here, arrows correspond to the excited sites, which oscillates with frequency ω, whereas the oscillators at rest are represented by *dots*. Only the site-centered (**a**) and bond-centered (**b**) solutions survive for $\epsilon \to \infty$, both of which approach the exact breather solution of the continuous sG equation, see Eq. (20). *Right*: Examples of discrete sG breathers for $\epsilon = 0.065$. The multibreathers in the *two bottom panels* collide in a fold bifurcation at $\epsilon \approx 0.152$. The labels correspond to those in the *left panel*

zero, so that the resonance is caused by the third harmonic ($m\omega = \omega_{\text{ph}}$, $m = 3$) and the first resonance is observed at $\epsilon_\pi = (9\omega^2 - 1)/4$.

Thus, if a smooth continuation is performed, the site- (or bond-) centered solution continuously transforms into a phonobreather (i.e., resonates with the linear modes) once the coupling reaches the value ϵ_π. The central site(s) will oscillate with frequency ω whereas the tail will oscillate with frequency $m\omega$. Nevertheless, if the step in the continuation method is large enough, the system is able to "jump" to a localized solution whose frequency lies in the gaps of the phonon band. Those solutions are the result of a hybridization of a breather with a phonon and are dubbed as *phantom breathers* [57].[2] Phantom breathers are in fact the discrete version of the nanopterons observed in the continuum ϕ^4 equation [52]. However, unlike the continuum models, they typically arise for on-site potentials (including e.g. the discrete sine-Gordon) as long as the coupling is finite and higher than ϵ_π. Similarly to nanopterons and the phonobreathers found by smooth continuation, phantom breathers are composed of a localized breather oscillating with frequency ω plus a low-amplitude background of frequency $m\omega$.

Let us mention that the bifurcation diagram for phantom breathers (see [57] for an extensive study of such solutions) resembles the Wannier–Stark ladders observed in a model of non-homogenous waveguide arrays [58] [see Fig. 8 (left)]; this implies the existence of three nonlinear modes for the same value of the coupling constant. Of course, the width of gaps in the phonon band decreases with the number of

[2] In the nonlinear Schrödinger equation framework those solutions have been called *hybrid lattice solitons* [58].

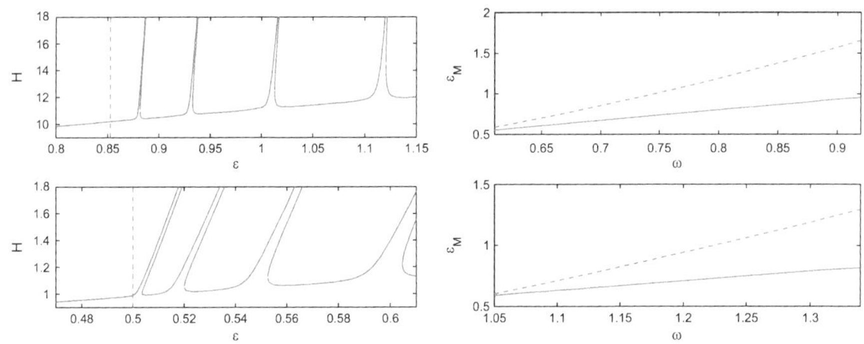

Fig. 8 (*Left panel*) Hamiltonian energy versus coupling for breathers with frequency $\omega = 0.7$ in the sine-Gordon model (*top*) and $\omega = 1$ for the ϕ^4 model (*bottom*). The *vertical dashed line* indicates the value ϵ_π. (*Right panel*) Exchange of stability bifurcation loci for breathers in the sG (*top*) and ϕ^4 potential (*bottom*); the *dashed line* indicates the value of $\epsilon_\pi(\omega)$

system particles N, and the gaps eventually vanish for $N \to \infty$. In other words, the hybridization of the breather with the background phonons is a finite-size artifact that disappears in the infinite lattice limit.

Similar to the kink situation described in Sect. 2.3, there are important details regarding the spectral properties of discrete breathers in the ϕ^4 and sine-Gordon lattices. In both models, the site- and bond-centered modes interchange their stability at a finite value of the coupling constant, which we denote as ϵ_M. This exchange-of-stability bifurcation (which is mediated by an intermediate breather) is closely related to the existence of *moving breathers* [59, 60]. This bifurcation is caused by a spatially anti-symmetric internal mode called *pinning mode*. For the site-centered breather mode, this internal mode detaches from the phonon band and becomes localized when ϵ is increased, colliding at $\theta = 0$ when the exchange of stability takes place. In addition, moving breathers not only exist close to the parameter values at which stability exchange takes place but also when the Floquet multiplier corresponding to the pinning mode is close to 1. In this case, the movement is possible because the height of the Peierls–Nabarro barrier is related to the distance of the unstable multiplier to 1. Figure 8 (right) shows the dependence of ϵ_M with respect to ω. In general, when ϵ becomes larger than ϵ_M the growth rate (distance of the unstable Floquet multiplier to 1) of the Sievers-Takeno mode grows up to a maximum value; beyond that point, there are many windows where stability exchanges again. Those windows correspond to the jumps into the phonon band and are difficult to control in the numerics. Nevertheless, as we approach $\epsilon \to \infty$, i.e., at the continuum limit, the relevant translational eigenvalue returns to the point $(1, 0)$ of the unit circle, signaling the restoration of the translational invariance in that limit.

4 A Recent Variant: sine-Gordon Equations with $\mathcal{PT}$ Symmetry

One of the interesting recent extensions that have emerged in the context of Klein–Gordon field theories is that of $\mathcal{PT}$-symmetric variants of these models. The analysis of Hamiltonians respecting the concurrent application of parity ($x \to -x$) and time-reversal ($t \to -t$) operations was initially proposed in a series of publications by Bender and collaborators [21–23]. While originally intended for Schrödinger Hamiltonians as a modification/extension of quantum mechanics, its purview has gradually become substantially wider. This stemmed to a considerable degree from the realization that other areas such as optics might be ideally suited not only for the theoretical study of this theme [61–63], but also for its experimental realization [64–66]. Recently, both at the mechanical level [67] and at the electrical one [68, 69] realizations of $\mathcal{PT}$ symmetry and its breaking have arisen, while a Klein–Gordon setting has also been explored theoretically for so-called $\mathcal{PT}$ symmetric nonlinear metamaterials and the formation of gain-driven discrete breathers therein [70].

These efforts have, in turn, prompted a number of works both at the discrete level [71, 72], as well as at the continuum one [72, 73] to identify prototypical settings of application of $\mathcal{PT}$-symmetry in nonlinear KG field theories. The model used in [72, 73] is of the form:

$$u_{tt} - u_{xx} + \epsilon\gamma(x)u_t + f(u) = 0. \tag{41}$$

Here, in order to preserve the $\mathcal{PT}$ symmetry, $\gamma(x)$, should be an antisymmetric function satisfying $\gamma(-x) = -\gamma(x)$. Physically, this implies that while this is an "open" system with gain and loss, the gain balances the loss, in preserving the symmetry.

Interestingly, for this class of models, the stationary states u_0, such as the kinks, are *unaffected* by the presence of the $\mathcal{PT}$ symmetry (since the relevant term involves u_t). On the other hand, the linearization problem becomes

$$\lambda^2 v + \epsilon\lambda\gamma v + Hv = 0. \tag{42}$$

where H is the Hamiltonian linearization operator of e.g. Eq. (9) or (12). Equivalently, the spectral problem can be written as:

$$\lambda \begin{pmatrix} v_1 \\ v_2 \end{pmatrix} = \begin{pmatrix} 0 & 1 \\ -H & -\epsilon\gamma(x) \end{pmatrix} \begin{pmatrix} v_1 \\ v_2 \end{pmatrix}. \tag{43}$$

Now, the key question becomes what will be the fate of the eigenstates associated with the linearization around the continuum kink, in the presence of the perturbation imposed by the $\mathcal{PT}$-symmetric gain/loss term. What has been argued in [72, 73] is that in both the continuum and the infinite lattice discrete problem, the continuous

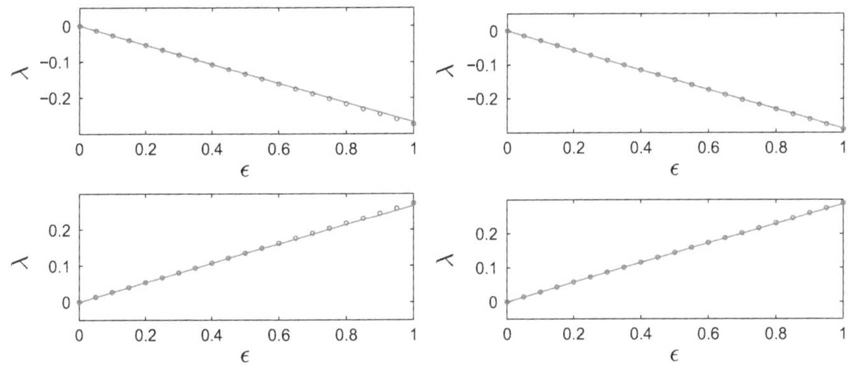

Fig. 9 A typical example of the comparison of the motion of the translational eigenvalue under a $\mathcal{PT}$-symmetric perturbation between numerics (points) and analytics of Eq. (44). The *left* is for sG and the *right* for ϕ^4, the *top* for a kink centered at the lossy side, while the *bottom* for a kink centered at the gain side

spectrum will not be altered. Hence, the question is what becomes of the point spectrum in the presence of this perturbation.

To determine what happens to the mode associated with translation, we project Eq. (42) to the eigenfunction of vanishing eigenvalue (in the unperturbed limit $v = u_{0,x}$), and observe that the last term in Eq. (42) disappears. Assuming a perturbative expansion of v in ϵ, which to leading order (when $\epsilon = 0$) has $v = u_x$, leads immediately to the leading order approximation

$$\lambda = 0 \quad \text{or} \quad \lambda = -\epsilon \frac{\int \gamma(x) u_{0,x}^2 dx}{\int u_{0,x}^2 dx}. \tag{44}$$

This explicit expression should yield a fairly accurate prediction of the fate of the former double 0 eigenvalue pair. Interpreting this result, we find that one of the two eigenvalues of the pair should remain at 0 (a result that persists to all orders in perturbation theory due to the remaining invariance of the location of the solution under translations). On the other hand, the fate of the second eigenvalue will depend on the relative position of the kink centered at x_0 with respect to the origin. In particular, for both the sine-Gordon and the ϕ^4, we find that if the kink is centered at the lossy side (say, $x_0 < 0$), then the numerator of the expression in Eq. (44) is negative, leading the second eigenvalue to the left hand plane. Reversing the sign of x_0 leads effectively to a reversal of the time flow (due to the nature of the system) and results into the corresponding eigenvalue moving by an equal amount to the right half of the spectral plane i.e., yielding instability. On the other hand, for the special case in which $x_0 = 0$, the parity of the relevant integral makes it vanish and the eigenvalue does not shift. A typical example showcasing the quality of the agreement of the theoretical prediction with full numerical results is shown in Fig. 9.

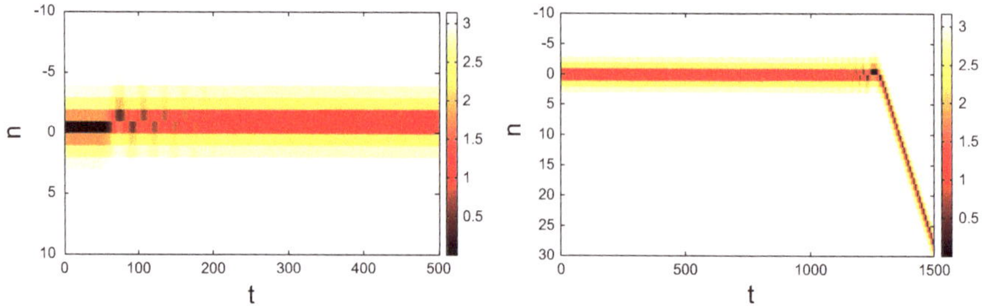

Fig. 10 Two canonical examples of the evolution of an unstable kink on the gain side of a discrete $\mathcal{PT}$-symmetrically perturbed sG model. The *left panel* shows the kink as settling into an intersite-centered structure on the lossy side, while the *right* forces it to propagate on the gain side, but without bearing a stable fixed point for the kink state. *Color bar* represents $|u_n - \pi|$

However, in the case of the ϕ^4 model, it is not sufficient to consider only the fate of the eigenvalue potentially bifurcating from the origin of the spectral plane. Recall that there is an eigenvalue at $\lambda^2 = -3$. One can consider a perturbative expansion of such an eigenvalue in the problem, according to $\lambda = \lambda_0 + \epsilon\lambda_1 + \ldots$, while $v = v_0 + \epsilon v_1 + \ldots$. Applying a solvability condition at the leading order of the perturbation theory then immediately yields the eigenvalue leading order correction as:

$$\lambda_1 = -\frac{1}{2}\frac{\int \gamma(x) v_0^2 dx}{\int v_0^2 dx}. \tag{45}$$

The interpretation of this result is that this isolated point spectrum eigenvalue (pair) will also move to the left half plane, for the case of the kink centered at the lossy side, while it will move (as a complex pair) to the right half plane when the kink is centered at the gain side. That is, overall stability of the kink's point spectrum is ensured when centered at the lossy region and instability when centered at the gain region. When the kink is at the interface, it should already be clear from the above considerations that all real contributions to the eigenvalue will vanish, allowing only imaginary ones that could shift the point spectrum along the imaginary axis, as proved earlier in the text. Again, very good agreement between this prediction and the full numerical result was found in [73].

In the discrete case, the same methods as explored above yield equally accurate predictions. Arguably, the only significant modification arising here is in the case of the intersite centered stable kink. For the onsite one, the perturbation shifts the eigenvalue along the real axis (weakly, for small perturbation strength), not changing the fate of the kink's (in)stability. However, in the marginal case of the intersite centered kink, once again, the lossy side leads to stability kicking the former translational (now imaginary in the discrete problem) eigenvalues to the left half-plane. Placing the kink on the gain side has the opposite effect with eigenvalues bifurcating on the right half plane (as a complex pair) and hence leading to instability. It is for that reason that the fate of the unstable discrete kink of the dynamical example of Fig. 10 leads to a static intersite centered structure on the lossy (left) side for the left panel of the figure. If, however, the perturbation pushes (intentionally) the kink toward the right hand (gain) side, it cannot find a stable equilibrium point and instead it continues to propagate as is shown in the right panel of the figure.

In the case of breathers, we can add that also based on the discussion given earlier for different variants of the sG equation and KG equations more generally, breathers are unlikely to exist under PT-symmetric perturbations. However, we do find that in the case example potentials that we considered, stationary breather solutions can only be found to exist at the delicate interface between gain and loss. Breathers seeded on the lossy side are found to become mobile. Perhaps more interestingly, such structures on the gain side appear to gain energy and transform themselves into a kink-antikink pair.

Acknowledgements PGK gratefully acknowledges support from grants NSF-CMMI-1000337, NSF-DMS-1312856, US-AFOSR FA-9550-12-1-0332, BSF-2010239 and from the ERC through an IRSES grant. The work of MCB was partially supported by the Deutsche Forschungsgemeinschaft (DFG) under the grant CH 957/1-1. We acknowledge Aslihan Demirkaya for her technical assistance.

References

1. R.K. Dodd, J.C. Eilbeck, J.D. Gibbon, H.C. Morris, *Solitons and Nonlinear Wave Equations* (Academic, New York, 1983)
2. T. Dauxois, M. Peyrard, *Physics of Solitons* (Cambridge University Press, Cambridge, 2006)
3. M. Cirillo, R.D. Parmentier, B. Savo, Physica D **3**, 565 (1981)
4. W.P. Su, J.R. Schrieffer, Phys. Rev. Lett. **46**, 738 (1981)
5. A.R. Bishop, S.E. Trullinger, J.A. Krumhansl, Physica D **1**, 1 (1980)
6. J. Frenkel, T. Kontorova, J. Phys. **1**, 137 (1939)
7. O.M. Braun, Yu.S. Kivshar, *The Frenkel-Kontorova Model: Concepts, Methods and Applications* (Springer, Berlin, 1998)
8. A.C. Scott, Am. J. Phys. **37**, 52 (1969)
9. A.C. Scott, *Nonlinear Science*, 2nd edn. (Oxford University Press, New York, 2003); A. Barone, F. Esposito, C.J. Magee, A.C. Scott, Riv. Nuovo Cim. **1**, 227 (1971)
10. J. Cuevas, L.Q. English, P.G. Kevrekidis, M. Anderson, Phys. Rev. Lett. **102**, 224101 (2009)
11. M. Kimura, T. Hikihara, Chaos **19**, 013138 (2009)
12. M.J. Ablowitz, H. Segur, *Solitons and the Inverse Scattering Transform* (SIAM, Philadelphia, 1981)
13. T.I. Belova, A.E. Kudryavtsev, Phys. Usp. **40**, 359 (1997)
14. P. Anninos, S. Oliveira, R.A. Matzner, Phys. Rev. D **44**, 1147 (1991)
15. D.K. Campbell, J.F. Schonfeld, C.A. Wingate, Physica D **9**, 1 (1983)
16. D.K. Campbell, M. Peyrard, Physica D **18**, 47 (1986)
17. R.H. Goodman, R. Haberman, SIAM J. Appl. Dyn. Sys. **4**, 1195 (2005)
18. R.H. Goodman, Chaos **18**, 023113 (2008)
19. A.J. Heeger, S. Kivelson, J.R. Schrieffer, W.P. Su, Rev. Mod. Phys. **60**, 781 (1988)
20. V.K. Wadhawan, *Introduction to Ferroic Materials* (Gordon and Breach, Amsterdam, 2000)
21. C.M. Bender, S. Boettcher, Phys. Rev. Lett. **80**, 5243 (1998)
22. C.M. Bender, S. Boettcher, P.N. Meisinger, J. Math. Phys. **40**, 2201 (1999)
23. C. M. Bender, Rep. Prog. Phys. **70**, 947 (2007)

24. P.G. Kevrekidis, Phys. Lett. A **285**, 383 (2001)
25. T. Sugiyama, Prog. Theor. Phys. **61**, 1550 (1979)
26. R.S. MacKay, S. Aubry, Nonlinearity **7**, 1623 (1994)
27. N.J. Balmforth, R.V. Craster, P.G. Kevrekidis, Physica D **135**, 212 (2000)
28. K.E. Atkinson, *An Introduction to Numerical Analysis*, 2nd edn. (Wiley, New York, 1989)
29. T. Kapitula, P. Kevrekidis, Nonlinearity **14**, 533 (2001)
30. J.M. Speight, R. Ward, Nonlinearity **7**, 475 (1994)
31. H. Segur, M.D. Kruskal, Phys. Rev. Lett. **58**, 747 (1987)
32. S. Kichenassamy, Comm. Pure Appl. Math. **44**, 789 (1991)
33. B. Birnir, H. McKean, A. Weinstein, Comm. Pure Appl. Math. **47**, 1043 (1994)
34. J. Denzler, Comm. Math. Phys. **158**, 397 (1993)
35. J. Denzler, Ann. Inst. H. Poincaré Anal. Non Linéaire **12**, 201 (1995)
36. M.D. Groves, G. Schneider, Comm. Math. Phys. **219**, 489 (2001)
37. J. Shatah, C. Zeng, Nonlinearity **16**, 591 (2003)
38. P. Kirrmann, G. Schneider, A. Mielke, Proc. R. Soc. Edinb. Sect. A-Math. **122**, 85 (1992)
39. D.W. McLaughlin, A.C. Scott, J. Math. Phys. **14**, 1817 (1973)
40. M. Chirilus-Bruckner, Ch. Chong, G. Schneider, H. Uecker, Math. Anal. Appl. **347**, 304 (2008)
41. M. Chirilus-Bruckner, G. Schneider, J. Diff. Equ. **253**, 2161 (2012)
42. C. Blank, M. Chirilus-Bruckner, V. Lescarret, G. Schneider, Comm. Math. Phys. **302**, 815 (2011)
43. A.J. Sievers, S. Takeno, Phys. Rev. Lett. **61**, 970 (1988)
44. J.B. Page, Phys. Rev. B **40**, 7835 (1990)
45. V. Koukouloyannis, Phys. Lett. A **377**, 2022 (2013)
46. V. Koukouloyannis, R.S. MacKay, J. Phys. A: Math. Gen. **38)**, 1021 (2005)
47. J. Cuevas, V. Koukouloyannis, P.G. Kevrekidis, J.F.R. Archilla, Int. J. Bif. Chaos **21**, 2161 (2011)
48. T. Cretegny, S. Aubry, Phys. Rev. B **55**, R11929 (2007)
49. T. Cretegny, S. Aubry, Physica D **113**, 162 (2008)
50. J. Cuevas, J.F.R. Archilla, F.R. Romero, J. Phys. A Math. Theor. **44**, 035102 (2011)
51. V. Koukouloyannis, J. Cuevas, P.G. Kevrekidis, V. Rothos, Physica D **242**, 16 (2013)
52. J. Boyd, Nonlinearity **3**, 177 (1990)
53. S. Aubry, Physica D **103**, 201 (1997)
54. J.F.R. Archilla, J. Cuevas, B. Sánchez-Rey, A. Álvarez, Physica D **180**, 235 (2003)
55. V. Koukouloyannis, P.G. Kevrekidis, Nonlinearity **22**, 2269 (2009)
56. M. Abramowitz, I.A. Stegun, *Handbook of Mathematical Functions* (Courier Dover Publications, New York, 1964)
57. A.M. Morgante, M. Johansson, S. Aubry, G. Kopidakis, J. Phys. A Math. Gen. **35**, 4999 (2002)
58. T. Pertsch, U. Peschel, F. Lederer, Phys. Rev. E **66**, 066604 (2002)
59. D. Chen, S. Aubry, G.P. Tsironis, Phys. Rev. Lett. **77**, 4776 (1996)
60. S. Aubry, T. Cretegny, Physica D **119**, 34 (1998)
61. Z.H. Musslimani, K.G. Makris, R. El-Ganainy, D.N. Christodoulides, Phys. Rev. Lett. **100**, 030402 (2008)
62. H. Ramezani, T. Kottos, R. El-Ganainy, D.N. Christodoulides, Phys. Rev. A **82**, 043803 (2010)
63. A. Ruschhaupt, F. Delgado, J.G. Muga, J. Phys. A Math. Gen. **38**, L171 (2005)
64. A. Guo, G.J. Salamo, D. Duchesne, R. Morandotti, M. Volatier-Ravat, V. Aimez, G. A. Siviloglou, D. N. Christodoulides, Phys. Rev. Lett. **103**, 093902 (2009)
65. C.E. Rüter, K.G. Makris, R. El-Ganainy, D.N. Christodoulides, M.Segev, D. Kip, Nat. Phys. **6**, 192 (2010)
66. B. Peng, S.K. Ozdemir, F. Lei, F. Monifi, M. Gianfreda, G.L. Long, S. Fan, F. Nori, C.M. Bender, L. Yang, Nonreciprocal light transmission in parity-time-symmetric whispering-gallery microcavities. arXiv:1308.4564
67. C.M. Bender, B.J. Berntson, D. Parker, E. Samuel, Am. J. Phys. **81**, 173 (2013)
68. J. Schindler, A. Li, M.C. Zheng, F.M. Ellis, T. Kottos, Phys. Rev. A **84**, 040101 (2011)
69. H. Ramezani, J. Schindler, F.M. Ellis, U. Günther, T. Kottos, Phys. Rev. A **85**, 062122 (2012)

70. N. Lazarides, G. P. Tsironis, Phys. Rev. Lett. **110**, 053901 (2013)
71. J. Cuevas, P.G. Kevrekidis, A. Saxena, A. Khare, Phys. Rev. A **88**, 032108 (2013)
72. A. Demirkaya, D.J. Frantzeskakis, P.G. Kevrekidis, A. Saxena, A. Stefanov, Phys. Rev. E **88**, 023203 (2013)
73. A. Demirkaya, M. Stanislavova, A. Stefanov, T. Kapitula, P.G. Kevrekidis, Spectral stability of kinks in $\mathcal{PT}$-symmetric Klein-Gordon. arXiv:1402.2942, Studies of Applied Mathematics, in press (2014)

Soliton Collisions

Sergey V. Dmitriev and Panayotis G. Kevrekidis

Abstract Soliton collisions constitute a fascinating topic of nonlinear science and its numerous applications. Solitary waves are well-known to be robust, being able to maintain their character upon mutual collisions and interactions with imperfections of the media. Therefore, they are efficient carriers of various physical quantities, e.g., energy, momentum, topological charge, etc. However, only in exactly integrable systems the soliton collisions are perfectly elastic. The addition of different types of perturbations to such models produces a number of intriguing effects such as radiation of small-amplitude wave packets, the possibility of excitation of the soliton internal modes, energy exchange between the solitons' internal and translational modes, radiationless energy exchange between colliding solitons, to name only some of the principal ones. The aim of this chapter is to present a brief historical overview of the theme of soliton collisions and a description of its current state-of-the-art within the realm of perturbed sine-Gordon and related Klein–Gordon equations.

Keywords ϕ^4 model • Collisions • Fractal soliton scattering • Internal modes • Radiation • Radiationless energy exchange • Solitary waves

S.V. Dmitriev (✉)
Institute for Metals Superplasticity Problems, Russian Academy of Science, Khalturin St. 39, Ufa 450001, Russia

National Research Tomsk State University, Lenin Prospekt 36, Tomsk 634050, Russia
e-mail: dmitriev.sergey.v@gmail.com

P.G. Kevrekidis
Department of Mathematics and Statistics, University of Massachusetts, Amherst, MA 01003-4515, USA
e-mail: kevrekid@gmail.com

J. Cuevas-Maraver et al. (eds.), *The sine-Gordon Model and its Applications*, Nonlinear Systems and Complexity 10, DOI 10.1007/978-3-319-06722-3_3,

1 Introduction

The study of the solitary wave collisions can provide an insight into the major physical effects supported by the considered system. The integrable sine-Gordon equation (SGE)

$$\phi_{tt} - \phi_{xx} + \sin\phi = 0 \tag{1}$$

supports two prototypical types of solitons, (anti-)kinks and breathers, and any number of them can be observed to collide purely elastically, maintaining their shapes and properties after the collision [1–4]. In many branches of physics, perturbed versions of SGE are considered. Integrability in such systems is lost and the soliton collisions depend, sometimes in a non-trivial way, on both the soliton parameters and on the type of perturbation. This chapter outlines the effect of perturbations on the soliton collisions in SGE. Kink collisions in the non-integrable ϕ^4 and ϕ^6 models are also discussed.

In Sect. 2 collisions in the integrable case are described. Various types of SGE perturbations are presented in Sect. 3. Effects of perturbations on SGE collisions in different non-integrable cases are classified in Sect. 4. As a special case example of intriguing collisional behavior, fractal soliton scattering is discussed in Sect. 4.9. A brief summary with the outline of some open problems is given in Sect. 5.

2 Soliton Collisions in the Integrable Case

We start with a list of several exact solutions to the integrable SGE (1) followed by a discussion relevant for understanding collision outcomes in various perturbed versions of SGE. The multi-soliton solutions can be obtained with the help of the Bäcklund transformation [3, 4] or of the Hirota method [2].

2.1 Exact Solutions to the Integrable SGE

The two basic soliton solutions to SGE (1) are the kink

$$\phi(x,t) = 4\arctan\{\exp[\delta_k(x - V_k t)]\}, \tag{2}$$

and the breather

$$\phi(x,t) = 4\arctan\frac{\eta\sin[\delta_b\omega(t - V_b x)]}{\omega\cosh[\delta_b\eta(x - V_b t)]}, \tag{3}$$

where V_k is kink velocity, V_b, ω are the breather velocity and frequency, and

$$\delta_{k,b} = \frac{1}{\sqrt{1 - V_{k,b}^2}}, \quad \eta = \sqrt{1 - \omega^2}. \tag{4}$$

The energy E and momentum P of one SGE (anti)kink are

$$E_k = 8\delta_k, \quad P_k = 8V_k\delta_k. \tag{5}$$

Energy, momentum, amplitude, wavelength and period of a breather are, respectively,

$$E_b = 16\eta\delta_b, \;\; P_b = 16\eta\delta_b V_b, \;\; A = 4\arctan\frac{\eta}{\omega}, \;\; \lambda = \frac{2\pi\delta_b|V_b|}{\omega}, \;\; T = \frac{2\pi\delta_b}{\omega}. \tag{6}$$

The breather solution Eq. (3) can be regarded as a kink-antikink bound state [5–7].

The collision between two kinks having velocities $\pm V_k$ is described by the following solution to (1)

$$\phi(x,t) = 4\arctan\frac{V_k\sinh(\delta_k x)}{\cosh(\delta_k V_k t)}. \tag{7}$$

For the collision between kink and antikink having velocities $\pm V_k$ one has the exact solution

$$\phi(x,t) = 4\arctan\frac{V_k\cosh(\delta_k x)}{\sinh(\delta_k V_k t)}. \tag{8}$$

A kink-breather collision is described by

$$\begin{aligned} \phi(x,t) &= 4\arctan(\exp B) + 4\arctan\frac{\eta Y}{\omega Z}, \\ Y &= 2\omega(\sinh D - \cos C\sinh B) + 2\delta_b\delta_k(V_k - V_b)\sin C\cosh B, \\ Z &= 2\eta(\cos C + \sinh D\sinh B) - 2\delta_b\delta_k(1 - V_k V_b)\cosh D\cosh B, \\ B &= \delta_k(x - x_k - V_k t), \quad C = -\omega\delta_b(t - V_b x), \quad D = \eta\delta_b(x - V_b t). \end{aligned} \tag{9}$$

Here V_k is kink velocity, x_k defines initial kink position, and V_b, ω are the velocity and frequency of the breather, respectively.

A collision between two breathers with velocities V_1, V_2 and frequencies ω_1, ω_2 is given by

$$\phi(x,t) = 4\arctan(S) - 4\arctan\frac{\eta_2(T\cosh B_1 + \sin C_1)}{\omega_2(\cosh B_1 + T\sin C_1)},$$

$$T = \varphi\frac{2\tau[(S-P)(1+SP) - Q^2S] - 2\beta Q(1+S^2)}{\varphi^2[(1+S)(1+SP) + Q^2S^2] + (\tau^2+\beta^2)[(S-P)^2+Q^2]},$$

$$P = \frac{\beta X + \kappa Y}{\varepsilon Z}, \quad Q = \frac{\beta Y - \kappa X}{\varepsilon Z}, \quad S = \frac{\eta_1 \sin C_1}{\cosh B_1 \omega_1},$$
$$X = \sinh B_2 \cos C_1 - \cos C_2 \sinh B_1, \; Y = \cosh B_2 \sin C_1 + \sin C_2 \cosh B_1,$$
$$Z = \cos(C_1 - C_2) + \cosh(B_1 + B_2),$$
$$B_{1,2} = \eta_{1,2}\delta_{1,2}(x - x_{1,2} - V_{1,2}t), \; C_{1,2} = \Delta_{1,2} - \omega_{1,2}\delta_{1,2}[t - (x - x_{1,2})V_{1,2}],$$
$$\delta_{1,2} = (1 - V_{1,2}^2)^{-1/2}, \; \eta_{1,2} = (1 - \omega_{1,2}^2)^{1/2}, \quad \alpha = \frac{\delta_2(1 + V_2)}{\delta_1(1 + V_1)},$$
$$\beta = \alpha - 1/\alpha, \quad \varepsilon = 2\alpha - \beta + 2(\omega_1\omega_2 - \eta_1\eta_2), \quad \tau = 2(\omega_1\eta_2 - \eta_1\omega_2),$$
$$\varphi = 2\alpha - \beta - 2(\omega_1\omega_2 + \eta_1\eta_2), \; \kappa = 2(\omega_1\eta_2 + \eta_1\omega_2). \tag{10}$$

Here $x_{1,2}$ and $\Delta_{1,2}$ define initial positions and initial phases of the two breathers, respectively.

2.2 Discussion of the Exact Solutions

The effect of collision between SGE solitons appears only in the relative phase shift calculated, e.g., in [2, 8, 9]. The sum of the relative phase shifts in multi-soliton collisions is equal to zero [2]. Imagine a sample kink embedded in the soliton gas of moving kinks and breathers. One can assume that the sample kink has zero velocity, taking into account the Lorentz invariance of SGE. Each collision with kinks and breathers of the gas produces a spatial shift of the sample kink, so that its coordinate changes in time, even though, between the collisions its velocity is equal to zero. In this sense, even for completely integrable nonlinear equations, one can derive a kinetic equation describing diffusion, internal friction and thermal conductivity of the soliton gas. Such calculations have been undertaken for different integrable models including the SGE [9, 10].

The robustness of solitons in integrable models is explained by the fact that their dynamics is constrained by an infinite number of conservation laws. Within the infinite number of conservation laws of the integrable SGE (1) there are only two that depend on the second-order terms, namely, the conservation of energy

$$E = \int_{-\infty}^{\infty} \left(\frac{1}{2}\phi_t^2 + \frac{1}{2}\phi_x^2 + 1 - \cos\phi\right)dx, \tag{11}$$

and that of linear momentum

$$P = -\int_{-\infty}^{\infty} \phi_t \phi_x dx. \tag{12}$$

All other conservation laws depend on fourth- and higher order terms [11, 12].

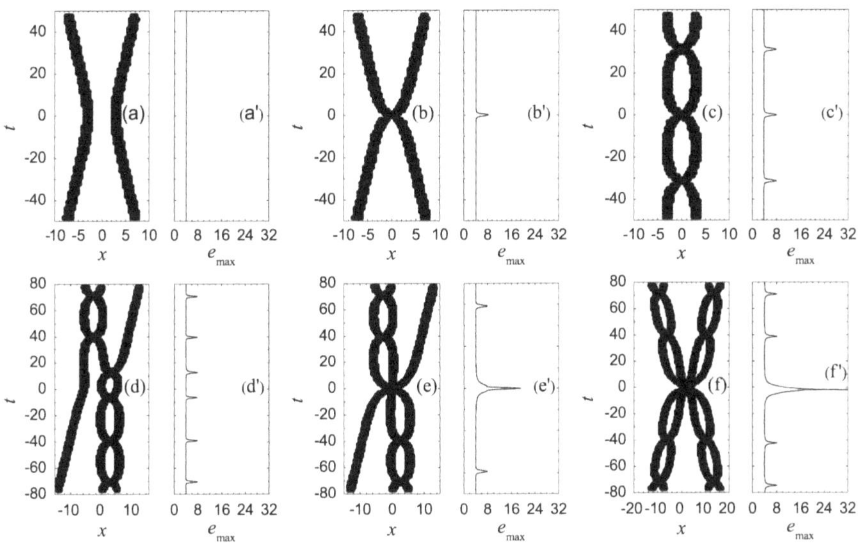

Fig. 1 Examples of elastic soliton collisions. (**a–f**) Regions of the (x, t)-plane with energy density $e(x, t) > 2$ revealing the cores of the solitons. ($\mathbf{a}'$–$\mathbf{f}'$) Maximum over spatial coordinate energy density $e_{\max}$ as a function of time. Examples are given for collisions between (**a**, $\mathbf{a}'$) two kinks, Eq. (7), $V_k = 0.1$; (**b**, $\mathbf{b}'$) kink and antikink, Eq. (8), $V_k = 0.1$; (**c**, $\mathbf{c}'$) kink and antikink bound state in the form of breather solution Eq. (3), $V_b = 0$, $\omega = 0.1$; (**d**, $\mathbf{d}'$) kink and breather, Eq. (9), $V_k = 0.1$, $x_k = -1$, $V_b = 0$, $\omega = 0.1$; (**e**, $\mathbf{e}'$) same as in (**d**, $\mathbf{d}'$) but for $x_k = 0$; and (**f**, $\mathbf{f}'$) two breathers, Eq. (10), $V_1 = -V_2 = 0.1$, $\omega_{1,2} = 0.1$, $x_{1,2} = 0$, $\Delta_1 = 0$, and $\Delta_2 = \pi$

The energy density of the SGE field is given by the integrand of Eq. (11),

$$e(x, t) = \frac{1}{2}\phi_t^2 + \frac{1}{2}\phi_x^2 + 1 - \cos\phi\,. \tag{13}$$

Examples of the multi-soliton solutions to SGE (1) are presented in Fig. 1, where in (a–f) the regions of the (x, t)-plane with energy density $e(x, t) > 2$ are shown to present the dynamics of the soliton cores and in (a′–f′) the corresponding time evolution of the maximum over spatial coordinate energy density, $e_{\max}$, is presented. Shown are: (a, a′) two kinks of Eq. (7) colliding with the velocities $V_k = \pm 0.1$; (b, b′) kink and antikink of Eq. (8) colliding with the velocities $V_k = \pm 0.1$; (c, c′) the breather solution of Eq. (3), $V_b = 0$, $\omega = 0.1$; (d, d′) kink and breather Eq. (9) colliding with $V_k = -0.1$, $V_b = 0$, $\omega = 0.1$, $x_k = -1$; (e, e′) same as in (d, d′) but for $x_k = 0$; (f, f′) breather-breather collision of Eq. (10) for $V_1 = -V_2 = 0.1$, $\omega_{1,2} = 0.1$, $x_{1,2} = 0$, $\Delta_1 = 0$ and $\Delta_2 = \pi$.

A few comments important for the discussion of soliton collisions in the perturbed SGE can be made.

Firstly, perturbations destroy higher order conservation laws more severely than the low order ones of Eqs. (11) and (12). Then, in a weakly perturbed SGE two nearly conserved quantities, E and P, will persist. From here it follows that no noticeable energy and momentum exchange is possible when two kinks collide in a weakly perturbed SGE because the two parameters of the two-soliton solution are constrained by the two weakly perturbed conservation laws. On the other hand, *strong* energy and momentum exchange is not forbidden for a collision between three and more solitons in a weakly perturbed SGE because there remain free parameters of the multi-soliton solution [13–16] (i.e., the number of parameters is larger than the number of effective constraints).

Secondly, separatrix solutions to SGE can be identified. For example, the breather solution (3) for $V_B = 0$ in the limit $\omega \to 0$, and the kink-antikink solution (8) in the limit $V_K \to 0$ come to the same separatrix two-soliton solution

$$\phi(x,t) = 4 \arctan \frac{t}{\cosh x}. \tag{14}$$

Qualitatively different near-separatrix solutions, in the presence of even weak perturbations, can transform one into another. For example, kink and antikink having small relative velocity can merge into a breather in presence of damping, or the reverse event can be observed in presence of energy gain, e.g., from a soliton internal mode. Even more spectacular collision outcomes are possible for multi-soliton near-separatrix collisions when the solitons can exchange energy and momentum between them. The separatrix nature of the following three-soliton wobble solution [14]

$$\phi(x,t) = 4 \arctan(\exp x) + 4 \arctan \frac{\eta(\sinh D - \cos C \sinh x)}{\eta(\cos C + \sinh D \sinh x) - \cosh D \cosh x},$$
$$C = -\omega t, \quad D = \eta x, \quad \eta = \sqrt{1-\omega^2}, \tag{15}$$

will be clarified later with the help of the three-particle model [15]. This solution describes the interaction of a kink with $V_K = 0$ and a breather with $V_B = 0$ placed at the same point. The breather frequency ω is the only parameter of the solution. Note that the solution (6) of the work [17] does not coincide with (15) because in that solution the kink and breather are not located at the same point.

Thirdly, kinks having the same topological charge repel each other and their cores do not merge during collisions, while for the kinks having opposite charge the situation is opposite, as can be seen from comparison of Fig. 1(a, a′) and (b, b′). Note that in (a′) at any time $e_{\max}$ is nearly equal to 4, which is the maximal energy density of a kink at rest. On the other hand, in (b′) at the collision point $e_{\max}$ rises up to a value close to 8, which is two times larger than that of a single kink. The same maximal value of $e_{\max} \approx 8$ can be observed in (c′) and (d′). In the latter case the kink-breather collision happens as a sequence of two-soliton collisions. In contrast, in (d′) soliton parameters are such that all three kinks collide at one point and $e_{\max}$ rises up to a value close to 20, which is not three but five times greater than that of a single kink. A special choice of parameters for the breather-breather collision presented in (f′) results in the collision of all four sub-kinks at one point with the maximal value of $e_{\max}$ close to 32. This is important because, in the presence of perturbations, collisions demonstrating higher energy density will be more affected by the perturbations. This is so because large values of $e_{\max}$ at the collision point mean large values of the partial derivatives ϕ_t and/or ϕ_x and, consequently, strong perturbation of the conservation laws of Eqs. (11) and (12).

The above three comments are not related to any particular type of perturbation of SGE and thus, they can be important for any type of perturbation.

3 Non-integrable Perturbations of SGE and Their Physical Significance

All types of perturbations can be divided into two wide classes, the Hamiltonian and non-Hamiltonian ones.

3.1 Modification of the On-Site Potential Profile

The SGE is a particular case of the Klein–Gordon equation

$$\phi_{tt} - \phi_{xx} + \frac{dV(\phi)}{d\phi} = 0, \tag{16}$$

where $V(\phi)$ is the on-site potential. This equation of motion follows from the Hamiltonian

$$H = \int \left[\frac{\phi_t^2}{2} + \frac{\phi_r^2}{2} + V(\phi) \right] dx. \tag{17}$$

For SGE the potential satisfies $V(\phi) = 1 - \cos\phi$.

Kink-antikink collisions and interaction of kinks with impurities have been studied extensively [18] (see also [19] for a recent detailed discussion) in the frame of the ϕ^4 model having non-periodic, double-well potential of the form

$$V(\phi) = \frac{1}{4}(\phi^2 - 1)^2. \tag{18}$$

Kinks of the ϕ^4 model support internal vibrational modes. In contrast, kinks of the following ϕ^6 model

$$V(\phi) = \frac{1}{2}\phi^2(\phi^2 - 1)^2, \tag{19}$$

do not support localized modes. Kink-antikink collisions in this model have been studied recently by Dorey et al. [20].

Peyrard and Remoissenet have proposed the following potential [21, 22]

$$V(\phi) = \frac{(1-r)^2(1-\cos\phi)}{1+r^2+2r\cos\phi}, \tag{20}$$

with a parameter $|r| < 1$. For $r = 0$ this reduces to the integrable SGE potential. For $r \neq 0$ the model is not integrable, though it supports kinks traveling without radiation [21]. What is important for the discussion of the kink-antikink collisions is that for $r < 0$ the kinks can bear the soliton internal mode [23], while for $r > 0$ they cannot [24]. Thus, for $r < 0$ the energy exchange between the translational and internal modes of the kinks can take place.

The double SGE with the potential

$$V(\phi) = -\frac{4}{1+|4\eta|}\left(-\cos\frac{\phi}{2} + \eta\cos\phi\right) \tag{21}$$

reduces to SGE potential in the limit $\eta \to \pm\infty$ and to the SGE potential with $\phi/2$ for $\eta = 0$. For $\eta < -1/4$ there are two potential wells of different height within a period of the potential. This gives rise to the existence of two types of kinks, the small and the large ones, interpolating across the small and high potential barriers, respectively. This model supports kink internal modes for $\eta > 0$, yet it does not support such modes for $-1/4 < \eta < 0$, and only small kinks can bear internal modes for $\eta < -1/4$. The resonance kink-antikink interactions can be observed in most ranges of η [25, 26]. Kink-kink interactions and excitation of the kinks' internal modes have been studied in [27]. Three-particle effects in the collision of a fast kink with two more slow ones have been studied in the weakly perturbed model to demonstrate the possibility of energy exchange between them [13].

To take into account local inhomogeneity (impurity) the following modification of the SGE potential has been proposed

$$V(\phi) = -[1 - \epsilon F(x)](1 - \cos\phi), \tag{22}$$

where one can take, e.g., $F(x) = 1/\cosh(\beta x)$ or $F(x) = \delta(x)$ [28–34]. A square potential well of width L can be simulated by taking $F(x) = 1$ for $-L/2 < x < L/2$ and $F(x) = 0$ elsewhere [35–38].

Another modification is the stepwise SGE $V(\phi) = a(1-\cos\phi)$ with $a = a_1$ (a_2) for $\phi < 0$ (> 0) [38–41].

3.2 Discreteness of Media

Taking into account the discreteness of media produces one of the most studied types of perturbation of SGE [42, 43]. The classical discrete version of SGE, which is the Frenkel–Kontorova model, reads

$$\frac{d^2\phi_n}{dt^2} - \frac{1}{h^2}\left(\phi_{n-1} - 2\phi_n + \phi_{n+1}\right) + \sin\phi_n = 0, \tag{23}$$

which stems from the Hamiltonian

$$H = \sum_n \left[\frac{1}{2}\dot{\phi}_n^2 + \frac{1}{2h^2}\left(\phi_{n+1} - \phi_n\right) + 1 - \cos\phi_n\right]. \tag{24}$$

The perturbation parameter here is $\epsilon = h^2$ with h being the lattice spacing. The lowest order correction to SGE due to the discretization can be quantified, upon a Taylor expansion of the second difference, as $(\epsilon/12)\phi_{xxxx}$.

Kinks in the discrete Klein–Gordon models typically experience the action of the Peierls–Nabarro potential [44–48] having period equal to h. A kink at rest can be stable only when centered at a high-symmetry position with respect to the lattice and moving kinks constantly radiate energy [39, 49, 50] until they get trapped and eventually pinned between two adjacent lattice sites [51–53].

Apart from the classical discretization Eq. (23) various discrete models having SGE in the continuum limit ($h^2 \to 0$) can be offered [54, 55].

A number of discrete Klein–Gordon equations free of the Peierls–Nabarro potential has been systematically constructed. The Hamiltonian set of such models was obtained by Speight and Ward [55, 56] using the Bogomol'nyi argument [57]. Then a series of momentum-conserving discretizations [58] was obtained. In [59,60] it was shown that PNp-free discrete Klein–Gordon models can be formulated by using a two-point discrete version of the first integral of a static continuum Klein–Gordon equation. Another Hamiltonian PNp-free model has been very recently discovered by Cooper et al. [61]. Standard discretizations of the Klein–Gordon models cannot support the energy and momentum conservation laws simultaneously [62]. A discrete ϕ^4 model supporting kinks traveling with arbitrary subsonic velocity has been derived by Oxtoby et al. [63].

The effects of discreteness on the soliton collisions have been studied in [15, 16, 64–66].

3.3 Radial SGE

The SGE is often used to model domain wall dynamics in different physical systems. When a domain has a shape of a circle (sphere) in a 2D (3D) problem the domain wall having width much smaller than the domain radius can be described approximately by the following 1D radial SGE to address domain wall dynamics in d-dimensional problem [67]

$$\phi_{tt} - \phi_{rr} - (d-1)\frac{\phi_r}{r} + \sin\phi = 0. \tag{25}$$

For large r this reduces to the integrable 1D SGE. The Hamiltonian of the d-dimensional radial SGE is [67]

$$H = \int_{r_0}^{r_1} \left[\frac{\phi_t^2}{2} + \frac{\phi_r^2}{2} + 1 - \cos\phi \right] r^{d-1} dr. \tag{26}$$

Solving Eq. (25) numerically on an interval $[r_0, r_1]$ the authors of [67] found that when a kink hits the r_0 boundary, for small r_0, a breather is reflected.

3.4 Driven SGE

Taking into account external forces and damping gives rise to non-Hamiltonian models except for the case of d.c. drive in the absence of damping. The equation of motion is in a general such case:

$$\phi_{tt} - \phi_{xx} + \sin\phi - \epsilon f(t) + \beta\phi_t = 0, \tag{27}$$

where in the case of harmonic a.c. drive $f(t) = \sin\Omega t$ with Ω being the driving frequency, or in the case of d.c. drive $f(t) =$ const. The damping term is introduced with the viscosity coefficient β.

The first attempts to describe SGE soliton dynamics in the presence of d.c. external force and damping have been done in the late 1970s [40, 68, 69]. McLaughlin and Scott developed a perturbative approach to describe the effect of d.c. drive and damping on the kink-antikink collisions and on the breather [69]. The critical values of the driving force and viscosity coefficient were found which separate the non-destructive kink-antikink collisions from the destructive ones (kink-antikink annihilation). The breather decay rate was estimated ignoring the loss of breather energy due to phonon radiation.

The main tool to describe dynamics of SGE kinks and breathers in the presence of perturbations is the collective coordinate approach [7, 70]. The problem of d.c. forces was given further consideration using the collective variable technique in [71, 72], where breather instabilities and pattern formation were the main subjects. A large driving force causes conversion of the breather into a kink-antikink pair. For small perturbations, the breather achieves a stationary state [71]. The initial phase of the breather φ was not taken into account and this was done later by Legrand and Reinisch [5]. Taking into account the initial phase of the breather is important as the energy gain of the system due to the d.c. driving force strongly depends on φ [5,71].

The effect of dissipation on breathers was studied in [73], where McLaughlin and Overman showed that damped breathers in the absence of driving were anihilated, their energy being dissipated into radiation. Subsequently, Lomdahl and Samuelsen [74,75] showed that the introduction of a.c. forces in the damped SGE compensated those losses and stabilized breathers, whose frequency became modulated by the driving. A nonlinear response of the SGE breather to an a.c. driver has been studied by Forinash and Willis in the frame of the collective variable approach [76]. The authors have demonstrated that the dynamics of the kink and antikink that make up the breather can be successfully described by the single collective variable $r(t)$, which is the separation between the center of masses of the kink and antikink. The ordinary differential equation for $r(t)$ describes dynamics of a point particle with r-dependent mass in a potential $V(r)$. In the frame of the model Eq. (27), for breathers, the force amplitude range for which they exist when dissipation is absent was found [77]. Analytical results were compared with numerical simulations to demonstrate the absence of d.c. motion of breathers under a.c. drive.

4 Effects of Perturbations on SGE Soliton Collisions

Phase shifts of colliding solitons are observed both in integrable and non-integrable models. All other effects discussed here are due to perturbations.

4.1 Radiation

Soliton collisions in non-integrable systems are typically accompanied by radiation of small-amplitude extended waves.

Radiation from a breather in the Frenkel–Kontorova model Eq. (23) has been studied numerically by Boesch and Peyrard [64]. The phonon spectrum of the model ranges from $\omega_{\min} = 1$ to $\omega_{\max} = \sqrt{1 + 4/h^2}$. Radiation of phonons results in gradual decrease of breather energy E_B and increase in frequency ω_B according to (6). When the frequency of the third harmonic $3\omega_B$ becomes greater than $\omega_{\max}$, the rate of radiation drops down and the breather lifetime sharply increases [64]. This effect can be seen only for sufficiently high degree of discreteness, $h^2 > 1/2$, otherwise the phonon spectrum is too wide and $3\omega_B$ cannot leave it because $0 < \omega_B < 1$.

Another intuitively clear effect is the *increase* of phonon emission rate when an odd breather harmonic enters the phonon spectrum, $(2m + 1)\omega_B = \omega_{\min}$, with an integer m [65]. This effect shows up not only for the Frenkel–Kontorova model studied in [65]. In Fig. 2 each dot shows the energy ΔE_n radiated by the breather in nth half-period of oscillation as the function of $\omega_{\min}/\omega_B$. Indeed the radiation rate increases when $\omega_{\min}/\omega_B$ approaches an odd integer. The most prominent resonance takes place for $m = 1$, i.e., when the third harmonic enters the spectrum. Panel (a) shows the results for the Frenkel–Kontorova model Eq. (23) with $h^2 = 0.04$. In panels (b) and (c) the results for SGE with the perturbed potential of Eq. (20) for $r = 0.003$ and $r = -0.003$, respectively, are presented. Higher order peaks can be seen only for weak perturbations, as in the examples given in Fig. 2. For relatively large perturbations the breather radiates energy rapidly and passes through the resonances too fast, so that only the $m = 1$ peak can be seen.

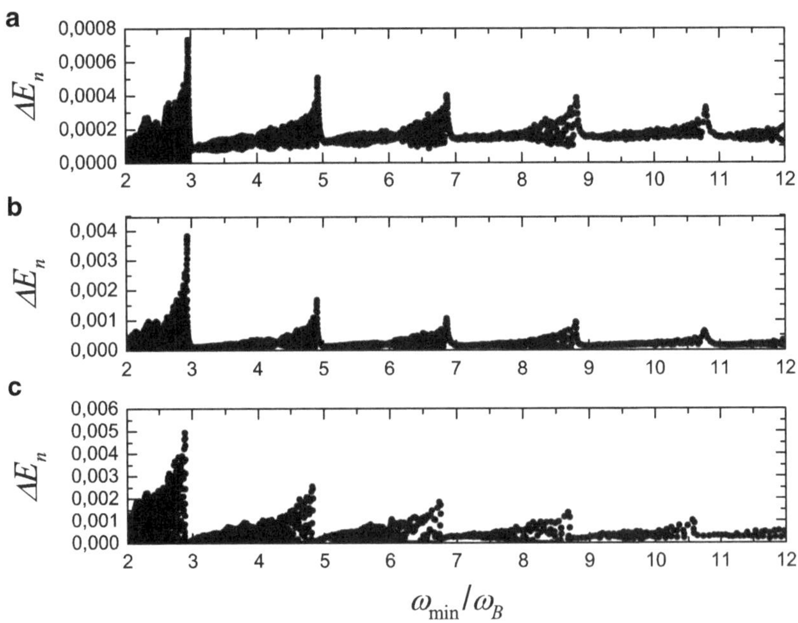

Fig. 2 *Dots* show the energy ΔE_n radiated by the breather during the nth half-period of oscillation as a function of ω_{min}/ω_B, where ω_{min} is the lower edge of the phonon spectrum of the considered model. (**a**) Results for the Frenkel–Kontorova model Eq. (23) with $h^2 = 0.04$. (**b**, **c**) Results for SGE with the perturbed potential Eq. (20) for $r = 0.003$ and $r = -0.003$, respectively

Radiation bursts are emitted by the breather in the form of small-amplitude wave packets when the kink and anti-kink constituting the breather collide and the energy density reaches a maximal value. Such radiation bursts can be seen in Fig. 4 of [65]. When the cores of more than two kinks and antikinks merge in a collision, even higher energy density is achieved and, consequently, the radiation bursts get larger.

4.2 *Energy Exchange Between Kink Translational and Internal Modes*

Kinks in non-integrable models such as the perturbed SGE or ϕ^4 model can support internal vibrational modes [23]. Such modes can be excited when traveling kinks collide with each other. In this process, a part of the kink's energy related to its translational motion is transformed into oscillatory mode energy localized on the kink [78, 79]. Interestingly, the reverse process is also possible when a kink bearing a vibrational mode, as a result of collision with another kink, gains additional momentum from the mode.

This possibility of energy transfer between the translational and vibrational modes was first detailed by Campbell and collaborators [25,78,79] for the ϕ^4 and the double SGE fields. The authors observed in numerical simulations that, as a result of the collision between a kink and an antikink, a part of their kinetic energy is spent to (1) excite soliton's internal modes and (2) emit a burst of radiation. Consequently, the velocity of the kinks after the collision V_{out} was always smaller than their initial velocity V_{in} and only a sufficiently fast kink and antikink could escape their mutual attraction after the first collision, while those moving with velocity smaller than the threshold value were returning to collide again forming a breather-like bound (so-called bion) state. Surprisingly, for particular collision velocities *below* the threshold value the kink and antikink can separate after the second collision gaining some energy back from the soliton internal modes excited at the first collision. The reflection windows in the range of collision velocity below the first collision escape threshold were described with the help of the collective variable approach [25, 78, 79]. Possibly the first numerical observation of such windows dates back to the work of Aubry [80].

Recently Goodman and Haberman have improved the collective variable theory developed in [25,78,79] to demonstrate that the kink and antikink can collide many times before they split, giving rise to a fractal structure of the $V_{\text{out}}(V_{\text{in}})$ dependence [34] (see also the detailed exposition of [19]). Importantly, this theory does not involve fitting parameters.

4.3 Energy Exchange Between Kink Translational and Impurity Modes

The reflection windows were also observed in the kink-impurity interactions by Fei, Kivshar and Vazquez in SGE [30] and ϕ^4 [31] models.

SGE kinks do not bear internal modes [24], however, when a kink hits an impurity, a part of its energy is trapped towards the excitation of the impurity mode [29, 30] and another fraction leads to the emission of radiation bursts [81]. For this reason the kink velocity decreases after passing through the attractive impurity and, if the initial kink velocity is smaller than a threshold value, the kink is trapped by the impurity and starts to oscillate in the potential well created by the impurity. While the kink is in the potential well, the internal mode energy can be given back to it, so that it can be reflected by the impurity or can pass it in the third half oscillation [31]. The threshold kink velocity and the resonance windows for SGE kink interacting with the localized impurity were described employing a collective variable method [29, 30].

ϕ^4 kinks support the presence of an internal mode and this opens the possibility to study the interaction of a kink bearing such a mode with an impurity (i.e., a richer framework). This possibility was realized in the work of [31]. The authors there developed an analytical approach to describe qualitatively the resonance windows observed numerically by the collective coordinate model taking into account the interaction of the kink with the internal mode and with the impurity mode.

4.4 Energy Exchange Between Kink Translational Mode and Phonons

Kinks in non-integrable models do not always support internal modes or may support them only in a part of the perturbation parameter domain, as was noticed in Sect. 3.1. The question then arises if kink-antikink collisions can (still) show non-trivial outcomes in this case and, if yes, what is the underlying mechanism. This problem has been addressed for the parametrically modified SGE with the potential Eq. (20) in [24] and for the ϕ^6 model with the potential Eq. (19) in [20].

As was mentioned above, the parametrically modified SGE Eq. (20) supports kink internal modes only for $r < 0$ with the frequency below the phonon band. For $r = 0$ (integrable case) the mode frequency lies exactly at the bottom edge of the phonon band and for $r > 0$ it enters the phonon band interacting with the odd parity modes close to the lower edge of the phonon band. These modes affect the kink's width (for $r < 0$ the internal mode also does so) [24]. It was found that for $r > 0$ a part of the translational energy of colliding kink and antikink can be given to the odd parity phonon modes with frequencies close to the lower phonon band edge. For kinks colliding with velocity smaller than the threshold value, the energy of these phonon modes can return to the kink translational mode. Although the efficiency of the energy storage by the phonons (for $r > 0$) seems to be smaller than that by the internal modes (for $r < 0$), instead of the reflection windows, the quasi-resonance pattern can be observed in the case $r > 0$ in the form of peaks of the curve $T_2(V_{\text{in}})$, where T_2 is the time between second and third collisions and V_{in} is the incoming kink velocity.

The ϕ^6 model of Eq. (19) supporting no kink internal modes has three vacua, $\phi = \{-1, 0, 1\}$. Collisions between kink and antikink $(0, 1) + (1, 0)$ do not show reflection windows, while the antikink-kink collisions $(1, 0) + (0, 1)$ demonstrate the two-bounce windows [20]. The authors present numerical evidence that this resonant structure is caused by the existence of bound states in the spectrum of small oscillations about a combined kink-antikink configuration. This work demonstrates that the existence of the (individual) kink internal modes is not a necessary condition for the appearance of the two-bounce windows.

4.5 Breather-Impurity Interactions

Scattering of a SGE breather on the square potential well has been studied numerically in the work of [35]. It has been found that the scattering process, apart from the depth and width of the scattering potential, depends on the incoming speed and phase of the breather. The breather can be reflected or trapped by the well. It can also pass through the potential well with a speed different, and sometimes larger, than the initial one. The increase of breather speed as a result of interaction with the impurity can be explained by the reduction of its amplitude (or increase of breather frequency) [see Eq. (6)].

4.6 Kink-Impurity Interactions in 2D Models

Exact solutions to 2D SGE of the form

$$\phi_{tt} - \phi_{xx} - \phi_{yy} + \sin\phi = 0, \tag{28}$$

describing moving kinks, moving breathers, colliding kinks, etc. have been reported (see [82,83] and references therein). The effect of perturbations on soliton dynamics in the 2D case is not well studied yet.

Gornostyrev et al. have studied the kink-impurity interactions in the 2D Frenkel–Kontorova model in the context of the theory of dislocations in crystals taking into account d.c. external force and thermal fluctuations of atoms [84]. Thermal fluctuations help the dislocation to overcome the Peierls–Nabarro barrier making dislocation glide possible even at relatively small external force. They have found that for small temperature the impurity initiates the kink-antikink pair formation on the dislocation line thus increasing the dislocation mobility which leads to solid solution softening. At higher temperatures the impurity traps moving kinks. This process reduces the dislocation mobility leading to the solid solution hardening.

The dynamics of a 180° domain wall in a magnet with two-dimensional inhomogeneities of the parameters of magnetic anisotropy and the conditions of excitation of nonlinear magnetization waves have been investigated using numerical methods by Ekomasov et al. [85]. Three types of localized magnetic inhomogeneities that appear in the region of anisotropic inhomogeneities have been revealed, namely, a pulson, a pulson transforming into a 2D soliton, and a 2D bell-shape soliton. Domains of model parameters that describe the magnetic-anisotropy inhomogeneities, where each of these solutions exists have been found.

4.7 Merger of Kink-Antikink Pairs into Breathers and Vice-Versa

Kink-antikink and breather solutions to the integrable SGE are linked by the separatrix solution of Eq. (14) having energy equal to 16, which is the energy of two kinks at rest. The center of mass of this solution does not move.

A merger of a colliding kink and antikink into a breather is possible in a non-integrable system when energy loss to radiation and/or excitation of the kinks' internal modes is sufficiently large [25, 34, 78, 79].

External d.c. driving force in the absence of damping [$f(t) = \text{const}$, $\beta = 0$ in Eq. (27)] for sufficiently large magnitude of the force causes the breather to split into a kink-antikink pair while for small driving force the breather excitations lead to stationary modes [71].

4.8 *Radiationless Energy and Momentum Exchange Between Colliding Solitons*

4.8.1 General Remarks

For a long time, the excitation of the soliton internal modes and the radiation losses were thought to be the two major effects observed in soliton collisions in the presence of perturbations. However, a qualitatively different effect was recently identified, namely, the radiationless energy exchange (REE) between colliding solitons [14–16, 86–92]. The energy transferred to soliton internal modes in soliton collisions, for a small perturbation strength ϵ, is typically proportional to ϵ^2, and the same is true for the radiation losses (here ϵ is the coefficient in front of a perturbation term, added to an integrable equation) [23, 49, 50, 64, 93]. On the other hand, the degree of inelasticity due to the REE effect, when the latter is present (see details below), grows proportionally to ϵ [14, 15, 87, 88]. This means that for weakly perturbed integrable systems, REE is a dominant effect, while the soliton's internal modes and radiation become increasingly important with stronger deviations from the integrable case.

We note that the terms proportional to ϵ^2 appear as the lowest-order correction terms in the collective variable approaches used to describe the soliton's internal modes [23], the radiation from the discrete SGE kink [49], the kink dynamics in the discrete ϕ^4 equation [50], and the kink and breather dynamics in the discrete SGE [64, 93].

A prototypical example of noticeable energy and momentum exchange between colliding solitons is the weakly-perturbed SGE, provided that three or more kinks collide simultaneously. This demand follows from the fact mentioned above that weak perturbations do not strongly affect the two low-order conservation laws, of the energy Eq. (11) and momentum Eq. (12). These conservation laws set two restrictions on the parameters of colliding solitons, and to have free parameters more than two solitons must participate in the collision.

Energy and momentum exchange is not possible if all colliding kinks have the same topological charge because in this case they repel each other and their cores do not merge during the collisions [cf. Fig. 1(a) for kink-kink and (b) for kink-antikink collisions]. On the other hand, the maximal value of the energy density Eq. (13) is achieved during kink-antikink collisions and can be used to predict the degree of inelasticity of collision; for larger $e_{\max}$ collisions are expected to be more inelastic. For example, the kink-breather collision in Fig. 1(d), (d′) happens in two successive two-soliton collisions with $e_{\max} \approx 8$, while in (e), (e′) all three kinks collide simultaneously and $e_{\max} \approx 20$. The latter collision in the presence of perturbations would be more inelastic.

The energy and momentum exchange is greatly enhanced for near-separatrix multi-soliton solutions [14, 15]. The latter fact will be illustrated below for the three-soliton solutions with the help of the three-particle model (see also [15]).

4.8.2 REE in Kink-Breather Collisions in Weakly Discrete SGE

Below we describe the numerical results for the kink-breather collisions in the weakly discrete ($\epsilon \equiv h^2 = 0.04$) SGE Eq. (23). The exact kink-breather solution to SGE Eq. (9) was employed for setting the initial conditions. The equations of motion Eq. (23) were integrated with the use of the Störmer method of order six with the time step $\tau = 0.005$. We register the parameters of quasi-particles after their collision and compare them with those before the collision. The larger the change in the parameters, the more inelastic the collision is.

Without loss of generality, we assume $P_K + P_B = 0$. Then we have two parameters, the breather momentum P_B and frequency ω. The outcome of the KB collisions is studied as a function of the initial separation between the kink and the breather controlled by the initial kink position $(x_0)_K$. Change of $(x_0)_K$ results in the change of breather phase at the collision point.

In Fig. 3 we plot the momenta of kinks after collision, $\tilde{P}_{K_j}$ (including the kinks constituting the breather, assuming that they share the breather's momentum equally), as a function of $(x_0)_K$ for (a) $P_B = 2.5$ (larger collision velocity) and (b) $P_B = 1.6$ (smaller collision velocity) with $\omega = 0.05$ in both cases. One can see that strong REE is possible in the KB collisions. Note that in Fig. 3 only a small part of one period of the output functions is shown for the region with strong REE. In (a) there is a range of $(x_0)_K$ where the breather obtains enough energy to split into a kink-antikink pair [example is shown in (a′)]. In (b), in addition to this possibility, there appears a region where the breather is reflected from the kink [example is shown in (b′)].

We point out again that the kink-breather collisions are strongly inelastic only in a narrow range of $(x_0)_K$ or, in other words, in a narrow range of the breather phase at the collision point, when the cores of all three kinks merge. Away from this narrow range collisions are nearly elastic. The ease of energy exchange between kinks in the presence of weak perturbation in a narrow range of soliton parameters suggests that such collisions happen in the vicinity of a separatrix solution that connects qualitatively different solutions. Indeed, it was demonstrated that two qualitatively different solutions to the integrable SGE can be very close to each other and thus, under the action of a weak perturbation, one of them can be transformed into another [94]. Particularly it was shown that two qualitatively different kink-breather

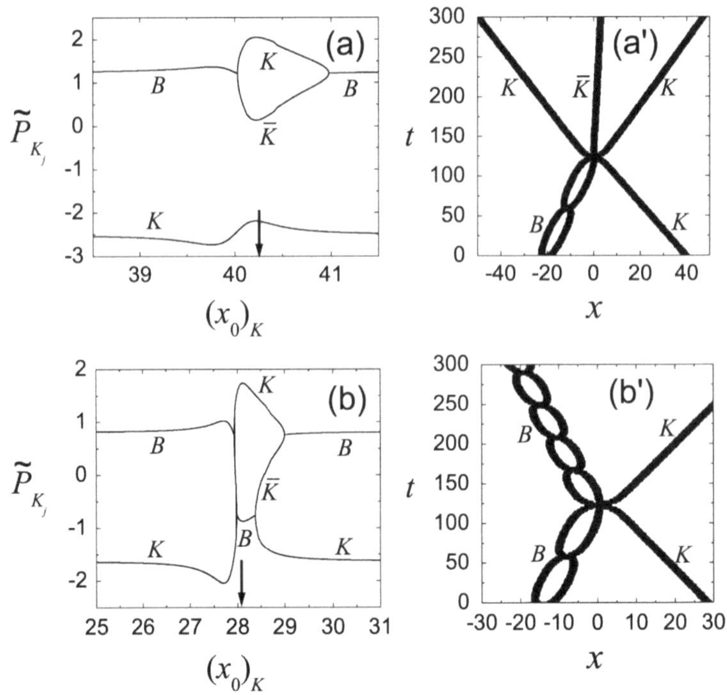

Fig. 3 Numerical results for the kink-breather collisions in weakly discrete SGE for (**a**, **a**′) $P_B = 2.5$ (larger collision velocity) and (**b**, **b**′) $P_B = 1.6$ (smaller collision velocity) and $\omega = 0.05$ in both cases. It is assumed that the kinks constituting the breather share momentum equally. The *left panels* show the momenta of the kinks after the collision $\tilde{P}_{K_j}$ as a function of the initial position of the kink, $(x_0)_K$. The *right panels* show examples of collisions on the (x,t) plane for (**a**′) $(x_0)_K = 40.25$ and (**b**′) $(x_0)_K = 28.1$

solutions $\phi(x,t)$ and $\phi^*(x,t)$ of the form of Eq. (9), with soliton parameters related as $V_K = -V_K^*$, $V_B = -V_B^*$, $\omega_B = \omega_B^*$ at the time of collision $t = t_c$ have exactly identical profiles $\phi(x,t_c) \equiv \phi^*(x,t_c)$ [94]. The velocity fields are related as $\phi_t(x,t_c) = -\phi_t^*(x,t_c)$. Under additional conditions $V_K \to 0$, $V_B \to 0$ one has $\phi_t(x,t_c) \to \phi_t^*(x,t_c)$ for any x. Separatrix three-soliton solutions to SGE can be visualized with the help of the three-particle model [15].

4.9 Fractal Soliton Scattering and Short-Lived Multi-soliton States

For some case example parameters (e.g. soliton velocities after collision), the collisional outcome as a function of the initial soliton parameters can exhibit self-similar patterns when observed at different scale levels [15, 16, 34, 90, 95–97].

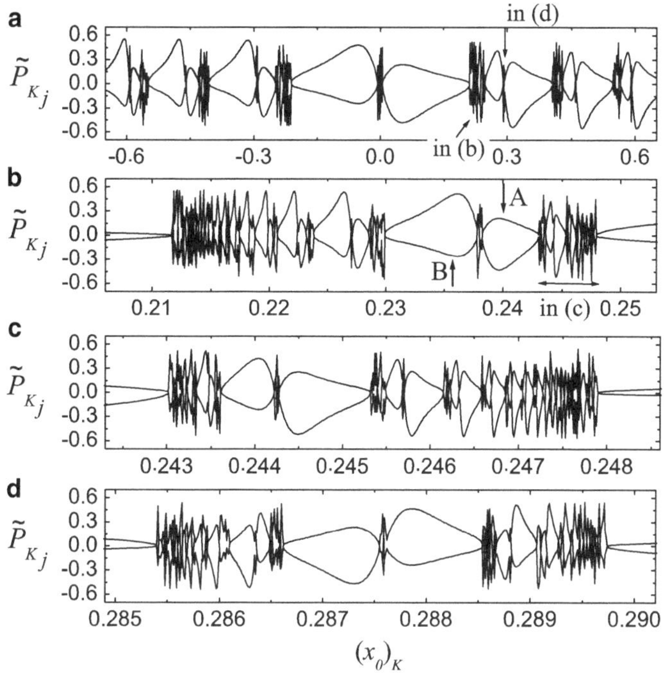

Fig. 4 Fractal kink-breather scattering observed for $\omega = 0.3$, $P_B = 0$ (kink and breather have zero initial velocities). The kinks' momenta after the collision $\tilde{P}_{K_j}$ are shown as the functions of the initial position of the kink, $(x_0)_K$, at different scales. At each scale smooth regions are separated by the apparently chaotic regions. Panels (**b**) and (**d**) present blowups of the regions indicated in panel (**a**); (**c**) is a blowup of the region indicated in (**b**)

Such self-similar structures, or fractals, can originate from the resonances with the internal shape mode oscillations [34, 95–97] or from REE [15, 16, 90]. Here we discuss in detail the latter source of fractal patterns in the soliton collision outcomes.

4.9.1 Kink-Breather Collisions

When a kink and a breather collide with sufficiently small relative velocity, their momenta after the collision as functions of the initial separation between them exhibit a fractal form. An example is presented in Fig. 4 for $\omega = 0.3$, $P_B = 0$ (kink and breather have zero initial velocities), where the soliton's momenta after collision $\tilde{P}_{K_j}$ are shown as the functions of initial distance between kink and breather, $(x_0)_K$.

The structure presented in Fig. 4 can be described as a chain of self-similar patterns. Panels (b) and (d) present blowups of the regions indicated in panel (a); (c) is a blowup of the region indicated in (b). At each scale smooth regions are separated by the apparently chaotic regions.

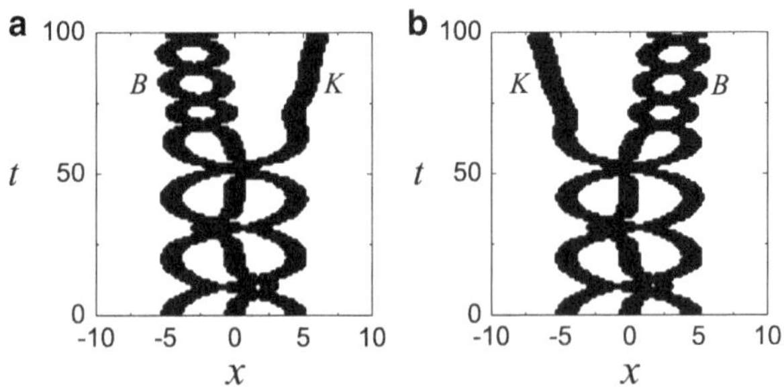

Fig. 5 Examples of the kink-breather dynamics for (**a**) $(x_0)_K = 0.24$ and (**b**) $(x_0)_K = 0.236$ [indicated in Fig. 4(b) by the arrows A and B, respectively]

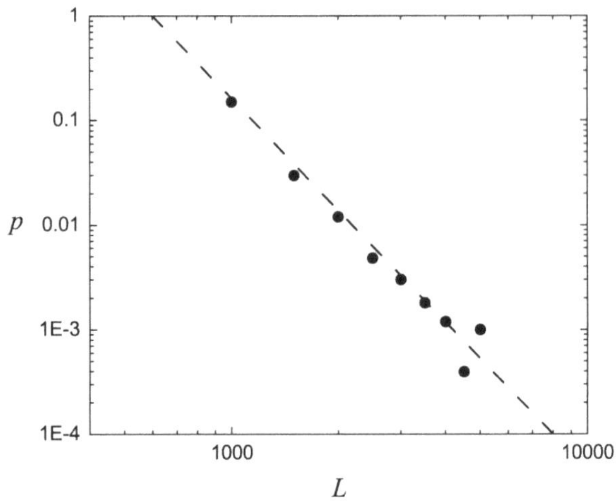

Fig. 6 Probability p to observe the kink-breather system with the lifetime L (in Fig. 5 we have $L \approx 55$). Numerical data is shown by *dots* and only the cases with $L > 1000$ were taken into account. The *dashed line* is a guide to the eye and it has slope -3.5

Two examples of the kink-breather dynamics are given in Fig. 5 for (a) $(x_0)_K = 0.24$ and (b) $(x_0)_K = 0.236$ [indicated in Fig. 4(b) by the arrows A and B, respectively]. The three-particle solution has a certain lifetime L (in this example $L \approx 55$) and then it splits into a kink and a breather. Similar dynamics has been reported, e.g., for the breather-breather system in the weakly discrete Frenkel–Kontorova model [16], in the weakly perturbed NLSE [90], and recently for the generalized NLSE [98]. Thus, this type of dynamics is rather general. For the two-soliton collisions in the weakly perturbed NLSE we have estimated numerically the probability p to observe the three-particle system with the lifetime L and found that $p \sim L^{-3}$ [90]. Here we carry out a similar estimation for the kink-breather solution in the Frenkel–Kontorova model and the result is shown in Fig. 6. The numerical data can be fitted as $p \sim L^{-3.5}$. There is evidence that for sufficiently small frequency of the breather the kink-breather system in the Frenkel–Kontorova model with a small h^2 never splits [89].

All the important features of the kink-breather fractal scattering including the existence of the two qualitatively different stochastic regions in the fractal structure will be clarified in the following with the help of the three-particle model.

4.9.2 Three-Particle Model

Attempting to explain the fractal soliton scattering observed in the kink-breather collisions in the weakly perturbed SGE reported above, we consider the solitary waves as effective point particles, and study the dynamics of three such particles in one-dimensional space. The particles have mass $m = 8$, which is the rest mass of SGE kink, and they carry topological charges $q_j = \pm 1$. Particles with $q_j = 1$ ($q_j = -1$) will be called kinks (respectively, antikinks) by analogy with the SGE solitons. We assume that particles i and j, having coordinates x_i and x_j, interact via the potential

$$U_{ij}(r_{ij}) = 16 + q_i q_j \frac{16}{\cosh(r_{ij})}, \quad r_{ij} = x_j - x_i, \tag{29}$$

which qualitatively approximates the interaction of two SGE kinks. The potential of Eq. (29) is attractive for $q_i = -q_j$ and repulsive for $q_i = q_j$. The binding energy of the kink-antikink pair is equal to 16, which is the energy of two standing SGE kinks. Note that for the kink and antikink at any finite distance r_{ij} the potential energy $U_{ij}(r_{ij})$ is less than 16. If the kinetic energy of relative motion of the particles is less than $16 - U_{ij}(r_{ij})$, then the particles cannot escape the mutual attraction and they form an oscillatory bound state, i.e., a breather.

The Hamiltonian of the three-particle system is

$$H = \frac{m}{2} \sum_{j=1}^{3} v_j^2 + U_{12}(r_{12}) + U_{13}(r_{13}) + U_{23}(r_{23}), \tag{30}$$

where $v_j = dx_j/dt$, and there is one more integral of motion, namely the conservation of momentum. Without loss of generality, we assume that the total momentum in the system is equal to zero, i.e., $m(v_1 + v_2 + v_3) = 0$. Introducing new variables

$$x_2 - x_1 \to \sqrt{3}\alpha + \beta, \;\; x_3 - x_1 \to 2\beta, \;\; t \to \sqrt{2m}t, \tag{31}$$

the Hamiltonian of Eq. (30) can be presented in the form

$$H = \frac{1}{2}\left(\dot{\alpha}^2 + \dot{\beta}^2\right) + U_{12}(\sqrt{3}\alpha + \beta) \\ + U_{13}(2\beta) + U_{23}(\sqrt{3}\alpha - \beta), \tag{32}$$

which is the Hamiltonian of a unit-mass particle moving in the two-dimensional scattering potential.

Now we solve numerically three equations of motion which can be derived from the Hamiltonian Eq. (30) and, inverting Eq. (31), present the three-particle dynamics by the trajectory of the particle in the (α, β)-plane.

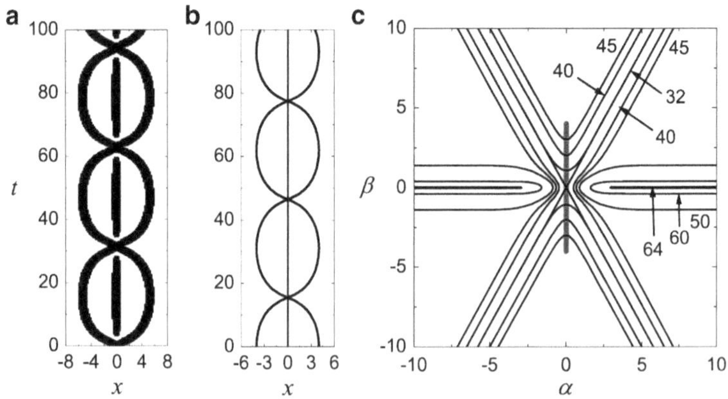

Fig. 7 (**a**) The SGE solution Eq. (15) for $\omega = 0.2$; (**b**) the three-particle dynamics with $q_1 = -q_2 = q_3 = 1$, $(v_0)_1 = (v_0)_2 = (v_0)_3 = 0$, $(x_0)_1 = -(x_0)_3 = -4$, and $(x_0)_2 = 0$ in the (x, t) space; (**c**) the *red line* shows the corresponding trajectory of the particle in the scattering potential in the (α, β)-plane (isopotential lines are shown in *black*). The particle in (**c**) oscillates along the potential ridge and this motion is, obviously, unstable. The picture in (**c**) gives a visual image of the separatrix kink-breather solution Eq. (15)

In Fig. 7 we plot (a) the SGE solution Eq. (15) for $\omega = 0.2$, (b) the three-particle dynamics in the (x, t) space for $q_1 = -q_2 = q_3 = 1$, $(v_0)_1 = (v_0)_2 = (v_0)_3 = 0$, $(x_0)_1 = -(x_0)_3 = -4$, and $(x_0)_2 = 0$, and in (c) the red line shows the corresponding trajectory in the (α, β)-plane. The particle in (c) oscillates along the potential ridge and this motion is, obviously, unstable. The picture presented in Fig. 7(c) gives a visual image of the separatrix kink-breather solution of Eq. (15).

When the red line passes the origin of the (α, β)-plane, from Eq. (31) one has $x_1 = x_2 = x_3$, i.e., all three particles meet at one point. In the SGE this corresponds to simultaneous collision of all three kinks.

Let us select the parameters of the three particles so as to simulate the collisions between a kink and a breather. In particular, we set the charges of the particles as $q_1 = -q_2 = q_3 = 1$, their initial velocities as $(v_0)_1 = (v_0)_2 = 0.3$, $(v_0)_3 = -0.6$; the initial positions of the particles constituting the "breather" are $(x_0)_1 = -16$, $(x_0)_2 = -13.5$, and the third particle initial position was varied. In Fig. 8 the results are shown for (a) $(x_0)_3 = 30.51$, (b) $(x_0)_3 = 23.398$, and (c) $(x_0)_3 = 23.391$. The top panels show the three-particle dynamics in the (x, t) space, while the bottom panels show the corresponding trajectory of the particle in the (α, β)-plane (red line).

Collisions in (a) and (b) are elastic but the difference is that while in (a′) the particle does not move along the separatrix line $\alpha = 0$, in (b′) it does, and a very small change in the initial conditions is sufficient to have a qualitatively different result of the collision, as presented in (c), (c′), where the breather reflects from the kink [compare (c) with actual kink-breather collision in the Frenkel–Kontorova model shown in Fig. 3(b′)].

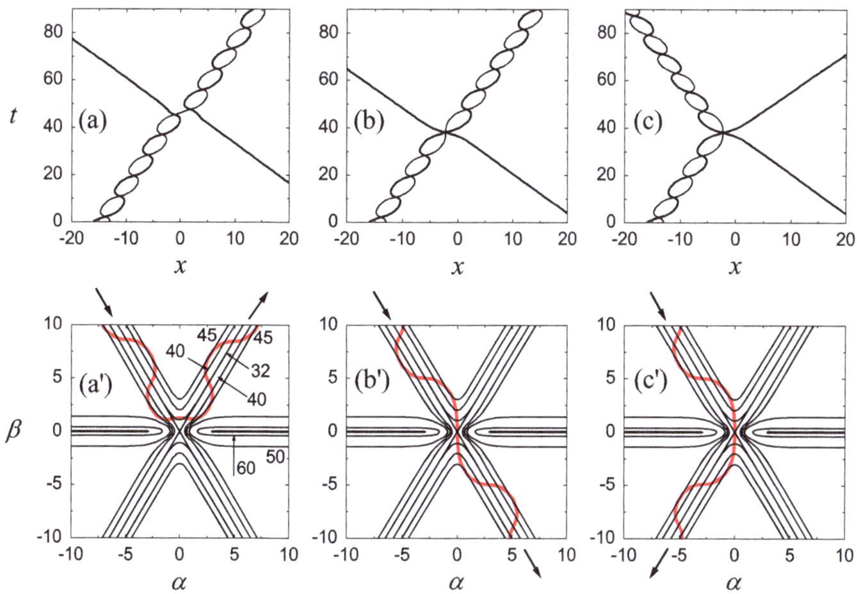

Fig. 8 Three-particle model simulating the kink-breather collisions. The *top panels* show the three-particle dynamics in the (x, t) space, while the *bottom panels* show the corresponding trajectory of the particle in the (α, β)-plane (*red line*). Only the initial position of the third particle is varied: (**a**) $(x_0)_3 = 30.51$, (**b**) $(x_0)_3 = 23.398$, and (**c**) $(x_0)_3 = 23.391$. Collisions in (**a**, **a**$'$) and (**b**, **b**$'$) are elastic but in the latter case it is close to the separatrix (see Fig. 7) resulting in a great sensitivity to variations in initial conditions, as demonstrated in (**c**, **c**$'$). The rest of the parameters are chosen as $q_1 = -q_2 = q_3 = 1$, $(v_0)_1 = (v_0)_2 = 0.3$, $(v_0)_3 = -0.6$, $(x_0)_1 = -16$, $(x_0)_2 = -13.5$

To reproduce the kink-breather fractal scattering described above for the Frenkel–Kontorova model we set the following parameters for the particles in the three-particle model: $q_1 = -q_2 = q_3 = 1$, $(v_0)_1 = (v_0)_2 = (v_0)_3 = 0$, $(x_0)_1 = -(x_0)_3 = -5$, and variable $(x_0)_2$. It was found that the velocities of particles after collision $\tilde{v}_j$ as the functions of $(x_0)_2$ reveal the qualitative similarity with the results presented in Fig. 4 for the kink breather collisions in the Frenkel–Kontorova model.

The three-particle model is useful because it gives the possibility to analyze the kink-breather fractal scattering from a different point of view, namely, by looking at the corresponding dynamics of the particle in the scattering potential in the (α, β) space.

It is well-known that the probability p of the time delay T for the particle interacting with the scattering potential *without* the periodic orbits decreases exponentially with T while in the presence of the periodic orbits it decreases algebraically [99]. The scattering potential in our case does have the periodic orbits (one of them is presented in Fig. 7) and the probability p to observe a bound state with the lifetime L (analogous to the time delay T) decreases algebraically,

$p \sim L^{-\alpha}$. This was found in [90] for the two-soliton collisions in the weakly perturbed NLSE, and in the present study this was also confirmed for the kink-breather system in the Frenkel–Kontorova model, as presented in Fig. 6.

5 Summary

We have discussed the major effects of perturbations on the soliton collisions and soliton-impurity interactions in Klein–Gordon systems including the perturbed sine-Gordon, ϕ^4, and ϕ^6 models (see Sect. 3).

The only consequence of soliton collisions in integrable systems is their phase shifts. On the contrary, soliton collisions in non-integrable systems are far more complex. In particular, they can result in (1) radiation of energy in the form of small-amplitude wave packets, (2) excitation of soliton's internal modes when such modes are supported by the solitons, and (3) radiationless energy exchange (REE) between colliding solitons. The energy spent to radiation and excitation of the soliton's internal modes increases as ϵ^2, while REE increases proportionally to ϵ, where ϵ is the perturbation parameter. This means that for weakly perturbed integrable systems REE is a dominant effect, while the soliton's internal modes and radiation become increasingly important with stronger deviation from the integrable case. On the other hand, REE in weakly perturbed sine-Gordon systems is possible only when more than two solitons collide, because the two low-order conservation laws, Eqs. (11) and (12), set two restrictions on the soliton parameters.

Analysis of the exact multi-soliton solutions to the unperturbed SGE has revealed the existence of narrow regions of soliton parameters when they collide with a large energy density (see Fig. 1). For such cases the effect of perturbations, regardless of their nature, is expected to be maximal. This is so because the energy density Eq. (13) is large only when ϕ_x and/or ϕ_t are large and then the conservation laws Eqs. (11) and (12) are largely affected, making the soliton collisions less elastic.

The excitation of the solitons' internal modes is responsible (but as discussed, not absolutely necessary) for the complex fractal structure of the reflection "windows" observed for the kink-antikink collisions with the velocity smaller than the escape threshold (Sect. 4.2). Similar effects are observed for the kink-impurity interactions due to the resonant energy exchange between the kink's translational mode and the impurity mode (Sect. 4.3). Much less explored is the energy exchange between the kink's translational mode and the phonon modes, that can also result in the fractal reflection "windows" or, at least, can modify the escape velocity of the kinks when they collide with large velocities (Sect. 4.4). REE can also produce fractal soliton scattering, as discussed in Sect. 4.9. In this case the separatrix multi-soliton solutions, e.g., Eq. (15), play an important role.

Despite the very substantial volume of studies on the field of soliton collisions, there is still a considerable volume of open problems. Among them, let us first mention the need to further investigate the energy exchange between the kink's translational mode and the phonon modes. The mechanisms (and rates) of energy transfer between these two types of states warrant a more substantial understanding. The role of higher dimensionality, and how it affects collisions of both quasi-one-dimensional and also genuinely higher dimensional (e.g. 2D breather) structures constitute another significant open problem. Although rather non-generic, the study of collisional examples with an increasing number of colliding solitons is expected to shed further light on the role and nature of phenomena such as the radiationless energy exchange and on how the maximal energy density and strength of the nonlinearity may affect the inelasticity of soliton collisions.

References

1. V.E. Zakharov, A.B. Shabat, Sov. Phys. JETP **34**, 62 (1972)
2. R. Hirota, J. Phys. Soc. Jpn. **33**, 1459 (1972)
3. R.K. Dodd, J.C. Eilbeck, J.D. Gibbon, H.C. Morries, *Solitons and Nonlinear Wave Equations* (Academic, London, 1982)
4. A.P. Fordy, in *Harmonic Maps and Integrable Systems*, ed. by A.P. Fordy, J.C. Wood (Vieweg, Wiesbaden, 1994), pp. 7–28
5. O. Legrand, G. Reinisch, Phys. Lett. A **35**, 3522 (1987)
6. O. Legrand, Phys. Lett. A **36**, 5068 (1987)
7. J.G. Caputo, N. Flytzanis, Phys. Rev. A **44**, 6219 (1991)
8. J.F. Currie, Phys. Rev. A **16**, 1692 (1977)
9. I.V. Baryakhtar, V.G. Baryakhtar, E.N. Economou, Phys. Rev. E **60**, 6645 (1999)
10. I.V. Baryakhtar, V.G. Baryakhtar, E.N. Economou, Phys. Lett. A **207**, 67 (1995)
11. A. Chodos, Phys. Rev. D **21**, 2818 (1980)
12. H. Sanuki, K. Konno, Phys. Lett. A **48**, 221 (1974)
13. Y.S. Kivshar, B.A. Malomed, Phys. Lett. A **115**, 377 (1986)
14. A.E. Miroshnichenko, S.V. Dmitriev, A.A. Vasiliev, T. Shigenari, Nonlinearity **13**, 837 (2000)
15. S.V. Dmitriev, P.G. Kevrekidis, Yu.S. Kivshar, Phys. Rev. E **64**, 046604 (2008)
16. S.V. Dmitriev, Yu.S. Kivshar, T. Shigenari, Phys. Rev. E **64**, 056613 (2001)
17. L.A. Ferreira, B. Piette, W.J. Zakrzewski, Phys. Rev. E **77**, 036613 (2008)
18. M. Peyrard, D.K. Campbell, Phys. D **9**, 33 (1983)
19. R.H. Goodman, Chaos **18**, 023113 (2008)
20. P. Dorey, K. Mersh, T. Romanczukiewicz, Y. Shnir, Phys. Rev. Lett. **107**, 091602 (2011)
21. M. Peyrard, M. Remoissenet, Phys. Rev. B **26**, 2886 (1982)
22. M. Remoissenet, M. Peyrard, Phys. Rev. B **29**, 3153 (1984)
23. Yu.S. Kivshar, D.E. Pelinovsky, T. Cretegny, M. Peyrard, Phys. Rev. Lett. **80**, 5032 (1998)
24. N.R. Quintero, P.G. Kevrekidis, Phys. D **170**, 31 (2002)
25. D.K. Campbell, M. Peyrard, P. Sodano, Phys. D **19**, 165 (1986)
26. V.A. Gani, A.E. Kudryavtsev, Phys. Rev. E **60**, 3305 (1999)
27. R. Ravelo, M. El-Batanouny, C.R. Willis, P. Sodano, Phys. Rev. B **38**, 4817 (1988)
28. B.A. Malomed, J. Phys. C Solid State Phys. **21**, 5163 (1988)
29. Y.S. Kivshar, Z. Fei, L. Vazquez, Phys. Rev. Lett. **67**, 1177 (1991)
30. Z. Fei, Yu.S. Kivshar, L. Vazquez, Phys. Rev. A **45**, 6019 (1992)
31. Z. Fei, Yu.S. Kivshar, L. Vazquez, Phys. Rev. A **46**, 5214 (1992)
32. B.A. Malomed, D.K. Campbell, N. Knowles, R.J. Flesch, Phys. Lett. A **178**, 271 (1993)
33. P. Woafo, T.C. Kofane, Solid State Commun. **89**, 261 (1994)
34. R.H. Goodman, R. Haberman, Phys. Rev. Lett. **98**, 104103 (2007)
35. B. Piette, W.J. Zakrzewski, Phys. Rev. E **79**, 046603 (2009)
36. E.G. Ekomasov, Sh.A. Azamatov, R.R. Murtazin, Phys. Met. Metallogr. **105**, 313 (2008)

37. E.G. Ekomasov, R.R. Murtazin, Sh.A. Azamatov, Phys. Solid State **54**, 1584 (2012)
38. E.G. Ekomasov, R.R. Murtazin, O.B. Bogomazova, A.M. Gumerov, J. Magn. Magn. Mater. **339**, 133 (2013)
39. J.F. Currie, S.E. Trullinger, A.R. Bishop, J.A. Krumhansl, Phys. Rev. B **15**, 5567 (1977)
40. M.B. Fogel, S.E. Trullinger, A.R. Bishop, J.A. Krumhansl, Phys. Rev. B **15**, 1578 (1977)
41. N. Riazi, A. Sheykhi, Phys. Rev. D **74**, 025003 (2006)
42. O.M. Braun, Yu.S. Kivshar, *The Frenkel-Kontorova Model: Concepts, Methods, and Applications* (Springer, Berlin, 2004)
43. Yu.S. Kivshar, B.A. Malomed, Rev. Mod. Phys. **61**, 763 (1989)
44. R. Boesch, C.R. Willis, Phys. Rev. B **39**, 361 (1989)
45. C. Willis, M. El-Batanouny, P. Stancioff, Phys. Rev. B **33**, 1904 (1986)
46. P. Stancioff, C. Willis, M. El-Batanouny, S. Burdick, Phys. Rev. B **33**, 1912 (1986)
47. S.P. Fitzgerald, D. Nguyen-Manh, Phys. Rev. Lett. **101**, 115504 (2008)
48. A.M. Dikande, Phys. Lett. A **304**, 143 (2002)
49. Y. Ishimori, T. Munakata, J. Phys. Soc. Jpn. **51**, 3367 (1982)
50. A. Combs, S. Yip, Phys. Rev. B **28**, 6873 (1983)
51. M. Peyrard, M.D. Kruskal, Phys. D **14**, 88 (1984)
52. P.G. Kevrekidis, M.I. Weinstein, Phys. D **142**, 113 (2000)
53. L.A. Cisneros, A.A. Minzoni, Phys. D **237**, 50 (2008)
54. M. Boiti, F. Pempinelli, B. Prinari, A. Spire, Inverse Probl. **18**, 1309 (2002)
55. J.M. Speight, R.S. Ward, Nonlinearity **7**, 475 (1994)
56. J.M. Speight, *ibid.*, Nonlinearity **10**, 1615 (1997); **12**, 1373 (1999)
57. E.B. Bogomol'nyi, J. Nucl. Phys. **24**, 449 (1976)
58. P.G. Kevrekidis, Phys. D **183**, 68 (2003)
59. S.V. Dmitriev, P.G. Kevrekidis, N. Yoshikawa, J. Phys. A Math. Gen. **38**, 7617 (2005)
60. I.V. Barashenkov, O.F. Oxtoby, D.E. Pelinovsky, Phys. Rev. E **72** 35602R (2005)
61. F. Cooper, A. Khare, B. Mihaila, A. Saxena, Phys. Rev. E **72**, 36605 (2005)
62. S.V. Dmitriev, P.G. Kevrekidis, N. Yoshikawa, J. Phys. A Math. Gen. **39**, 7217 (2006)
63. O.F. Oxtoby, D.E. Pelinovsky, I.V. Barashenkov, Nonlinearity **19**, 217 (2006)
64. R. Boesch, M. Peyrard, Phys. Rev. B **43**, 8491 (1991)
65. S.V. Dmitriev, T. Shigenari, K. Abe, A.A. Vasiliev, A.E. Miroshnichenko, Comput. Mater. Sci. **18**, 303 (2000)
66. I. Roy, S.V. Dmitriev, P.G. Kevrekidis, A. Saxena, Phys. Rev. E **76**, 026601 (2007)
67. J.-G. Caputo, M.P. Soerensen, Phys. Rev. E **88**, 022915 (2013)
68. M.B. Fogel, S.E. Trullinger, A.R. Bishop, J.A. Krumhans, Phys. Rev. Lett. **36**, 1411 (1976)
69. D.W. McLaughlin, A.C. Scott, Phys. Rev. A **18**, 1652 (1978)
70. A. Sánchez, A.R. Bishop, SIAM Rev. **40**, 579 (1998)
71. P.S. Lomdahl, O.H. Olsen, M.R. Samuelson, Phys. Rev. A **29**, 350 (1984)
72. J.C. Ariyasu, A.R. Bishop, Phys. Rev. B **35**, 3207 (1987)
73. D.W. McLaughlin, E.A. Overman, Phys. Rev. A **26**, 3497 (1982)
74. P.S. Lomdahl, M.R. Samuelsen, Phys. Rev. A. **34**, 664 (1986)
75. P.S. Lomdahl, M.R. Samuelsen, Phys. Lett. A. **128**, 427 (1988)
76. K. Forinash, C.R. Willis, Phys. D **149**, 95 (2001)
77. N.R. Quintero, A. Sánchez, Eur. Phys. J. B **6**, 133 (1998)
78. D.K. Campbell, J.S. Schonfeld, C.A. Wingate, Phys. D **9**, 1 (1983)
79. D.K. Campbell, M. Peyrard, Phys. D **18**, 47 (1986)
80. S. Aubry, J. Chem. Phys. **64**, 3392 (1976)
81. B.A. Malomed, Phys. D **15**, 385 (1985)
82. S. Johnson, P. Suarez, A. Biswas, Comput. Math. Math. Phys. **52**, 98 (2012)
83. A.M. Wazwaz, J. Appl. Math. Inform. **30**, 925 (2012)
84. Yu.N. Gornostyrev, M.I. Katsnelson, A.Y. Stroev, A.V. Trefilov, Phys. Rev. B **71**, 094105 (2005)
85. E.G. Ekomasov, R.R. Murtazin, Sh.A. Azamatov, A.E. Ekomasov, Phys. Met. Metallogr. **112**, 213 (2011)

86. S.V. Dmitriev, L.V. Nauman, A.A. Ovcharov, M.D. Starostenkov, Russ. Phys. J. **39**, 164 (1996)
87. H. Frauenkron, Yu.S. Kivshar, B.A. Malomed, Phys. Rev. E **54**, R2244 (1996)
88. S.V. Dmitriev, L.V. Nauman, A.M. Wusatowska-Sarnek, M.D. Starostenkov, Phys. Status Solidi B **201**, 89 (1997)
89. S.V. Dmitriev, T. Miyauchi, K. Abe, T. Shigenari, Phys. Rev. E **61**, 5880 (2000)
90. S.V. Dmitriev, T. Shigenari, Chaos **12**, 324 (2002)
91. S.V. Dmitriev, D.A. Semagin, A.A. Sukhorukov, T. Shigenari, Phys. Rev. E **66**, 046609 (2002)
92. S.V. Dmitriev, P.G. Kevrekidis, B.A. Malomed, D.J. Frantzeskakis, Phys. Rev. E **68**, 056603 (2003)
93. R. Boesch, C.R. Willis, M. El-Batanouny, Phys. Rev. B **40**, 2284 (1989)
94. S.V. Dmitriev, T. Shigenari, A.A. Vasiliev, A.E. Miroshnichenko, Phys. Lett. A **246**, 129 (1998)
95. P. Anninos, S. Oliveira, R.A. Matzner, Phys. Rev. D **44**, 1147 (1991)
96. J. Yang, Y. Tan, Phys. Rev. Lett. **85**, 3624 (2000)
97. Y. Tan, J. Yang, Phys. Rev. E **64**, 056616 (2001)
98. Y. Zhu, R. Haberman, J. Yang, Phys. Rev. Lett. **100**, 143901 (2008)
99. E. Ott, T. Tel, Chaos **3**, 417 (1993)

Effects of Radiation on sine-Gordon Coherent Structures in the Continuous and Discrete Cases

Luis A. Cisneros-Ake and A.A. Minzoni

Abstract We present a modulation approach coupled with radiation to study the behavior of one and two space dimensional coherent structures in the sG equation. We show how the radiation induced motion can be coupled to the collective coordinates which describe the soliton. It is also shown how the usual geometric optics can be used to calculate the effect of two dimensional spiral waves on solitons. In turn this allows to study the stabilization of coherent solutions by internal degrees of freedom or the so-called PN effect. We point out some open questions in the last section.

Keywords Averaged Lagrangian • Born-Infeld equations • Coherent structure • Discrete WKB • Kink • Moyal product • Nonlinear Schrödinger • Peierls–Nabarro potential • Poisson summation formula • Radiation • Skyrme model

1 The sG Equation in One Space Dimension

The solution for the sG equation can be obtained using the inverse scattering [1] for both the line and periodic problem. However the relation of the initial values to the spectral data involves solving an eigenvalue problem which in general can only be solved approximately. Also in order to obtain the solution one must solve an integral equation which again must be solved approximately. Because of this reason

L.A. Cisneros-Ake (✉)
Department of Mathematics, ESFM, Instituto Politécnico Nacional, Unidad Profesional Adolfo López Mateos Edificio 9, 07738 México, D.F., México
e-mail: cisneros@esfm.ipn.mx

A.A. Minzoni
FENOMEC, Department of Mathematics and Mechanics, IIMAS, Universidad Nacional Autónoma de México, Apdo. 20-726, 01000 México, D.F., México
e-mail: tim@mym.iimas.unam.mx

J. Cuevas-Maraver et al. (eds.), *The sine-Gordon Model and its Applications*, Nonlinear Systems and Complexity 10, DOI 10.1007/978-3-319-06722-3_4,

it is useful to construct directly for a family of physically relevant initial conditions, an asymptotic solution independently of the inverse scattering ideas which could be used for sG like problems which are not only one space dimensional.

The main question in the construction of an asymptotic solution is the study of the interaction between the coherent structure; the kink soliton and the shed radiation. This involves the coupling of the finite dimensional degrees of freedom of the kink with the infinite dimensional superposition of radiative modes. In this chapter we will show how this can be formulated using appropriate trial functions and conservation laws for the equation. It will be shown that this approach provides a unified understanding of the solution and a very good method for a quantitative study of this evolution.

1.1 Asymptotics for the Adjustment of sG Kink Solutions Due to Emission of Radiation

Since the inverse scattering solution for the continuous sG equation is based on a similar eigenvalue problem as for the NLS equation. We expect for the sG equation a similar behavior for the evolution of the radiation. For the NLS equation this asymptotic study was performed by Kath and Smyth in [2]. They showed that a soliton like initial condition develops a pedestal (shelf) structure around the core. This is due to the long wave radiation, whose dispersion relation has very small group velocity for long waves. This radiation couples to the core of the soliton and provides the mechanism for mass shedding. Thus the initial value settles to the steady solution by emitting radiation as it oscillates. This oscillation is produced by the coupling between the soliton width and the long wave radiation.

Since the dispersion relation for the linear waves in the sG equation has also small group velocity for long waves we expect a similar mechanism to control the evolution of the kink solution. This was explored in detail in [3]. To study this problem the sG equation is taken in the form:

$$u_{tt} - u_{xx} + \sin u = 0, \tag{1}$$

with initial conditions $u(x,0)$ and $u_t(x,0)$ are derived from the kink like solution

$$u = 4 \arctan \exp\left(-\frac{x - U_0 t}{w_0}\right). \tag{2}$$

The width w_0 of the kink and its velocity U_0 satisfy the nonlinear dispersion relation

$$w_0 = \frac{1}{\sqrt{1 - U_0^2}}, \tag{3}$$

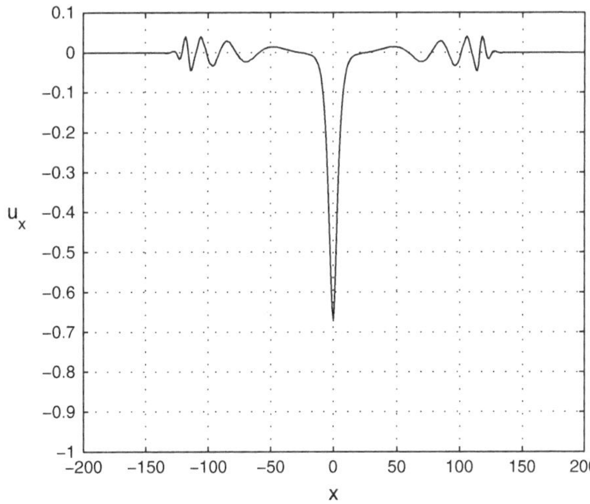

Fig. 1 Space derivative of coherent structure (2) and linear dispersed radiation from the full numerics of (1) at $\xi(0) = 0$, $\dot{\xi}(0) = U_0 = 0$, $w(0) = w_0 = 0.8$ and $t = 45$

for the exact kink solution. We assume that in (3) this is not satisfied. The equations are then solved numerically. The evolution of u_x in the core is displayed in Fig. 1. It is clear that the long wave radiation deforms the core and adjusts the solution to the final state. In this case the shelf in u_x is responsible for the readjustment of the core and of the velocity. To analyze this we proceed as in [3] using the Lagrangian:

$$L = \int_{t_1}^{t_2} \int_{-\infty}^{\infty} \left(\frac{1}{2} u_t^2 - \frac{1}{2} u_x^2 + \cos u - 1 \right) dx dt, \tag{4}$$

and the trial function

$$u(x,t) = 4 \arctan \exp \left(- \frac{x - \xi(t)}{w(t)} \right), \tag{5}$$

in the averaged Lagrangian to obtain the appropriate modulation equations [4]. The averaged Lagrangian is:

$$L = \frac{\pi^2}{3} \frac{\dot{w}^2}{w} + 4 \frac{\dot{\xi}^2}{w} - \frac{4}{w} - 4w. \tag{6}$$

The modulation equations, obtained from the Euler-Lagrange equations for L, take the form:

$$\frac{2\pi^2}{3w} \ddot{w} - \frac{\pi^2}{3w^2} \dot{w}^2 + 4 \frac{\dot{\xi}^2}{w^2} - \frac{4}{w^2} + 4 = 0, \tag{7}$$

$$\frac{d}{dt} \frac{\xi}{w} = 0. \tag{8}$$

Since the system is Hamiltonian it has only one stable (elliptic) fixed point. Thus to describe the evolution to the steady state we need to couple the modulational equations (7)–(8) to the motion of the linear waves which take away mass and momentum from the soliton, providing the desired radiation damping for Eq. (6). Clearly the total mass and energy are conserved but the excess is radiated away from the coherent soliton. To include this effect in [3] it is assumed that the linear radiation is emitted by the oscillating edge of the shelf (core) of u_x. Thus the linear radiation satisfies

$$u^r_{tt} - u^r_{xx} + u^r = 0, \tag{9}$$

subjected to the signaling condition

$$u^r_x(l,t) = g(t) \text{ and } u = u_t = 0 \text{ for } |x| \geq l, \tag{10}$$

where $g(t)$ is the vertical motion of the shelf. The solution is readily obtained using the Laplace transform in the form:

$$u = -\int_{x-t}^{t} J_0\left(\sqrt{\tau^2 - (x-l)^2}\right) g\,(t-\tau)\, d\tau \tag{11}$$

and

$$u_t = -g(t) + \int_0^t J_1(\tau) g(t-\tau) d\tau, \tag{12}$$

since $g(0) = 0$. The equation for w will now be modified taking into account the radiation. Energy conservation gives

$$\frac{d}{dt}\int_{-\infty}^{\infty}\left(\frac{1}{2}u_t^2 + \frac{1}{2}u_x^2 + 1 - \cos u\right) dx = 0. \tag{13}$$

Now we assume that in the core region $|x| \leq l$ the solution is the soliton with modulated parameters described in (5). Away from the soliton we have linear radiation. To start we assume zero initial velocity. The interval of integration is divided in the region $|x| \leq l$ and the radiation region $|x| \geq l$, this gives:

$$\begin{aligned}\frac{d}{dt}\int_{-l}^{l} \frac{1}{2}u_t^2 + \frac{1}{2}u_x^2 + (1-\cos u)\, dx = &-\frac{d}{dt}\int_{-\infty}^{-l} \frac{1}{2}u_t^2 + \frac{1}{2}u_x^2 + (1-\cos u)\, dx\\ &-\frac{d}{dt}\int_{l}^{\infty} \frac{1}{2}u_t^2 + \frac{1}{2}u_x^2 + (1-\cos u)\, dx\end{aligned}$$

the infinite integrals contribute to leading order only flux terms [3], we obtain

$$\frac{d}{dt}\int_{-l}^{l}\frac{1}{2}u_t^2+\frac{1}{2}u_x^2+(1-\cos u)\,dx=[u_tu_x]_l^t-[u_tu_x]_{-t}^{-l}. \tag{14}$$

This exact identity is now approximated in the core region by the trial kink and in the radiation region by the linear solution generated by the oscillating shelf in the derivative. This gives:

$$\frac{d}{dt}\left(\frac{\pi^2}{3}\dot{w}^2+\frac{4}{w}+4w\right)=-2g^2(t)+2g(t)\int_0^t J_1(\tau)\,g(t-\tau)\,d\tau \tag{15}$$

and no equation for $\dot{\xi}$ since $\dot{\xi}=U_0=0$. It remains to obtain g in terms of the soliton parameter w. At this stage we recall the main assumption about the behavior of the radiation. All the excess energy from the fixed point is radiated for the kink to settle to the steady state.

The total energy E_r contained in the radiation region is given by:

$$E_r=\frac{1}{2}\int_{|x|\geq l}u_t^2+u_x^2+u^2dx. \tag{16}$$

Since the solution $u=0$ at the fronts $|x|=t$ we can use the trapezoidal rule to evaluate the integral (16) using the boundary values at $|x|=l$, $u_x=g$ and $u_t=\dot{g}$ and obtain the approximation

$$E_r=g^2t+\frac{1}{2}\dot{g}^2t. \tag{17}$$

Now we calculate the excess energy when the kink is away from the fixed point w_f. We use $w=w_f+\tilde{w}$ where $\tilde{w}<<1$ to obtain the excess energy at the fixed point:

$$\frac{\pi^2}{3}\frac{\dot{w}^2}{w_f}+\frac{4\tilde{w}^2}{w_f^3}\text{ since }\dot{\tilde{w}}=\dot{w}. \tag{18}$$

Since we are assuming $\tilde{w}$ to be slowly varying and since to the linear approximation $\tilde{w}\approx\ddot{\tilde{w}}$ we neglect this term compared to the first one. Finally we obtain:

$$E_r=\frac{\pi^2}{3}\frac{\dot{w}^2}{w_f} \tag{19}$$

which gives

$$g^2=\frac{\pi^2}{3}\frac{\dot{w}^2}{w_ft}\approx\frac{\pi^2}{3}\frac{\dot{w}^2}{wt}. \tag{20}$$

This closes the equation for w in the form:

$$
\begin{aligned}
'\dot{w}\left(\frac{2}{3}\pi^2\ddot{w} - \frac{\pi^2}{3w^2}\dot{w}^2 - \frac{4}{w^2} + 4\right) \\
= \dot{w}\left(-\frac{2}{3}\frac{\pi^2}{wt}\dot{w} + \frac{2\pi^2}{3\sqrt{wt}}\int_0^t J_1\left(t-\tau\right)\frac{\dot{w}\left(\tau\right)}{\sqrt{\tau w(\tau)}}d\tau\right)
\end{aligned} \tag{21}
$$

which gives the final form of an oscillator equation with the nonlocal radiation damping just as for the NLS equation [2].

The argument is now extended to the moving soliton. The radiation calculation proceeds as before but we now neglect the acceleration and calculate the radiation of momentum as emitted by a source moving at constant speed. The final equations take the form:

$$
\frac{2\pi^2}{3w}\ddot{w} - \frac{\pi^2}{3w^2}\dot{w}^2 + 4\frac{\xi'^2}{w^2} - \frac{4}{w^2} + 4 = -\frac{2\pi^2}{3wt}\dot{w} + \frac{2\pi^2}{3\sqrt{wt}}\int_0^t J_1\left(t-\tau\right)\frac{\dot{w}\left(\tau\right)}{\sqrt{\tau w\left(\tau\right)}}d\tau \tag{22}
$$

and

$$
\frac{d}{dt}\left(\frac{\dot{\xi}}{w}\right) = \frac{\pi^2\dot{\xi}}{12}\dot{w}\left[\frac{\dot{w}}{wt} - \frac{1}{\sqrt{wt}}\int_0^t J_1\left(t-\tau\right)\frac{\dot{w}\left(\tau\right)}{\sqrt{\tau w\left(\tau\right)}}d\tau\right]. \tag{23}
$$

It was shown in [3] that the solution of the system of modulated equations (22)–(23) gives a very good approximation to the evolution of the kink. It is to be remarked that the role of the radiation is fundamental to understand the evolution of the kink towards its final state.

The balance of energy and momentum give now the appropriate coupling of the kink degrees of freedom with the radiation. This produces a nonlocal damping in the Hamiltonian which accounts in simple form for the radiation effects.

1.2 Radiation Analysis for the Discrete Kink Propagation: Asymptotics for Radiation Bursts Using Discrete Analogue of the W.K.B. Solutions for Linear Problems

We now consider the discrete sG equation in the form:

$$
\ddot{u}_n = u_{n-1} - 2u_n + u_{n+1} - \frac{1}{d^2}\sin u_n. \tag{24}
$$

We are interested in studying the evolution of a kink type solution generated by the initial condition:

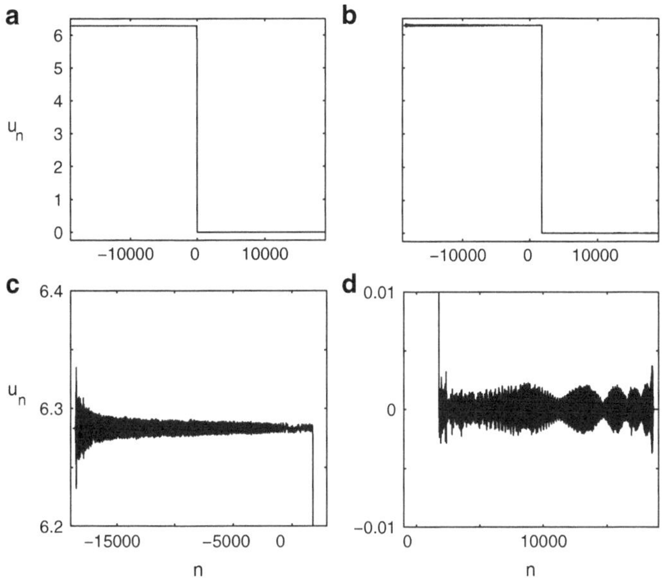

Fig. 2 Full numerical solution of system (24) for $2N + 1$ particles with $\dot{\xi}(0) = 0.4$, $\xi(0) = 0$, $w(0) = 1$, $d = 1$, $N = 20000$ and $\delta t = 0.025$. (**a**) Initial condition (25), (**b**) numerical solution at $t = 30000$, (**c**) backward radiation and (**d**) forward radiation

$$u_n(t) = 4 \arctan \exp\left(-\frac{n - \xi(t)}{w(t)}\right) \tag{25}$$

This problem has been considered for example in [5, 6] without taking into account the role of radiation in the dynamics of the kink motion. In [7] the radiation was studied for the first time and a new behavior was found. Namely the motion of the kink resonates with the linear waves of the lattice. Since the lattice is discrete, it can be considered to leading order as a periodic medium thus different velocities will resonate with the different conduction bands of the linear equation. Thus as the velocity diminishes, as the wave speed decreases due to the loss of kink momentum to radiation momentum, different resonances will appear. This in turn causes radiation bursts to further slow the kink motion on longer and longer time scales [7]. This mechanism explained qualitatively the numerical result of Fig. 2. In [8] the results were used to identify the mechanisms responsible for the bursting of radiation.

We begin by averaging the Lagrangian:

$$L = \sum_{n=-\infty}^{\infty} \frac{\dot{u}_n^2}{2} - \frac{1}{2}(u_{n+1} - u_n)^2 - \frac{1}{d^2}(1 - \cos u_n), \tag{26}$$

using the trial function (25). The sums in (26) are evaluated using the Poisson summation formula in the form:

$$\sum_{n=-\infty}^{\infty} f(n) = \sum_{n=-\infty}^{\infty} \hat{f}(n) \tag{27}$$

where $\hat{f}(n)$ is the Fourier transform of $f(x)$ at $x = n$. We thus obtain:

$$L_K = \left(\frac{2\dot{\xi}^2}{w^2} - \frac{2}{d^2}\right)\left(2w + \frac{4\pi^2 w^2}{\sinh(\pi^2 w)}\cos(2\pi\xi)\right) + \frac{\pi^2 \dot{w}^2}{3w} - h(w), \tag{28}$$

where $h(w)$ can be approximated in the form

$$h(w) = 8\int_{-\infty}^{\infty} \arctan^2\left(\sinh\frac{1}{2w}\operatorname{sech}\frac{x}{w}\right)dx \approx \frac{28 + 120w}{1.4 + 6w + 30w^2}. \tag{29}$$

There are two major differences with the continuum case. There is a term of the form $\frac{4w^2}{\sinh(\pi^2 w)}\cos(2\pi\xi)$ which depends on the position of the kink. This is the so-called PN potential term which accounts for the discrete nature of the lattice. We note that as the kink becomes wider (w large) this term vanishes recovering the continuum approximation. The term $h(w)$ behaves as $2/w$ for large w and the Lagrangian recovers the continuum limit. To study the leading order behavior of the kink we obtain the modulation equations in the form:

$$\ddot{\xi} = \left(\dot{\xi}^2 + \frac{w^2}{d^2}\right)\frac{2\pi^3 w}{\sinh(\pi^2 w)}\sin(2\pi\xi), \tag{30}$$

$$\ddot{w} = \frac{\dot{w}^2}{2w} - \frac{6w}{\pi^2 d^2} - \frac{6\dot{\xi}^2}{\pi^2 w} - \frac{3w}{2\pi^2}h'(w). \tag{31}$$

We have from the first equation that to this order steady propagating kink solutions are not possible since we need $\ddot{\xi} = 0$ giving $\sin(2\pi\xi)$ which fixes the steady kinks. From the width equation we obtain:

$$\dot{\xi}^2 = -\frac{w^2}{4}\left(\frac{4}{d^2} + h'(w)\right) \tag{32}$$

which determines the width to leading order as a function of the velocity. We note however that unlike the continuum case there are two possible widths. The smaller one gives an unstable solution while the second one gives a traveling kink with an exponentially small acceleration as it jumps from site to site. The question is then to study the motion of the kink as it progress along the lattice.

As the kink accelerates it will radiate in two different modes as can be seen in Fig. 2. On the one hand there is the long wave radiation which is also present in the continuum limit which readjusts the width of the kink by radiating the excess mass.

The long wave radiation damping is calculated as in the continuum case balancing the excess of energy around the fixed point with the energy contained in the radiation region. This gives the energy balancing as in the continuum case

$$\frac{d}{dt}\sum_{n=-\infty}^{\infty}\frac{\dot{u}_n^2}{2}+\frac{1}{2}\left(u_{n+1}-u_n\right)^2+\frac{1}{d^2}\left(1-\cos u_n\right)=\left[\dot{u}_n^c\frac{du_n^c}{dn}\right]_{\xi-l}^{\xi+l} \tag{33}$$

where u_n^c is the continuum approximation to the linear radiation and $\frac{du_n^c}{dn}$ is the continuous limit approximation of the backward difference . The function g of Sect. 1 is determined as before but now the radiation is confined to the region bounded by the group velocity lines $l \pm c_g t$ where c_g is the group velocity of the waves produced by the oscillating width. We thus obtain the new equation for the width w:

$$\ddot{w}-\frac{\dot{w}^2}{2w}+\frac{6w}{\pi^2 d^2}+\frac{6\dot{\xi}^2}{\pi^2 w}+\frac{3w}{2\pi^2}h'\left(w\right)=-\frac{\dot{w}}{c_g t}. \tag{34}$$

Equations (30) and (34) are similar to the ones obtained in the first section. However it was observed by Peyrard and Kruskal in [7] that a new type of radiation is emitted in bursts as the kink jumps the PN potential. This radiation is shown in Fig. 3 and is produced by the asymmetric motion of the peak of the forward divided difference $u_{n+1}-u_n$ of the solution. We now examine in detail the behavior of this new effect which is a feature of the discrete nature of the lattice.

For the Peyrard and Kruskal radiation y_n^{PK} we need to solve the moving boundary value problem

$$\ddot{y}_n^{PK}-\left(y_{n-1}^{PK}-2y_n^{PK}+y_{n+1}^{PK}\right)+\frac{y_n^{PK}}{d^2}=0, n<q(t) \tag{35}$$

with an appropriate matching condition to the core of the kink. The numerical solution shows that the radiation is produced by the change in width and a flip flop in the back of the kink. We will relate this structure to the shape modes of the kink later in the argument. Meanwhile we will assume it to be a given function $r\left(n-\xi\left(t\right)\right)$ to be at the boundary to be matched by y_n^{PK} at some point $q(t)$ to be determined. It is clear that this problem has no exact analytic solution and we now find an approximation.

To study (35) we take the time Laplace transform of r and denote it by $\tilde{r}$. We can only calculate it approximately. In fact the function $\xi\left(t\right)$ is unknown and slowly varying we can approximate $\xi\left(t\right)=\dot{\xi}\left(t\right)t$ and since $\dot{\xi}$ is also slowly varying we take $\dot{\xi}=v$ to be a constant in the integration to obtain:

$$\tilde{r}\left(s,n\right)=\frac{e^{-ns/v}}{v}\int_{-n}^{\infty}e^{-su/v}r\left(-u\right)du=e^{-ns/v}B\left(s,n\right) \tag{36}$$

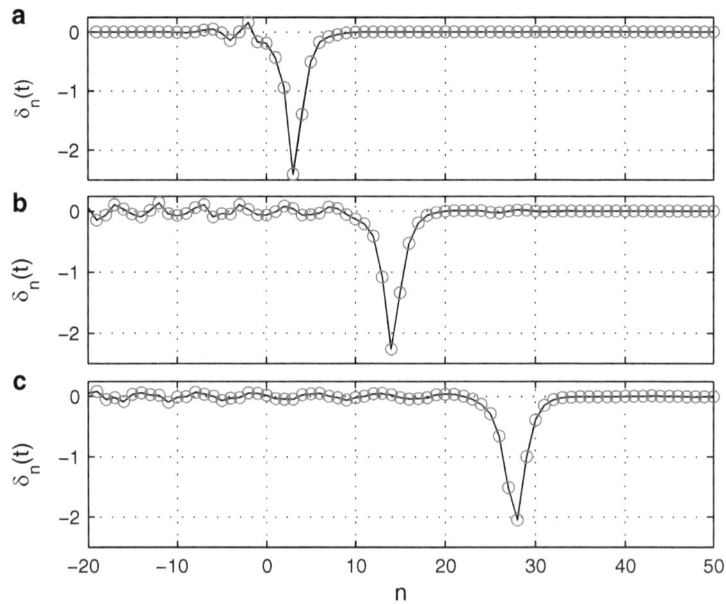

Fig. 3 Numerical forward divided difference $\delta_n(t) = u_{n+1} - u_n$ showing the dynamics of an asymmetry in the peak at (**a**) $t = 10$, (**b**) $t = 45$ and (**c**) $t = 100$

which gives a forcing term in the W.K.B. form as a function of the variable n. Now in the radiation region we also assume a W.K.B. form for the transform of the solution $A(s,n)\, e^{\phi(s,n)}$. This form is substituted in the Laplace transform or the discrete wave equation to obtain:

$$\left(s^2 + 2 + \frac{1}{d^2}\right) - e^{\phi(s,n+1)-\phi(s,n)} + e^{\phi(s,n-1)-\phi(s,n)} = 0 \tag{37}$$

which is the discrete Eikonal equation analogue.

To find the value of the amplitude A we need to match the waves in the region $n \leq q(t)$ to the function $r(n - \xi(t))$. Since the function r is a difference w will match the difference of the radiation solution. The matching is as follows: we take $N = \xi(t)$ and the nodes head and behind of $\xi(t)$. Then the difference of the radiation must satisfy:

$$\tilde{y}^{PK}_{N+1}(s) - \tilde{y}^{PK}_{N-1}(s) = 2Be^{-Ns/v}. \tag{38}$$

We now use this equation to determine the amplitude $A(s,n)$. Since A is slowly varying we have:

$$A(s,n)\left(e^{\phi(s,N+1)} - e^{\phi(s,N-1)}\right) = e^{-Ns/v}2B. \tag{39}$$

For $n \leq N$ the phase ϕ satisfies the Eikonal equation. The matching condition is $\phi = ns/v$ for $n \geq N$ which is the phase of the forcing at the moving boundary. Then

$$\phi(s, n) = \lambda(s)(n - N) - Ns/v \tag{40}$$

for $n \leq N$. Using these expressions we obtain:

$$A(s, n) = \frac{2B}{e^{-s/v} - \lambda^{-1}(s)}. \tag{41}$$

Then the radiation solution

$$y_n^{PK}(t) = \frac{1}{2\pi i} \int_C \lambda^n(s) \frac{2B(s)}{e^{-s/v} - \lambda^{-1}(s)} e^{st} ds \tag{42}$$

provides the desired approximate solution. Note that the velocity $v = \dot{\xi}(t)$ is a slowly varying function. The term $\lambda^n(s)$ for imaginary s provides the outgoing wave. the integrand has the branch points of $\lambda^{-1}(s)$. Their contribution vanishes as $t \to \infty$. We are left with the contribution of the poles which are only the imaginary roots (which come in the complex conjugate pairs) of the denominator. The integral can be evaluated using the residues in the form:

$$y_n^{PK}(t) = \sum_k e^{inw_k/v} e^{iw_k t} \frac{2B(iw_k)}{-e^{-iw_k/v}/v + \lambda'(iw_k)/\lambda^2(iw_k)} + c.c. \tag{43}$$

Since $B(s, n)$ is a Laplace transform it becomes smaller as w_k increases. With this we can analyze the behavior of the radiation.

It can be shown that for subsonic waves the dispersion relation starts with one root. This produces a burst of waves which slows the kink. As the kink slows the roots becomes larger as v decreases leading to a decreasing B. Thus a smaller radiation is emitted. At the same critical speed two more roots giving radiated waves appear. However the first wave dominates and the kink keeps slowing down. At a critical speed of $v \approx 0.15$ the first and middle roots are lost. There is just a larger root which emits an even smaller radiation. Thus the deceleration due to radiation becomes small and the velocity has a slower decay than in the first phase of the motion. The process then repeats. We will show that as $t \to \infty$ the deceleration becomes smaller and it is not clear if the kink will eventually stop. This question will be resolved by a rigorous analysis.

Thus the final expression for $y_n^{PK}(t)$ is given by

$$y_n^{PK}(t) = -\sum_k \frac{8\pi}{v} \exp\left(-\frac{\pi w_k}{2v} w\rho\right) R(w_k) \cos\left(\frac{w_k}{v} n + w_k t + \psi_k\right) \tag{44}$$

where

$$R(w_k) = \frac{\left|\lambda^2(iw_k)\right| v}{\left|\lambda'(iw_k) - \lambda^2 e^{-iw_k/v}\right|}, \tag{45}$$

$$\psi_k = \arg\left(1/\left(\lambda'(iw_k)/\lambda^2(iw_k) - e^{-iw_k/v}\right)\right). \tag{46}$$

This expression is now used to calculate the energy flow away from the kink. Since the radiation oscillates with frequency w_k and the acceleration is small we only need the average flow produced by the radiation $y_n^c(t) + y_n^{PK}(t)$ in the form

$$\text{Energyflux} = \left[\dot{y}_n^c(y_{n+1}^c - y_n^c)\right]_{\xi-l}^{\xi+l} - \sum_k 128\pi^2 |R(w_k)|^2 |B(iw_k)|^2$$
$$\frac{w_k^2}{v} \sin^2\left(\frac{w_k}{v}(\xi + l) + w_k t + \psi_k\right)$$

the average of the cross terms is zero due to the slowly time varying nature of $y_n^c(t)$.

Using the energy identity into the last equation we have:

$$\frac{d}{dt}\left[\left(\frac{2\dot{\xi}^2}{w^2} + \frac{2}{d^2}\right)\left(2w + \frac{4\pi^2 w^2}{\sinh(\pi^2 w)}\cos(2\pi\xi)\right) + \frac{\pi^2 \dot{w}^2}{3w} + h(w)\right]$$
$$= -\frac{\pi^2 \dot{w}^2}{3c_g w t} - \sum_k 64\pi^2 |R(w_k)|^2 |B(iw_k)|^2 \frac{w_k^2}{v}.$$

Finally the energy loss is split to avoid singularities to obtain:

$$\ddot{w} - \frac{\dot{w}^2}{2w} + \frac{6w}{\pi^2 d^2} + \frac{6\dot{\xi}^2}{\pi^2 w} + \frac{3w}{2\pi^2} h'(w) = -\frac{\dot{w}}{2c_g t}, \tag{47}$$

$$\ddot{\xi} - \left(\dot{\xi}^2 + \frac{w^2}{d^2}\right)\frac{2\pi^3 w}{\sinh(\pi^2 w)} \sin(2\pi\xi) = -\dot{\xi}\sum_k 8\pi^2 w |R(w_k)|^2 |B(iw_k)|^2 \frac{w_k^3}{\dot{\xi}}. \tag{48}$$

The first equation is analogous to the one for the continuum model and controls the approach to a kink of steady width due to mass shedding with the group velocity. The second equation decelerates the soliton in steps since the roots w_k are complicated functions of time [7]. The mechanism of soliton deceleration due to radiation relies on the resonance between the phase velocity of the linear lattice waves and the kink velocity.

To close the equations it remains to find the amplitude $|B(iw_k)|$. We recall that the function $B(iw_k)$ is given in the form:

$$B(iw_k) = \frac{1}{v}\int_{-N}^{\infty} e^{iw_k u/v} r(-u)\, du \approx \frac{1}{v}\int_{-\infty}^{\infty} e^{iw_k u/v} r(-u)\, du \qquad (49)$$

since we assume that the function $r(u)$ as a narrow width as discussed from the numerical solution. To determine $r(u)$ we recall that the static kink solutions have two so called shape modes (see [9] for a complete study). They are eigenfunctions of the linearized problem around the kink and they provide internal oscillations [9]. Thus to leading order we expect these shape modes will be excited during the motion. One of the shape modes is even with respect to the kink position and has only one sign. It is the discrete counterpart of the zero mode eigenfunction in the continuum problem which is the derivative of the kink with respect to the phase. The second shape mode is odd and was taken into account in the previous analysis as the one producing the long wave radiation. We thus assume a new trial function in the form:

$$y_n(t) = 4\arctan\exp\left(-\frac{n-\xi(t)}{w(t)}\right) + A(t)\,\mathrm{sech}\left(\frac{n-\xi(t)}{\lambda(t)}\right) \qquad (50)$$

where we assume a shape mode which is also modulated. Thus from the last equation it is clear that the flip flop motion of the core, as Fig. 3 shows, responsible for the radiation is just the difference of the shape mode which can be approximated by:

$$\frac{A(t)}{\lambda(t)}\mathrm{sech}\left(\frac{n-\xi(t)}{\lambda(t)}\right)\tanh\left(\frac{n-\xi(t)}{\lambda(t)}\right). \qquad (51)$$

Taking only the right side of the previous equation as the source of radiation, we thus have:

$$B(iw_k) = A(t)\left(-\frac{2\pi}{v}\right)\exp\left(-\frac{\lambda w_k}{2v}\right). \qquad (52)$$

In order to close the equations we need modulation equations for A and λ. We refer to [8] for the details and we just reproduce and explain the main steps.

Substitution of (50) into the Lagrangian (26) and summing using the Poisson formulas in the n variable we obtain a Lagrangian involving the self interactions of the kink, the self interaction of the shape mode and the cross interaction between the kink and the shape mode. The Lagrangian L_A for the self interaction of the shape mode with itself has the form:

$$L_A = \lambda\dot{A}^2 + \frac{\dot{\lambda}^2 A^2}{36\lambda}\left(12+\pi^2\right) + A^2 Q_1\left(\lambda, \dot{\xi}, \cos(2\pi\xi)\right) + A^2 R_1\left(\lambda, w, \cos(2\pi\xi)\right)$$
$$+ A^4 Q_2\left(\lambda, w, \cos(2\pi\xi)\right).$$

The interaction Lagrangian L_{int} in the continuum limit gives zero since the shape mode has zero eigenvalue. The contribution to leading order is then periodic in ξ and takes the form:

$$L_{int} = \frac{4A\sin(2\pi\xi)}{\lambda w}\left(\dot{\xi}^2 - 1\right) V(\lambda, w) - \frac{4A}{d^2}\sin(2\pi\xi) V(\lambda, w) + \frac{2A^3}{3d^2}\sin(2\pi\xi) S(\lambda, w).$$

The final average Lagrangian is just the Lagrangian $L_k + L_A + L_{int}$. The modulation equations are obtained taking appropriate variations and are given in [8]. The equation for the position of the kink is modified in the form:

$$\ddot{\xi} - 4w\pi^3\exp(-\pi^2 w)\left(\dot{\xi}^2 + \frac{w^2}{d^2}\right)\sin(2\pi\xi) + \frac{\pi}{\lambda}A\left(\dot{\xi}^2 + 1\right)V(\lambda, w)\cos(2\pi\xi) + \frac{\pi w}{d^2}AV(\lambda, w)\cos(2\pi\xi) - \frac{\dot{\xi}\dot{w}}{w} = l_\xi$$

and the equation for the radiation damping takes the form:

$$l_\xi = -\dot{\xi}\sum_k 8\pi^2 w\exp\left(-\frac{\rho\pi w w_k}{v}\right)\frac{w_k^2}{v^3}. \qquad (53)$$

We can now describe the qualitative behavior of the solutions for the system modulation equations.

As the kink begins to move it will accelerate or decelerate according to the PN potential. This will in turn imply a readjustment of the width w and an oscillation of the shape mode amplitude. This in turn produces the radiation which continuously decelerates the kink. As this happens the radiation will abruptly change character and become much smaller. This gives an evolution with almost constant average velocity. The sudden change in radiation acceleration will show as a kink in the velocity versus time evolution.

In Fig. 4 we present a detailed comparison between the modulational and numerical solution for a special value of the initial conditions. We observe that all the qualitative features are well reproduced by the modulational solution which includes the interaction and feed back between the kink, the shape mode and the radiation. The quantitative agreement is remarkable and is of a similar nature for a wide range of initial conditions.

The results strongly suggest that the evolution of the kink is strongly influenced by radiation. Moreover the asymptotic solution displays a mechanism with few degrees of freedom which explains all the observed features with very good quantitative accuracy. The question of ultimate stopping of the kink will have to be resolved with a rigorous study of the existence of a traveling kink which should be proved to be attractive in a suitable sense. It is to be remarked that the

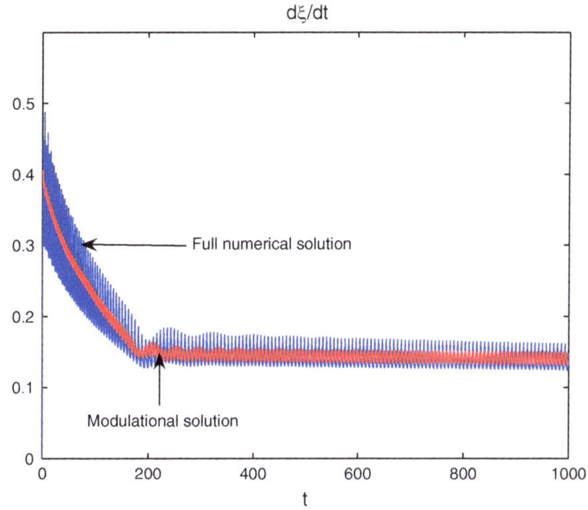

Fig. 4 Numerical comparison between the full numerics and the modulational solution including radiation damping and shape modes interactions

modulation theory predicts successive knees due to the coalescing roots. Those pinning occurs over very large distance, which depends on the width of the soliton, but only after successive slowing down. It has been shown in other problems [10] that parametric excitation could help the kink to escape the pinning, it will be of interest to explore this mechanism in the present context. The problem will be to determine if parametric resonance could widen the soliton to diminish the PN barrier and avoid the pinning.

2 The sG Equation in Two Space Dimensions

The problem for standing and traveling waves in the two dimensional sG equation was considered in [11, 12]. It was shown using a scaling argument in the Lagrangian that there were no traveling localized waves of permanent form. On the other hand it was shown numerically that standing lump initial conditions collapsed in finite time emitting a considerable amount of radiation. On the other hand the question of lump solution in more than one dimension is related to the Skyrme model which is a system of non linear equations to describe fundamental particles. In this model the topological constraints give, in the appropriate local coordinates, a system of sG like equations. In this case there is an interplay between the constraints and the radiation which has not been studied. We now study the effect of radiation for a single equation.

We now describe how the radiation plays a fundamental role in two space dimensions to destabilize coherent structures and how they can be stabilized by non topological effects.

2.1 Standing Coherent Structures

We now revisit the problem of stabilizing lump solutions by adding an internal degree of freedom as in [11].

In [11] a solution with an oscillating boundary for the lump was assumed. It was shown that the boundary layer satisfies a Born-Infeld type equations which led to non collapsing asymptotic solutions. Thus the problem of stability of these solutions needs consideration. To this end we study the two space dimensional sG equation in the form:

$$\frac{\partial^2 u}{\partial t^2} - \nabla^2 u + \frac{\sin u}{\epsilon^2} = 0 \tag{54}$$

with the initial profile

$$u = 4 \arctan \exp \left(-\frac{r - R(t) - ar \cos n (\theta - \xi(t))}{\epsilon} \right) \tag{55}$$

where r and θ are polar coordinates at the center of the lump. The cos term provides the boundary modulation and $R(t)$ is the basic radius of the lump.

We then average the Lagrangian

$$L = \int_{t_1}^{t_2} \int_{-\infty}^{\infty} \int_{-\infty}^{\infty} \left(\frac{1}{2} u_t^2 - \frac{1}{2} u_x^2 - \frac{1}{2} u_y^2 + \frac{\cos u - 1}{\epsilon^2} \right) dx dy dt \tag{56}$$

with the trial function (55) to obtain:

$$L = \frac{8\pi}{\epsilon} R \dot{R}^2 + \frac{4\pi}{\epsilon} a^2 n^2 \dot{\xi}^2 - \frac{16\pi}{\epsilon} R - \frac{4\pi n^2}{\epsilon R} a^2. \tag{57}$$

The equations of motion take the form:

$$8\pi a n^2 \left(\dot{\xi}^2 R - \frac{1}{R} \right) = 0, \tag{58}$$

$$\frac{d}{dt} \left(a^2 \dot{\xi} R \right) = 0, \tag{59}$$

$$\frac{d}{dt} (4 R \dot{R}) - \frac{2a^2 n^2}{R^2} + 4 = 0. \tag{60}$$

Equations (58) and (59) show that as in the Kepler problem the angular momentum of the wave running along the boundary of the lump is conserved. Moreover in Eq. (60) is the term involving the angular motion $n^2 a^2$ prevents the collapse. To study the stability of these waves we need to consider their interaction with the radiation.

We now show that the stability of these solutions is controlled by their interaction with the radiation. To this end we now consider the effect of the radiation emitted by the lump. As it was shown in the first section the mass emitted in the radial direction will produce a damping for the radial motion of the boundary of the lump giving a modified equation for the radial motion in the form:

$$2R\ddot{R} + \dot{R}^2 - \frac{a^2n^2}{R^2} + 2 = -\frac{an}{\sqrt{2t}}\dot{R} + \epsilon\frac{\sqrt{anR}}{\sqrt{2t}}\int_0^t \frac{\sqrt{na(\tau)}}{\sqrt{R(\tau)}}\dot{R}(\tau)\, J_1\left(\frac{t-\tau}{\epsilon}\right)d\tau. \tag{61}$$

There is however a new effect to be considered which is due to the conservation of angular momentum since we have a spinning solution. The total angular momentum is given by:

$$L = \int_0^{2\pi}\int_0^{\infty} u_t u_\theta r dr d\theta \text{ and } \dot{L} = 0. \tag{62}$$

As before the angular momentum balance of the lump and the radiation is considered in the form:

$$\frac{d}{dt}\int_0^{2\pi}\int_0^{\infty} u_t u_\theta r dr d\theta = -\int_0^{2\pi} R v_r v_\theta d\theta \tag{63}$$

where v is the solution of the linearized equation

$$v_{tt} - \nabla^2 v + v = 0 \tag{64}$$

with signaling boundary values at the lump edge of the form:

$$\begin{aligned} v_r = -u_r &= -\frac{2}{\epsilon}\operatorname{sech}\left(\frac{\mu - Ra\cos\left(n\left(\theta - \xi(t)\right)\right)}{\epsilon}\right) \\ &= -\frac{2}{\epsilon}\left[\frac{\epsilon_0}{2} + \sum_{k=0}^{\infty} c_k(\mu, R, a)\cos\left(kn\left(\theta - \xi(t)\right)\right)\right]. \end{aligned} \tag{65}$$

The first term was taken into account in the radial equation which produced the damping due to energy shedding. This term does not give angular momentum loss since it is independent of the angle.

It is remarkable that the boundary value problem can be solved with geometric optics obtaining good quantitative agreement. The details are now sketched.

We use the geometric optics approximation in the form,

$$v = \sum_{k=1}^{\infty} A_R(r, \theta, t)\, e^{inS_k(r,\theta,t)} \tag{66}$$

to obtain the Eikonal equations for the phases S_k in the form.

$$S_{kt}^2 - S_{kr}^2 - \frac{1}{r^2} S_{R\theta}^2 + \frac{1}{n^2 k^2 \epsilon^2} = 0. \tag{67}$$

with boundary conditions $S_k\left(R\left(t\right)+\mu,\theta,t\right) = \theta - \xi\left(t\right)$ and we need to take the outgoing solution for the quadratic equation for S_r obtained from the Eikonal equation.

From this equation we have always a real solution for the outgoing wave number in the radial direction in the form:

$$S_k = \frac{R\left(t\right) - r}{\epsilon k n} + \theta - \xi\left(t\right) \tag{68}$$

which gives the phase of a spiral wave. Thus we see that the angular momentum is always lost into the spiral wave radiation. This loss is given by:

$$\frac{d}{dt} a^2 = -\frac{R\left(t\right)}{2\pi^2 n} \sum_{k=1}^{\infty} k^2 c_k^2\left(a\right). \tag{69}$$

This shows how lump solutions even if long lived are ultimately dissolved into radiation. Presumably for the full Skyrme model is stabilized due to topological constraints which isolate the nontrivial topological soliton in 2 or 3 space dimensions from the topologically trivial radiation. However to our knowledge this has not been explained.

2.2 *Stabilization of Lump Solutions in a Lattice: Description of the Stabilizing Effect of the Peierls–Nabarro Self-Induced Potential*

As we discussed in Sect. 2.2 one of the effects of the lattice is to produce stable steady kink solutions at the minima of the PN potential induced on the kink by the lattice in a self consistent way. On the other hand there are two other mechanisms which stabilize lump solutions for a single equation which we now describe.

We now show how a circularly symmetric kink induces an approximately circularly symmetric PN potential which will prevent the collapse of the kink. Unlike the angular momentum stabilization which was radiated away this stabilization due to the discreteness of the lattice is persistent giving a full family of standing two dimensional lump solutions.

Again we use an average Lagrangian to calculate the desired solutions and describe the dynamics. The discrete sG equation to be considered is:

$$\ddot{u}_{nm} = u_{n-1,m} + u_{n+1,m} + u_{n,m-1} + u_{n,m+1} - 4u_{n,m} - \frac{1}{d^2} \sin u_{nm}, \tag{70}$$

with a lump like initial condition in the form:

$$u_{nm}(t) = 4 \arctan \exp\left(-\frac{\sqrt{n^2+m^2} - R(t)}{w}\right) \tag{71}$$

and $\dot{u}_{nm} = 0$ at $t = 0$. We recall that for large d we have the continuum limit and the usual one dimensional width velocity relation is satisfied in the form $w = d\sqrt{1-\dot{R}^2}$. To study the evolution of the lump we proceed as in the continuum case assuming the width w to be constant. We use the trial function (71) to average the Lagrangian:

$$L = \sum_{m=-\infty}^{\infty} \sum_{n=-\infty}^{\infty} \left[\frac{1}{2}\dot{u}_{nm}^2 - \frac{1}{2}\left(u_{n+1,m} - u_{nm}\right)^2 - \frac{1}{2}\left(u_{nm} - u_{n,m+1}\right)^2 - \frac{1}{d^2}\left(1 - \cos u_{nm}\right)\right], \tag{72}$$

to obtain

$$\begin{aligned} L &= \frac{2}{w^2}\left(\dot{R}^2 - 1 - \frac{w^2}{d^2}\right) \sum_{m=-\infty}^{\infty} \sum_{n=-\infty}^{\infty} \operatorname{sech}^2\left(\frac{\sqrt{n^2+m^2} - R(t)}{w}\right) \\ &= \frac{2}{w^2}\left(\dot{R}^2 - 1 - \frac{w^2}{d^2}\right) f\left(\frac{R}{w}, w\right). \end{aligned}$$

In the continuum limit $w \to \infty$ and the sum is independent of R. The Lagrangian is the same as the one found in the previous section when the angular momentum term is omitted. This slows collapse as is also observed in the numerical solutions.

To evaluate the sum we proceed as in Sect. 2.2 using the Poisson sum formula. The spatial Fourier transforms which arise are evaluated using the stationary phase formula for the angular integrals and using the boundary layer nature of the solution to approximate the radial integral. In this way we obtain to leading order in the exponentially small PN terms:

$$f(R) = \sum_{m=-\infty}^{\infty} \sum_{n=-\infty}^{\infty} f_{nm} \approx 4\pi w R + \frac{16\pi w^2 \sqrt{R}}{\sinh(\pi^2 w)} \cos\left[\pi\left(2R - \frac{1}{4}\right)\right]. \tag{73}$$

Now there is an oscillatory potential which becomes larger as R increases. In Fig. 5 we plot the level lines of the Hamiltonian which show an infinite family of centers which correspond to stable lump solutions. However the centers only appear for sufficiently large radius. Below a threshold in the radius there are always collapsing lumps. In [13] quantitative comparisons between modulational and numerical solutions is given and good quantitative agreement is found. For the minima of the PN potential there is no collapse since the appropriate minima of the angular average PN potential stop the lump from collapsing, we actually show an

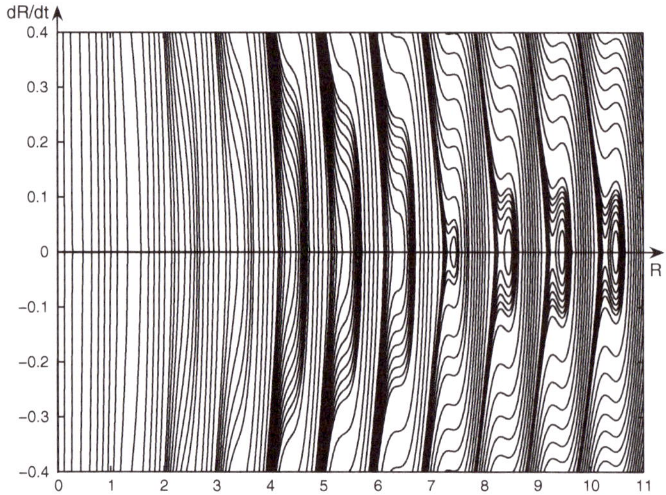

Fig. 5 Phase plane for R obtained from the potential (73)

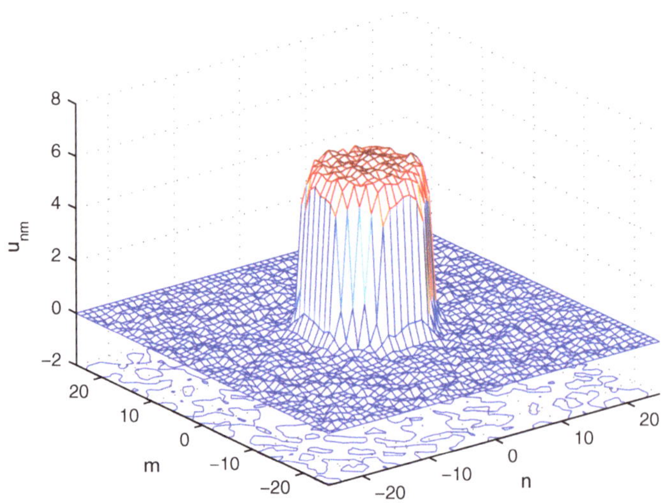

Fig. 6 Non collapsing evolution for $R = 8.5$

example in Fig. 6 for $R = 8.5$ according to Fig. 5. Thus we see that due to the PN self consistent potential, the lattice stops the pulse from collapse. We thus see that lattices have a stabilizing effect. In the next section we examine another mechanism for stabilization which is based on internal degrees of freedom but unlike the angular momentum mechanism does not radiate.

2.3 Stabilization of Lump Solutions Due to Internal Breathing

We consider again the sG equation

$$u_{tt} - \nabla^2 u + \sin u = 0 \tag{74}$$

with initial values corresponding to

$$u(x,t) = -4 \arctan \left[a(t) \sin \phi(t) \operatorname{sech}(\lambda(t) x) \operatorname{sech}(\lambda(t) y) \right]. \tag{75}$$

We also take $a(t) = \frac{\lambda(t)}{1-\lambda^2(t)}$ which is the correct relation for the one dimensional breather solution of the one space dimensional sG equation. It is to be noted that the trial function in (75) is just an one dimensional modulated breather with a local transverse dependence [12].

Again the modulation equations for $\lambda(t)$ and $\phi(t)$ are obtained again the Lagrangian to obtain:

$$L = A(\lambda)\dot{\lambda}^2 + B(\lambda)\dot{\phi}^2 - V(\lambda) \tag{76}$$

where $A(\lambda) = \frac{2\lambda^2+(1-\lambda^2)^2}{(1-\lambda^2)(1-2\lambda^2/3)^2}$, $B(\lambda) = \frac{2(1-\lambda^2)}{(1-2\lambda^2/3)^3}$ and $V(\lambda) = \frac{2(1-\lambda^2)(1+2\lambda^2/3)}{(1-2\lambda^2/3)^2}$.
The variational equations give:

$$\frac{d}{dt}\left(B(\lambda)\dot{\phi}\right) = 0, \tag{77}$$

$$H = A(\lambda)\dot{\lambda}^2 + B(\lambda)\dot{\phi}^2 + V(\lambda) = E \tag{78}$$

where E is the initial energy. The first equation gives $\dot{\phi} = \frac{D}{B(\lambda)}$ while the second one

$$\dot{\lambda}^2 = \frac{1}{A(\lambda)}\left(E - \left(\frac{D^2}{B(\lambda)} + V(\lambda)\right)\right). \tag{79}$$

We note again that the internal degree of freedom $\phi(t)$ provides an angular momentum like term which stabilizes the "radial" λ equation. If there is no internal breathing $D = 0$. The equation for λ has a minimum for the potential at $\lambda = 0$. Thus there is no breather as predicted by Derricks' argument. This situation persists up to $D = 1$. As D increases beyond $D = 1$ $\lambda = 0$ becomes an unstable maximum and between $\lambda = 0$ and the repelling singularity at $\lambda = 1$ (corresponding to total collapse) there is a minimum at λ^*. This gives again a standing breather. In [12] a comparison with other purely numerical approaches is given.

We now consider the effect of radiation. Again the linearized equation must be solved subject to the appropriate signaling boundary condition this is

$$v_{tt} - \nabla^2 v + v = 0,$$
$$v_t = h(\phi(t), t) \text{ on } r = \frac{1}{\lambda(t)}$$

where the function is given in terms of the boundary value of the lump solution. Now we expand the periodic function h in Fourier Series in the form:

$$h(\phi(t), t) = \sum_{n=0}^{\infty} c_n(t) e^{in\phi(t)} + c_0(t). \tag{80}$$

The solution corresponding to $c_0(t)$ gives as before the radial energy shedding. $c_0(t)$ can be also calculated from the excess energy between the standing breather and the deformed breather. However there are new periodic terms which need to be calculated. Again we seek a geometric optics solution in the form:

$$v = \sum_{n \neq 0} A_n(t) e^{inS_n(r,\theta,t)}. \tag{81}$$

The phases satisfy the Eikonal equation in the form:

$$S_{nt}^2 = S_{nr}^2 + 1 \tag{82}$$

with boundary condition $S_n(R + \mu, \theta, t) = \phi(t)$. Now it is clear that for $\dot{\phi}(t)$ small enough, this is when $\dot{\phi} = \frac{D}{B(\lambda)}$ with $D > 1$ but $\lambda \approx 0$ the wave number S_r is complex. Thus no waves are shed, the radiation is trapped and only the low phases contribute to the energy shedding leading to the readjustment of the breather. On the other hand as $\lambda \to 1$ due to the choice of specific initial conditions we have a real wave number and waves are shed leading to a decrease in $\dot{\phi}$. When the total energy is balanced with the energy lost by the waves there are terms involving $\dot{\lambda}$ and $\dot{\phi}$ which are then matched separately to the energy flux caused by the waves. The details are documented in [12] and the final equations take the form:

$$A(\lambda, \dot{\phi}) \ddot{\lambda} + \dot{A}(\lambda, \dot{\phi}) \dot{\lambda} + P(\lambda, \dot{\phi}) = -Q(\lambda) \dot{\lambda} \tag{83}$$

where the term $Q(\lambda)$ is again obtained by matching the excess energy around the coherent structure to the energy in the radiation. The phase equation now takes the nonconservative form:

$$B(\lambda) \ddot{\phi} + B(\dot{\lambda}) \dot{\phi} = -2\pi \lambda^{-1} \left(\dot{\phi}^2 - 1\right)^{1/2}. \tag{84}$$

Now we see from (84) that as $\lambda \to 1$ the phase begins to collapse. However (84) sheds radiation and the pulse has a smaller frequency which stabilizes again the structure. This behavior is described in [12]. However a complete analysis of (83) and (84) is not available.

The same ideas are applied to a traveling pulse and again a propagating breather structure is obtained.

3 Conclusions and Perspectives

We have shown how to extend the average Lagrangian ideas (or the collective coordinates approach) to include the coupling between the finite dimensional approximate description of the soliton with the radiation produced by the distortion of the soliton. In turn the equations are closed using the conservation equations, which are exact, together with the assumption that the excess energy between the approximate solution at the approximate steady state is radiated entirely into linear waves.

We have shown in very different continuum and discrete equations how the radiation determines the ultimate stability or instability of the soliton solutions and plays a major role in the dynamics of distortion of the coherent structure.

We have also shown that even if the moving boundary value problem which determines the radiation which is not exactly solvable (even if linear) it can be solved to a sufficient accuracy using geometric optics or a discrete version of the same idea. It is to be remarked the accuracy of the asymptotics when compared with the full numerical solutions. These ideas explain in a simple and unified way the evolution of two dimensional coherent structures.

To some extent the approximations used to determine the radiation in the vicinity of the core of the soliton are related to the detailed knowledge of the spectrum of the linearization around the exact coherent structure. However a precise comparison has still eluded us. It is to be noted that in the cases studied there we do not have an exact solution but only an approximation which makes the spectral analysis one step more complicated. We believe this clarification is very important in order to make further progress in more complicated problems where is not clear how to describe the transition region core-linear radiation.

We have shown that resonant radiation in lattices plays a major role in the evolution of the solutions. Moreover we also described lattices and non resonant internal degrees of freedom is stabilizing effects for two space dimensional sG coherent structures.

Finally there is a new set of sG type problems which arise when point wise products are substituted by non local Moyal type products. This substitution arises as an attempt to study the effects of non commutativity in the spatial coordinates [14]. It has been shown for the non local Moyal NLS approximation that the Moyal product induces in the dynamics a PN like potential which avoids the collapse [14]. It is also known [15] that non commutative approximations to the sG equations have coherent lump solutions which are stable. Moreover it is also known that modulation like solutions of two colliding solitons show 90° scattering due to the fact that the coordinates of the centers of the soliton move on a Khaler manifold whose conical singularities control the scattering process. It is presumed that the addition of radiation may force the fusion of these objects in a toroidal solution [15]. It will be of interest to develop a modulation approach which takes into account the radiation and understand from a dynamical view how the Moyal product, which prevents angular momentum from being conserved in a central potential forces the scattering process at 90°. This problem seems to be of a very difficult nature from other non local problems [16] where the non locality arises from the Green function of an elliptic equation. For the Moyal product as far as we know there is no representation in terms of a Green function of an explicitly defined operator.

The behavior of systems such as the full Skyrme model has not yet been attempted from a modulation point of view.

We close this section by mentioning that sG like equations describe the torsional modes in backbones of protein chains. By enlarging the subject of the role of the kinks in the dynamics of these structures is just an emerging subject. From a modulational point of view the presence of bends and defects in the chain likely to produce substantial radiation which will affect the further dynamics of the soliton.

Acknowledgements L.A. Cisneros-Ake gratefully acknowledges support from COFAA-IPN and project IPN-CGPI-20130803. L.A. Cisneros-Ake and A.A. Minzoni acknowledge support from Conacyt-177246. We acknowledge Elsevier for having given permission to reproduce Figs. 2, 4 and 5 of this chapter, which were included in [8].

References

1. S.P. Novikov, S.V. Manakov, L.P. Pitaevskii, V.E. Zakarov, *Theory of Solitons. The Inverse Scattering Transform* (Consultants Bureau, New York, 1984)
2. W.L. Kath, N.F. Smyth, Phys. Rev. E **51**, 1484 (1995)
3. N.F. Smyth, A.L. Worthy, Phys. Rev. E **60**, 2330 (1999)
4. G.B. Whitham, *Linear and Nonlinear Waves* (Wiley-Interscience, New York, 1999)
5. Y. Ishimori, T. Munakata, J. Phys. Soc. Jpn. **51**, 3367 (1982)
6. R. Boesch, C.R. Willis, M. El-Batanouny, Phys. Rev. B **40**, 2284 (1989)
7. M. Peyrard, M. Kruskal, Phys. D **14**, 88 (1984)
8. L.A. Cisneros, A.A. Minzoni, Phys. D **237**, 50 (2008)
9. P.G. Kevrekidis, M.I. Weinstein, Phys. D **142**, 113 (2000)
10. G. Assanto, L.A. Cisneros, A.A. Minzoni, B.D. Skuse, N.F. Smyth, A.L. Worthy, Phys. Rev. Lett. **104**, 053903 (2010)
11. A.A. Minzoni, N.F. Smyth, A.L. Worthy, Phys. D **159**, 101 (2001)
12. A.A. Minzoni, N.F. Smyth, A.L. Worthy, Phys. D **189**, 167 (2004)
13. L.A. Cisneros, J. Ize, A.A. Minzoni, Phys. D **238**, 1229 (2009)
14. A.A. Minzoni, L.R. Juarez, M. Rosenbaum, Phys. Lett. A **373**, 1510 (2009)
15. R. Gopakumar, S. Minwalla, A. Strominger, J. High Energy Phys. **05**, 020 (2000)
16. A. Fratalocchi, G. Assanto, Phys. Rev. E **72**, 066608 (2005)

Experimental Results for the sine-Gordon Equation in Arrays of Coupled Torsion Pendula

Lars Q. English

Abstract The pendulum chain with torsional spring coupling represents a physical system that is described exactly by the spatially discrete sG equation. In order to study its solutions experimentally, however, we have to drive the system to counteract the inevitable dissipation we face in experiments. In fact, it is important to model the main sources of energy dissipation realistically; here we find that both on-site and intersite contributions to dissipation are relevant. We show that lattice solitons, also known as discrete breathers (DB) or intrinsic localized modes (ILMs), can be produced and stabilized in this system in the presence of driving (forcing) and damping. One way to do this is by exploiting the modulational instability of the uniform mode. Once a discrete soliton has been phase-locked to the driver, it persists indefinitely, and its properties can be studied, as can the range of its stability in parameter space and the types of instability outside of it. We also explore in some detail the interesting effect of which there exists no analogue in continuous media: the existence of two types of discrete solitons of different symmetry—the one-site centered and the two-site centered ILM. Furthermore, we find an exchange of stability between those two breather types as certain system parameters are varied. Comparison to analytical work based on realistic model parameters will be discussed as well.

Keywords Breather symmetry • Breather mobility • Driven-damped ILM • Lattice soliton • Mechanical DB • Naimark-Sacker bifurcation • Peierls–Nabarro barrier • Pendulum chain • Stability exchange • Torsional spring coupling

---

L.Q. English (✉)
Department of Physics and Astronomy, Dickinson College, Carlisle, PA 17013, USA
e-mail: englishl@dickinson.edu

J. Cuevas-Maraver et al. (eds.), *The sine-Gordon Model and its Applications*,
Nonlinear Systems and Complexity 10, DOI 10.1007/978-3-319-06722-3_5,

1 Introduction: The sine-Gordon Equation and Its Approximate Physical Realization in the Pendulum Chain

Imagine an array of equally-spaced pendula rotating about a common axis and connected together via springs. When one pendulum rotates away from vertical, it twists the springs connecting it to its neighbors on either side, and this twist causes a torque on them. Newton's second law (in rotational form) then dictates that the angular acceleration of any given pendulum depends on the gravitational restoring torque and the spring-elastic torque, and we arrive at an equation of the form, $I\theta_{tt} = \beta(\theta_{n+1}-\theta_n)+\beta(\theta_{n-1}-\theta_n)-mgL_{com}\sin(\theta_n)$. Here time is a continuous variable, but space is discrete and indexed by the subscript n. Exact solutions cannot be found analytically. However, if we go into the continuous spatial limit (and re-dimensionalize), we arrive at the standard sG equation:

$$\theta_{tt} + \sin\theta - \theta_{xx} = 0 \tag{1}$$

This equation is integrable and closed-form solutions for solitons, anti-soliton, and breathers exist (see, for instance, [1]). The soliton and anti-soliton solutions are topological; essentially they represent traveling kinks, where pendula on one side start out at $\theta = 0$ and on the other side end up at $\theta = 2\pi$, and where the 2π phase shift is bridged over a narrow spatial region. This region can then travel along the chain at a speed that is tied to its width—the faster the kink travels, the sharper it is (with the relationship being captured by the Lorentz contraction formula). Finally, the breather solution can be understood as a bound state of a soliton and anti-soliton.

Experiments were performed early on to generate such solutions in physical systems (see, for instance, [2] and references therein). In these studies, the aim was to be close to the continuous limit, which favored stiff spring, and short, light, closely spaced pendula. Here, a kink could be launched from one end of the line, by taking the first pendulum and rotating it by a full 360° where it is again at a local minimum. Subsequently the disturbance can travel down the line in the form of a topological kink. The bound state of kink and anti-kink—the breather—was harder to see experimentally in these chains. Here the best strategy was to set up a very slowly moving kink and anti-kink, but even then, the stationary breather turned out to be a fairly fragile mode. Due to the presence of small dissipation in the system, it would inevitably disintegrate within a few oscillations.

In this study, we take a different approach. We are interested in discrete breathers, or ILMs, and therefore choose a system sufficiently far away from the continuous limit. Furthermore, instead of launching disturbances from the ends of a long line of pendula and seeing the ensuing solitons propagate across it, we will attempt to produce and stabilize a stationary discrete breather in a smaller chain by external sinusoidal driving. We will then be able to investigate the ILM properties as a function of driving/forcing conditions, and explore its stability regime. We find that two types of driven ILMs can exist in this system, and that a stability-exchange leads from one to the other [3, 4]. Analytical results will corroborate the experimental observations.

2 The Experimental System

2.1 A Detailed Description of the Pendula and Their Coupling

Several immediate requirements of a physical realization went into the initial design considerations. We wanted to be able to drive the pendulum chain sinusoidally with as little environmental noise as possible, as well as with precise control of the driving conditions (frequency and amplitude). Secondly, we wanted to keep the effect of friction as low as possible. This second constraint led to the use of fairly heavy and long pendula suspended on thin piano wire. The pendula were made from threaded rods (mass of 13 g, length of 25.4 cm) as well as end masses on the order of a few grams; this (together with our choice of springs) also ensured that the system would be far away from the continuous limit. The first constraint (i.e., precise driving) suggested a platform on wheels on which the pendulum array would rest and which would be driven by a high-torque electric motor (in order to eliminate feedback from the system to the driver).

Figure 1 illustrates the larger experimental setup in more detail [3]. The pendulum array (i) rests securely on a frame (ii) with low-friction wheels (iii). A high-torque electric motor (90 V, 190 W) is attached to the cart via a crankshaft (iv) and moves it horizontally at a constant but voltage-adjustable frequency. This sinusoidal horizontal motion of the pendulum-platform translates into a uniform driving torque on each pendulum. The frequency and amplitude of the driver is independently determined using a Vernier linear motion sensor tracking the pendulum platform. While the driving frequency can be dynamically tuned (or chirped) during a run, the amplitude of platform motion can only be adjusted between different runs. This is done by changing the radius of the crankshaft connection to the driving wheel. Finally, the pendulum motion can be captured by an overhead wide-angle web-cam (v) which was fixed to the frame of the pendulum array (vi).

Let us now zoom in a bit and examine the pendulum chain more closely. Again, the desire to reduce friction and maximize the Q-value of the pendulum oscillation dictated certain mechanical design features. The pendulum array is schematically shown in Fig. 2. The array consisted of 15 pendulums connected by springs which couple them torsionally. Each pendulum was comprised of a brass base piece near the suspension point, a threaded rod screwed vertically into the base, and a nut or a disc attached to the bottom end of the threaded rod, which served as the end-mass, M. The pendula were suspended on taut piano wire of 0.041 in. ($\approx$ 1 mm) diameter which runs through the base of each pendulum and was held up by the outer frame. Springs of diameter 0.4 in. (1.02 cm) slipped tightly over sleeves at each end of the

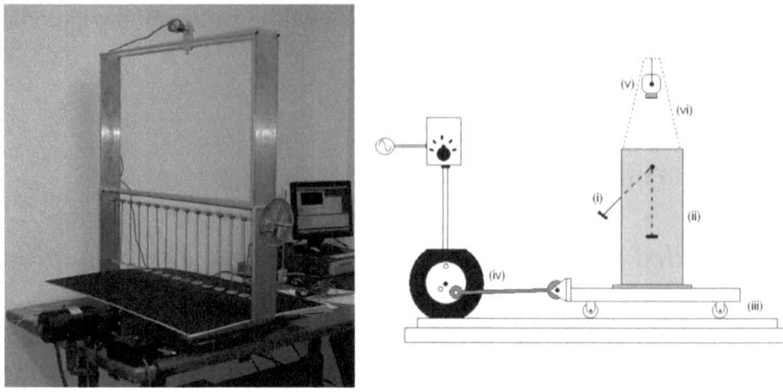

Fig. 1 A picture (*left*) and a schematic overview (*right*, from [3]) of the experimental setup. The pendulum array and its frame sits on a platform that can freely move horizontally; this platform is driven by a crankshaft and electric motor

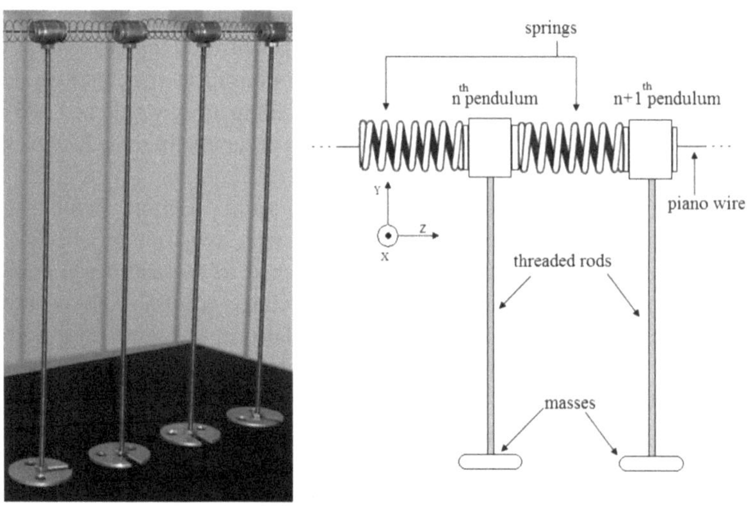

Fig. 2 Picture (*left*) and schematic diagram (*right*, from [3]) of the pendulum and coupling springs

base pieces and were then glued onto that sleeve for additional support. No slippage of the spring with respect to the pendulum base was ever observed. Care needed to be taken to ensure that the springs were completely relaxed when the pendula were in their vertical position.

The first and last pendula were coupled only to their one neighbor, the other end resting against a spacer piece to keep it at a fixed distance from the frame; this results in open boundary conditions. Fixed boundary conditions could also be achieved by simply keeping the outer two pendula locked in a vertical position, in which case the number of pendula reduced effectively to 13. Due to the use of a threaded rod, the effective length of the pendulums is adjustable, but this experimental parameter was chosen to be fixed at 10 in. (25.4 cm). (This adjustability was nonetheless convenient in improving spatial homogeneity.) The end-masses were initially calibrated 5 g discs, but in later studies we reduced it by using small nuts that screwed onto the threaded rods down to omitting end-masses altogether.

The coupling between pendula occurs via the restoring torque in the torsional springs . This spring restoring torque was confirmed experimentally to be quite linear up to an angle difference of 90°. The linearity was confirmed by taking a pendulum triplet and holding the two end pendula fixed in a vertical position, while applying a force to the bottom of the middle pendulum via a force sensor; the net torque increases linearly.

2.2 Damping and Driving Considerations, and Their Model Description

Four main torques act on each pendulum: gravitational, spring, and frictional torques, as well as a centrifugal driving torque by virtue of the non-inertial reference frame. The gravitational torques, of course, are on-site and try to enforce vertical alignment on pendula; spring torques are inter-site and try to enforce relative alignment between neighboring pendula. The driving torque has to be examined a little more closely.

It comes about due to the horizontal motion of the pendulum array by an outside motor. To obtain the expression for the driving torque, we stay within the non-inertial reference frame of the pendula. In this frame, the fictitious centrifugal force acts on each pendulum, namely, $F_c = -ma_{base}$. This force acts on the rod at its center of mass in a horizontal direction, and it thus gives rise to the torque, $\tau_d = A\omega_d^2 \cos(\omega_d t)\frac{mL}{2}\cos(\theta)$. For the end-mass, the torque expression is almost identical, except that the force acts at a distance L from the pivot. Thus, we obtain for the driving torque the expression,

$$\tau_d = A\omega_d^2 \cos(\omega_d t)\cos(\theta_n)(\frac{mL}{2} + ML) = \left(\frac{I\omega_0^2}{g}\right) A\omega_d^2 \cos(\omega_d t)\cos(\theta_n), \quad (2)$$

where I is the pendulum's moment of inertia, $I = ML^2 + \frac{1}{3}mL^2$, and ω_0 is the pendulum's natural frequency of oscillation with $\omega_0^2 = \frac{1}{I}(mgL/2 + MgL)$.

Two types of friction are present in the experiment—there is an on-site component and an intersite one. The on-site component comes from air drag as well as contact friction at the suspension points. The intersite component has its origin in the energy dissipation due to the twisting of the springs, and it should be related to angle differences between neighboring springs. The on-site contribution to the frictional torque is well approximated simply by $\gamma_1\dot{\theta}_n$. The intersite contribution depends on how quickly the angle difference varies, and should thus be given by, $\gamma_2\frac{d}{dt}(\theta_{n+1} - \theta_n)$. The γ constants here are empirical damping coefficients. The way

Table 1 Experimental constants

β	γ_1	γ_2	L	m	A
0.0165 N m/rad	284 g cm^2/s	70 g cm^2/s	25.4 cm	13 g	1.12 cm

they were experimentally determined was via the following procedure: The on-site γ_1 was measured using just one isolated pendulum, tracking its amplitude decay over time, which yields $\frac{\gamma_1}{I}$ directly from the exponential decay constant. Incidentally, γ_1 was measured for a number of end masses, M, and found to be approximately independent of end mass for $M < 10$ g. The damping parameter γ_2 was estimated by starting with three pendula connected in a line via springs; the two outer ones are held fixed vertically, and the amplitude decay of the center pendulum was tracked. The exponential decay constant thus obtained is then divided by the angular velocity ω for this mode (in the linear regime), which can be measured independently. This number corresponds to an effective α, and it is larger than $\alpha_1 = \frac{\gamma_1}{I\omega_0}$. From the difference, the value of γ_2 can be extracted.

We thus obtain the governing equations for the system as,

$$\ddot{\theta}_n + \omega_0^2 \sin\theta_n - \left(\frac{\beta}{I}\right) \Delta_2\theta_n + \left(\frac{\gamma_1}{I}\right) \dot{\theta}_n + \left(\frac{\gamma_2}{I}\right) \Delta_2\dot{\theta}_n + F\omega_d^2 \cos(\omega_d t) \cos\theta_n = 0, \tag{3}$$

where the driving strength is given by $F = \frac{A\omega_0^2}{g}$, β is the torsional spring constant, and Δ_2 represents the discrete Laplacian. We see that nonlinearity in the undriven array enters only through the sine-function in the gravitational term, but not through coupling springs (in contrast to the MEMS cantilever arrays, for instance). Comparing Eqs. (3) and (1), it is clear that, absent damping and driving, the dynamics of this physical system is exactly governed by the discrete sG-equation.

Table 1 summarizes the important experimental constants of our system. The torsional spring constant, β, was actually varied a number of times. The value shown in the table represents the largest value used for the study of stability exchange discussed in Sect. 3.4. Our discussion of mechanical ILMs begins with a chain of much weaker coupling, where β was about a factor 10 lower (at 0.0015 Nm/rad). Also notice that while the intersite contribution is smaller than the on-site contribution, it is by no means negligible. This means, for instance, that the magnitude of dissipation is mode-dependent. The uniform mode, $k = 0$, experiences the least energy dissipation and should therefore have the largest Q-value, with every higher-order standing-wave mode facing successively larger damping.

The dispersion relation for small-amplitude plane-waves is given by

$$\omega(k) = \sqrt{\omega_0^2 + 2\left(\frac{\beta}{I}\right)(1 - \cos k)}, \tag{4}$$

and so for the more weakly coupled chain, $\omega_0 = 6.95$ rad/s and $\omega(\pi) = 7.6$ rad/s, resulting in a fairly narrow dispersion band. Later, when the value of β is increased, the band will broaden as well by shifting up the zone boundary mode to 12.6 rad/s. In both cases, there is an interval below ω_0 for which the first overtone is still above the top of the linear dispersion curve.

2.3 Recording and Analyzing Pendula Dynamics

The easiest way to extract pendula angles θ_n is by analyzing the video clip recorded from a wide-angle web-cam mounted at the top of the frame. This web-cam looks vertically down on the pendula and moves with them in their accelerated reference frame. This has the benefit that the pendulum pivot points never move from one frame to another but always appear stationary. The frame rate of 30 per second is enough to capture the pendulum positions, especially at the turning points. At the maximum speed through $\theta = 0$, some blurring does occur.

The individual frames can then be analyzed in any video analysis software package; we have used Logger Pro for the purpose of obtaining coordinates for all of the pendulum positions. These coordinates can then be converted into an angle, provided the proper calibration was carefully performed. For highly excited and localized ILMs, the center pendulum can sometimes reach angles exceeding 180°, meaning that the pendulum swings past the vertically up position. Here, of course, care has to be taken, since the relationship between lateral coordinate and angle is no longer one-to-one. Finally, this technique also means that the error in determining angle is lowest for small angles and highest for angles near the maximum in the calibrated relationship between lateral coordinate and angle (near 115°).

3 Experimental Results

3.1 Modulational Instability Within the Chain as a Precursor to Soliton Formation

As is the case in many experimental systems, ranging from anti-ferromagnetic spin lattices [5], and micro-mechanical cantilever arrays [6], to granular crystals [7], modulational instability of the uniform mode of oscillation can be used to create ILMs in lattices. This pathway towards energy localization was first proposed in theoretical work (see e.g. [8, 9] for relevant recent discussion). Not surprisingly, therefore, this strategy can be successfully employed here as well.

The way to initiate the modulational instability in this damped-driven system of pendula is to select a driving frequency very close to the frequency of the uniform mode, $\omega_d \cong \omega_0$. In this way, energy is quickly transferred into the uniform mode,

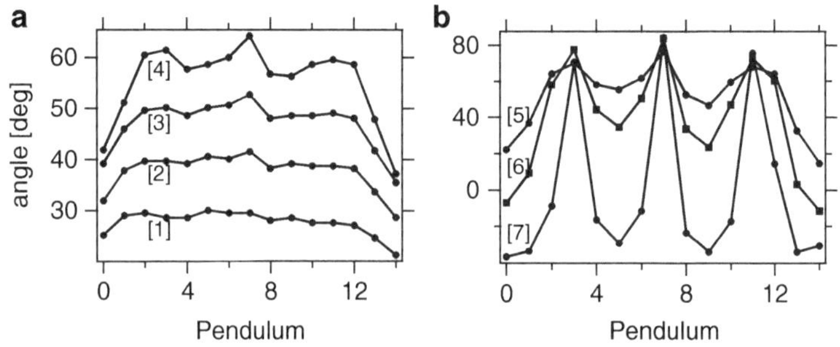

Fig. 3 The modulational instability of the uniform mode as the driver feeds energy into the mode. Between panels (**a**) and (**b**) the driver frequency is lowered. The times shown are 0.33, 2.3, 4.2, 6.2, 8.3, 9.4, and 10.6 s (reproduced from [3])

and as its amplitude grows the driving frequency has to be chirped down to stay in resonance with it due to the system's soft nonlinearity. This gradual downward chirping of the frequency usually continues until the uniform mode has become visibly unstable and ILMs begin to form. At that point, the frequency is kept steady at the lower value. Typically, the final driving frequency at this point will have been lowered from ω_0 by about 10 %.

Figure 3 shows the modulational instability (MI) of the uniform pendulum mode. The traces depict the positions of the 15 pendulums at the instant the central pendulum reaches its turning point. The first trace in Fig. 3a shows the pendulums soon after the driver has been turned on. The subsequent traces in Fig. 3a show the turning points of the pendula at later times as they acquire more energy from the driver. In traces 3 and 4, an emerging spatial variation can already be observed. This spatial variation quickly escalates with time, as seen in Fig. 3b. Traces 5 through 7 show the position of the pendulum chain at three successive periods of the center pendulum. In trace 7—taken 11 s after the driver is turned on—we see that three pendula (including the center one) are now out of phase relative to the rest of the chain.

The MI essentially amplifies the noise sufficiently for the system to eventually settle into one or more ILMs [9]. Observationally, the most accurate way of describing the process is that the MI breaks up the uniform mode where all the pendula swing in lock-step. A period ensues where the phases and amplitudes of the pendula are widely varying and the energy distribution in the lattice is seemingly chaotic. Out of this chaotic phase, eventually some coherent structures assert themselves, reinforced by the driver. Other features that are not in phase with the driving signal over sufficiently long a time window, gradually lose energy. Eventually, only one or two solitons persist in the lattice of 15 pendula.

It should be emphasized that in order to produce an ILM using the MI, it is essential that the frequency be decreased during the run. In this run, the frequency was first set to resonance and then lowered within 3–4 s by about 10 % to ω_d = 6.1 rad/s; a steady decrease of frequency over this time interval works well. In this

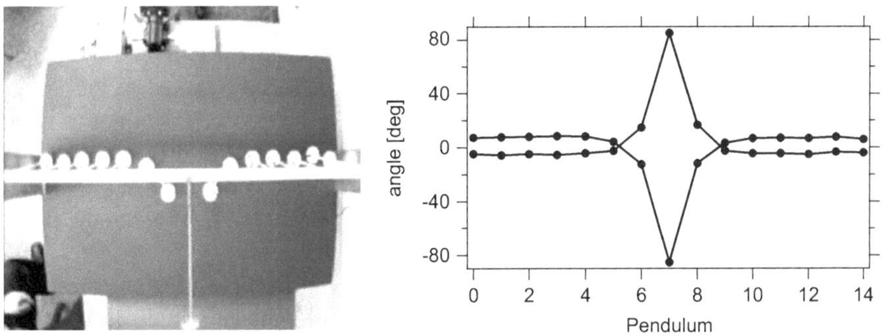

Fig. 4 The profile of the discrete soliton entrained to the driver. The driver amplitude was 2.5 mm, and driver angular frequency 6.1 rad/s

way, a stable driven ILM was generated in this particular run centered at pendulum 7. The location of the ILM within the chain and even the number of ILMs produced cannot be predicted apriori and is different from run to run, as expected for a uniform system of discrete nonlinear elements; the sites near the boundaries are, however, avoided.

3.2 The Driven Lattice Soliton-Dependence on Driving Conditions and Observed Instability

The lattice soliton (or ILM) that is eventually obtained after the MI has broken up the uniform mode and after the ensuing chaotic dynamics has died down is sharply localized. This is depicted in Fig. 4. We see that the central pendulum has a fairly large amplitude of oscillation hitting just below 90°. The nearest neighbors also swing in phase with the center, but their amplitude is reduced to about 20°. The next nearest pendula almost represent nodes in that they hardly move at all. If we move even farther away from the center, we are now in the wings of the driven soliton. The small value for the torsional spring constant, β, produces ILMs that are effectively localized only on three lattice sites. It is clear that the continuous approximation must fail in this system, and that we are in fact closer to the anti-continuous limit.

As we mentioned, the driven-ILM wings are seen to oscillate out-of-phase with the center. Subsequent comparison to the driver phase reveals that the center is out of phase with the driver, and so the wings are in-phase with the driver. Theoretical studies [9–11] have predicted the existence of ILMs with a 180° phase difference between center and wings. This creates what [10] calls a "phase-domain wall" of just one or two pendulums in width. In systems with soft nonlinearity, the low-amplitude wings are in phase with the driver (as the driver frequency is below the low-amplitude resonance frequency), and the ILM center is out-of-phase. A second type of ILM is also predicted where both center and wings are in-phase with the driver. We have not been able to stabilize this solution experimentally over long times.

Once a stable discrete breather has formed that is firmly locked to the driver, the driver frequency can be adiabatically varied. Care has to be taken not to move the frequency too rapidly, or else the ILM de-phases from the driver and decays away. Even small fractional changes in frequency have to occur over multiple ILM periods. The question we can explore is: what happens to the ILM when the driver frequency is adiabatically reduced?

Since we are moving away from the linear dispersion curve, it is not surprising that the ILM has to become even more nonlinear. The swing amplitude of the center pendulum has to be larger to accommodate the lower frequency. The spatial profile of the discrete soliton becomes even sharper. At some lower cutoff frequency, the ILM can no longer stay locked to the driver, decouples from it, decays in amplitude (due to damping), and eventually disintegrates into a plane-wave mode. The location of this frequency cutoff, i.e., the lowest possible driver frequency, is (as expected) sensitive to the driver amplitude. For instance, the cutoff is at around $0.9\omega_0$ for $A = 2.5$ mm, and $0.77\omega_0$ for $A = 7.5$ mm. At this second cutoff-frequency (for the larger driving amplitude), the center pendulum reaches a maximum deflection angle of 120°. That angle can exceed even 180° in cases where stiffer coupling springs and larger driver amplitudes are selected.

Next, we can explore the higher frequency limit. What happens when the driven ILM, once produced and stabilized, is made to lock to ever-higher frequencies, approaching ω_0 from below? Here, too, a threshold of soliton stability is encountered. Beyond an upper frequency cutoff, the soliton is seen to undergo an instability that eventually destroys it. This cutoff is lower for larger driver amplitudes. For $A = 2.5$ mm, the threshold is at 0.94 ω_0, but for 7.5 mm is has moved down to 0.91 ω_0. We will characterize the type of instability more fully in Sect. 4; here let us simply describe its experimental signature in greater detail. For this purpose, we switch to a fixed-boundary chain where the results are somewhat cleaner, although the basic instability occurs for both boundary conditions.

Figure 5 shows a picture taken by the overhead web-cam that illustrates the onset of this instability. The wings are still out-of-phase with the ILM center, but they are gaining energy. In fact, to appreciate the full dynamics, Fig. 6 shows what happens to the oscillation amplitude of a pendulum in the wings and one at the center.

An energy pulsation between wings and center is evident in Fig. 6a; energy is transferred back and forth in time at a time-scale exceeding the oscillation period. It is also evident that the magnitude of those pulsations grows with time, leading to the eventual destruction of the soliton. Examining the full soliton profile at certain times (indicated in panel (a) by numbers) suggests an emerging resonance with a nonlinear standing-wave mode. The following qualitative interpretation of the breather-instability mechanism thus emerges. At sufficiently high driver frequencies, there is enough energy in the ILM wings that its presence pulls down the third allowed plane wave of wavelength $k_n = \frac{\pi}{L}n, n = 3$ to the same frequency. At that point—their frequencies matched—energy can be transferred from the ILM to this plane-wave mode and back (see panel (c), where the dotted lines indicates

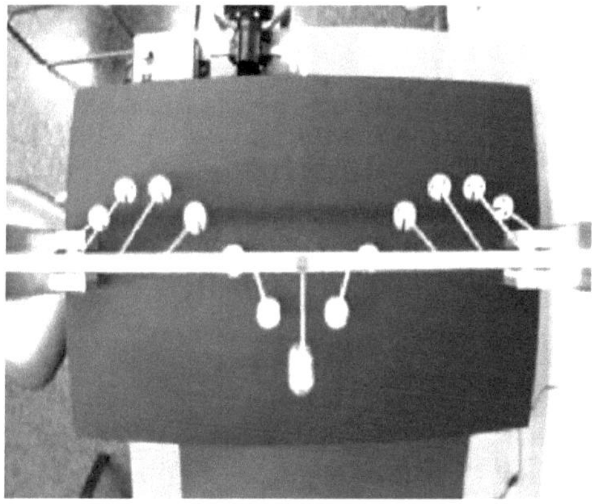

Fig. 5 The profile of the discrete soliton where the driver frequency slightly exceeds the upper threshold, and an instability sets in

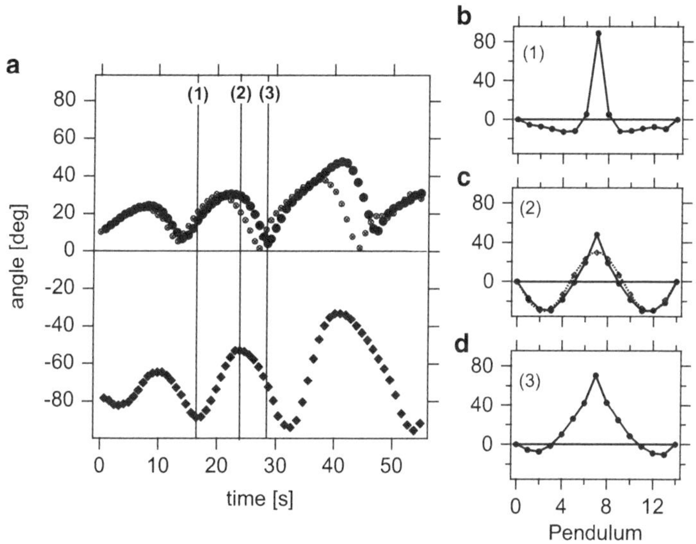

Fig. 6 The evolution of the instability. The lower trace shows the maximum angle attained by the center pendulum in each period; the upper two traces are for the two pendula in the wings. The panels on the right show the ILM profiles at select times (1), (2), and (3) (reproduced from [3])

the standing mode). The nonlinear coupling and decoupling between ILM and third allowed plane-wave mode, as well as both modes interaction with the driver, appears to give rise to this pulsating breather instability. We will revisit this instability in the analytical section, where it can be shown to be a Naimark-Sacker (NS) bifurcation (a generalized Hopf or Hopf bifurcation of periodic orbits).

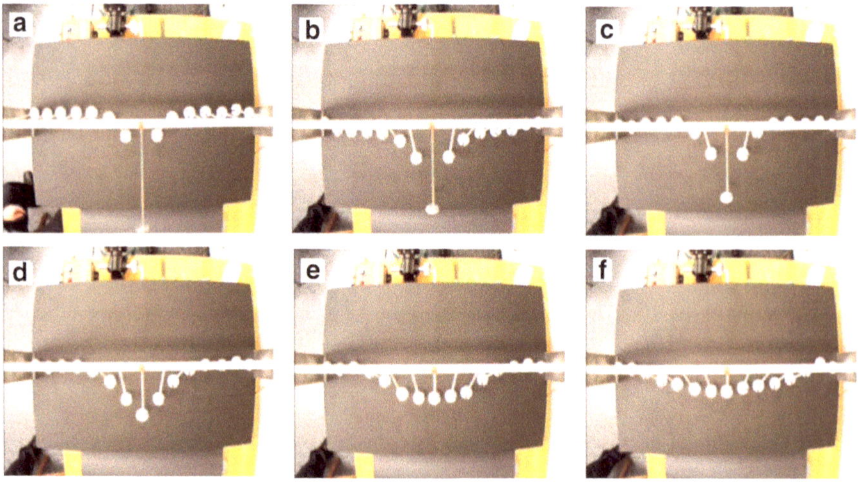

Fig. 7 Turning off the driver—the decay of the discrete soliton. (**a**) Just before the driver is turned off. (**b**) After 7 periods, (**c**) 18 periods, (**d**) 24 periods, (**e**) 30 periods, and (**f**) 38 periods (note that the periods are decreasing in duration)

3.3 Turning off the Driving/Forcing: A Glimpse of the 'Bare' Soliton and Its Eventual Decay

In this high-Q array, we can turn off the driver and have a chance to see the undriven soliton over a number of periods before it disintegrates. This is illustrated in Fig. 7, where a chronological sequence of snapshots taken by the overhead web-cam are shown. The first panel depicts the driven soliton just before the driver is turned off; the characteristic π-phase shift between ILM-center and wings is evident. The second panel (b) shows the situation after 7.1 seconds have passed (or 7 periods); by this time, the wings have come into phase with the ILM center. After another 10.8 seconds (or 11 periods), the discrete soliton has lost energy but is still intact. Notice that the period is decreasing in duration, which means that the frequency is increasing towards ω_0, as expected. Panel (d) depicts the soliton after 24 periods. In panels (e) and (f), at 30 and 38 periods, respectively, the soliton has disintegrated into the standing wave mode.

3.4 Two Types of Discrete Solitons: From Onsite to Inter-Site Symmetry

There are basically two sets of accessible parameters in the experiment. One relates to the driver (amplitude and frequency), and we have already investigated the effects on the soliton profile and its stability. We now turn to the other set of parameters

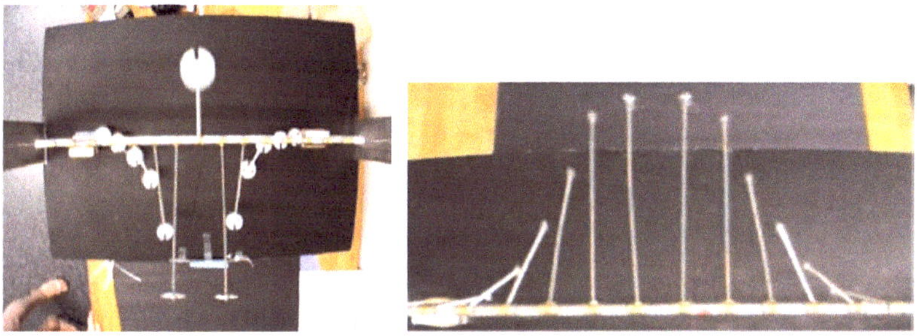

Fig. 8 Two snapshots taken by the overhead web-cam of discrete solitons in a system with increased spring coupling; as the mass is lowered, symmetry changes to two-site. *Left*: The end mass is 5.0 g. The center pendulum swings past the vertical orientation ($\theta > 180°$). *Right*: the two-site lattice soliton just after the transition (here the end-mass is decreased to $M = 0.9$ g)

related to the undriven mechanical system—the coupling strength of the springs and the moment of inertia of the pendula.

The situation becomes more transparent upon nondimensionalizing Eq. (3) by introducing the following parameters: $\omega_b = \frac{\omega_d}{\omega_0}$, $C = \frac{\beta}{I\omega_0^2}$, and $\alpha = \frac{\gamma}{I\omega_0}$ (F is already dimensionless). F and ω_b are driving (forcing) parameters, α refers to damping, and C encapsulates the mechanical properties of the pendulum chain. The definition of C suggests two experimental ways of adjusting this parameter. For gross adjustment, one can change the springs between the pendula modifying the torsional stiffness and coupling. A finer (and more convenient) way of varying C is via I and ω_0. This can be done most easily by changing the end-mass, M, which affects I and ω_0. A larger M reduces the effective coupling constant C. Note that by changing ω_0, ω_b and F are also affected.

Upon increasing the stiffness of springs (by swapping out the original springs with ones made from larger-diameter wire), the solitons become much broader in width. This is, of course, expected, as we are moving further away from the anti-continuous limit. We are, finally, left with springs characterized by a tenfold increase in torsional spring constant ($\beta = 0.0165$ Nm/rad). Having increased the value of C quite substantially in this way, we now proceed to raise C in much finer increments by reducing the end mass, M, down from 5 g all the way to 0 g. During this gradual increase in C, we observe an interesting soliton transition (see Fig. 8).

The symmetry of the ILM abruptly changes from on-site (centered on one site) to intersite (centered on two sites). This transition happens at an M between 1.8 and 0.9 g [4]. Furthermore, it is observed that for masses in that range, the breather becomes very mobile . This can be ascertained in the following manner: bring a fixed boundary into the wings of the ILM, for instance by holding a particular pendulum in place in the vertical orientation. In most situations, such an interference will either not significantly disturb the ILM (if the fixed pendulum is sufficiently far away from the ILM center), or it will destroy the ILM. However, in the regime near the stability exchange, the ILM will move away from the newly introduced fixed boundary. The ILM will move fairly quickly at first, alternating between one-site and two-site symmetry, and in the limited lattice size of the experiment it will slow down and finally come to rest three sites over.

4 Analytical Results and Comparison to Experimental Data

It has been known that, unlike solitons in continuous media, ILMs/discrete breathers come in two principal varieties in one-dimensional chains: the on-site centered (or one-site) modes [12] and the intersite centered ones (two-site modes) [13]. In the simplest ones among the relevant models supporting such structures, either the former, or the latter modes are stable, and this property is maintained when relevant parameters, such as the lattice spacing, are varied [14]. However, more recently it has been illustrated that stability exchanges are theoretically possible between these modes [15], which result either in a true or approximate vanishing of the so-called Peierls–Nabarro (PN) barrier, namely, the energy difference between them. This vanishing is, in turn, relevant in that it enables the localized modes to move along the lattice without facing such an energy barrier, thereby enhancing their mobility [15, 16].

The experimental results indicate that this system exhibits just such a stability exchange, together with a corresponding vanishing of the PN barrier. We now investigate the matter analytically to confirm this picture [4]. Our computations start at the anticontinuous (AC) limit ($C = 0$), using the two (different amplitude) attractors of an uncoupled pendulum. Breathers close to the AC limit only exist if there are two coexisting attractors of the single driven and damped pendulum—both phase-locked to the external driving—but this bi-stability is a fairly generic feature of the resonance curve of a nonlinear oscillator of sufficient Q-value and nonlinearity. The highest amplitude attractor will then correspond to the excited breather site(s) whereas the smallest-amplitude one will be chosen for the low-amplitude background. Then, the breather can be thought of as a collective oscillation at the driving frequency, ω_b. Upon creation at the AC limit, such solutions are continued for arbitrary $C > 0$, through a path-following scheme in the real space, as proposed in [17]; this method allows us to identify both stable and unstable solutions. The idea is that the non-vanishing coupling now mediates (or interpolates) between the two solutions at ω_b. The breather profiles thus obtained compare very well with the experimental results, as shown in Fig. 9. The good agreement is maintained across the transition from one-site to two-site breathers.

The linear stability of the discrete breathers obtained in this way is studied by means of a Floquet analysis. Here the basic idea is to imagine adding a perturbation, $\eta_n(t)$, to the breather profile. Upon substituting $\theta_n + \eta_n$ back into Eq. (3) (or its non-dimensional versions) and linearizing, a differential equation in η, governing the evolution of the perturbations, can be extracted. If we look at Poincaré sections of this differential equation at intervals of the driver period, T, we can think of it as a map of the form, $(\eta_n(T), \dot{\eta}_n(T)) = M\ (\eta_n(0), \dot{\eta}_n(0))$. Here the matrix M represents

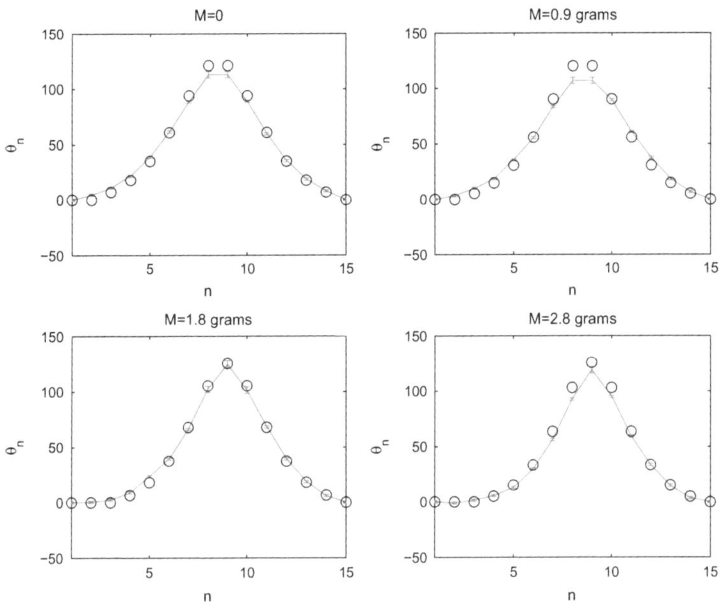

Fig. 9 The numerical profiles (*circles*) of the discrete soliton compared to the experimental results (*lines*)

the map from one period to the next, and linear stability of the original unperturbed solution now requires that its eigenvalues have a magnitude less than or equal to 1.

Figure 10 depicts the results. Panel (a) shows the stability regime in the ω_b-M plane for fixed driver amplitude. The dots in the figure correspond to the experimental conditions shown in Fig. 9. The analysis confirms that between $M = 0.9$ and 1.8 g, there occurs an exchange of stability between the two types of breathers. Conversely, for a given M, an exchange of stability can be induced by varying ω_b. Turning back to the experiment for a moment, for $M = 1.8$ g, the one-site ILM is found to be stable down to $\omega_b \cong 0.83$. At that point, a transition to two-site ILM appears to be initiated in the experiment. The resulting two-site ILM is, however, difficult to stabilize in the system of 13 pendula, and it usually decays within 5–10 periods. The observed transition frequency agrees reasonably well with the predicted value of 0.805. At $M = 1.2$ g, the transition can be observed more clearly. The on-site breather prevails at $\omega_b = 0.96$. Below $\omega_b = 0.91$ the intersite breather is stable; the predicted value is, in this case, 0.852, as seen in the figure.

Above a critical value (≈ 0.875), the system experiences a Naimark-Sacker (NS) bifurcation leading to the eventual breather disappearance. The NS-bifurcation at larger ω_b is also clearly observed experimentally, as discussed earlier, but for end masses above $M = 6.8$ g. For driver frequencies ω_b below a critical lower value, we observe the nonexistence of any kind of breather (even unstable ones), again in agreement with the experiment.

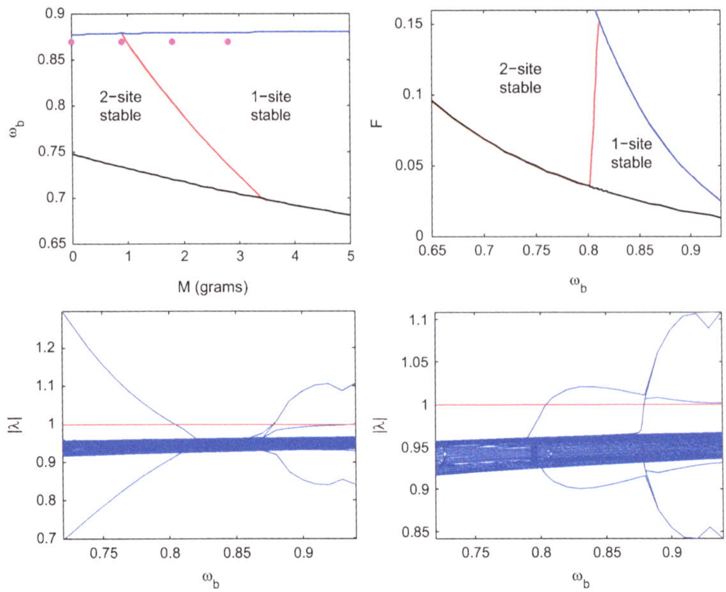

Fig. 10 *Top*: Numerical two-parameter planes for (*left*) fixed $A = 1.2$ cm and (*right*) fixed $M = 1.8$ g. Full circles in the left panel correspond to the solutions shown in Fig. 1. *Bottom*: numerically computed moduli of Floquet multipliers (indicating stability, when they are all smaller than 1) for (*left*) one-site and (*right*) two-site breathers. In this case, $F = 0.06$ and $M = 1.8$ g are fixed, while ω_b is variable. An exchange of stability bifurcation is observed for $\omega_b \cong 0.804$. In both cases, the bifurcation arising for higher ω_b is a Naimark-Sacker one

In Fig. 10b, the breather stability is shown in the F-ω_b plane, and again a sharp delineation between one-site and two-site breather is found. To see how such phase diagrams are constructed, Fig. 10c, d show a typical parametric dependence of the Floquet exponents for two different configurations. The left panel refers to the one-site breather solution, and the right panel to the two-site breather. We observe that as ω_b increases, the two-site breather loses stability (a Floquet multiplier crosses above 1) just as the one-site breather gains stability (multiplier crosses below 1).

5 Other Experimental Work on Pendulum Arrays

As mentioned earlier, pendulum arrays were first constructed to realize the analytical solutions for solitons of the continuous sG equation. The early work in [18, 19] thus focused on characterizing traveling kinks (or topological solitons) and their velocity-dependent shape. Kinks were usually launched at the end of the pendulum chain. In [19], the setup allowed for adjustable damping coefficient and for the inclusion of a constant (and spatially uniform) external torque. The results indicated that larger torques gave rise to faster (and therefore narrower) kinks, as did lower damping.

A significant modification of these original experiments came with the inclusion of periodic driving, initially by means of shaking the pendulum system vertically. Denardo et al. [20] used masses held up by two cords in a V-shape; neighboring masses were coupled because the V from adjacent masses overlapped and were tied together at that point. This coupling yielded an approximately linear restoring force for small angle differences. Placing this assembly on a shake-table allowed certain wave-modes to be excited parametrically by a driver at twice the mode's frequency. For instance, the zone-boundary mode (where each pendulum is exactly π out-of-phase with its neighbors) can be excited in this way, but the novel observation was the appearance of a stable kink separating two realizations of this mode in different halves of the lattice. Kink-like domain walls that transitioned between modes of different k-values were also seen. Again, the domain wall was found to have a fairly narrow spatial width ranging only over a few lattice sites.

Chen [21] extended these findings in a very similar experimental system by directly producing discrete solitons via parametric driving. At lower frequencies of the vertical shake, the driven soliton would exhibit complete phase coherence between all pendula, with the amplitude envelope closely matching a hyperbolic secant fit. At somewhat higher driver frequencies, the soliton featured π phase-shifts between neighboring pendula, and it thus resembled a zone-boundary envelope soliton. In di-element pendulum lattices, gap-solitons (residing between the acoustic and optic branches) were also generated via parametric driving [22]. In a more recent study [23], coupled pendula were arranged in a ring configuration, which naturally realizes periodic boundary conditions. Using parametric (vertical) driving of this system, certain plane-wave modes could be excited. Upon exciting these extended modes beyond a critical amplitude, modulational instability was shown to lead to excitations localized to only a small region of the ring. This emergence of a localized vibrational mode in the ring visually demonstrates in a dramatic manner that symmetry was broken in the process.

6 Conclusion

The experimental results in an array of coupled pendula, driven uniformly, reveal a dynamical richness inherent in the discrete sG model. ILMs/discrete breathers can phase-lock to the driver in such a pendulum lattice and thereby sustain themselves indefinitely, if the driver frequency is tuned below the bottom of the dispersion curve. The soliton profile is characterized by sharp localization of energy on just a few pendula, which oscillate out-of-phase with respect to the other pendula in the wings. Furthermore, for a given driver amplitude, the ILM can only exist within a certain interval of frequencies below the dispersion curve. At lower frequencies, no breather solution exists, and at higher frequencies, the breather undergoes a dynamic instability akin to a Hopf bifurcation. Underlining the importance of discreteness in this system, an interesting phenomenon occurs when the coupling strength between neighboring pendula is sufficiently increased. We observe a transition from a one-site centered to a two-site centered ILM; at a critical coupling strength, an exchange of stability occurs between these two types of ILMs. At this point, the PN-barrier vanishes and the ILMs exhibit a large enhancement in mobility both in the experiment as well as in simulations. The experimental determination of the critical point was accurately corroborated by Floquet analysis of breather stability. Other experimental studies have shown that stationary lattice solitons can also be generated and stabilized via parametric driving.

A few immediate possibilities for future work suggest themselves. Since parametric and direct driving have now been shown experimentally to generate stable lattice solitons, the question is if subharmonic driving (in the form of a horizontal force near twice the natural frequency) could also be employed. In other (more highly damped) nonlinear lattices [24], subharmonic ILMs were experimentally generated and identified as a new type of driven localized mode recently, so the question arises if the sG system would also admit them? Could the number of ILMs (in a lattice of a given size) similarly be controlled via the driver frequency? Finally, we can ask if there is any straightforward physical realizations of a higher-dimensional sG system. For instance, could a two-dimensional sG lattice somehow be constructed. The difficulty with pendula is that their coupling proceeds along their common rotation axis. Coupling along the second lattice dimension would likely have to utilize something other than torsional springs; perhaps regular springs connecting the ends of neighboring pendula may be feasible (at least for angle differences not too large). It would be interesting to see if true two-dimensional localization could be achieved in such a system.

Acknowledgements I thank Richard Lindsey (technician at Dickinson College) for much help in constructing the pendulum system. He will be missed.

References

1. T. Dauxois, M. Peyrard, *Physics of Solitons*, Chap. 2 (Cambridge University Press, Cambridge, 2006),
2. M. Remoissenet, *Waves Called Solitons*, 3rd edn., Chap. 6 (Springer, Berlin/Heidelberg/ New York, 1999)
3. R. Basu Thakur, L.Q. English, A.J. Sievers, J. Phys. D Appl. Phys. **41**, 015503 (2008)
4. J. Cuevas, L.Q. English, P.G. Kevrekidis, M. Anderson, Phys. Rev. Lett. **102**, 224101 (2009)
5. R. Lai, A.J. Sievers, Phys. Rep. **314**, 147 (1999); L.Q. English, M. Sato, A.J. Sievers, J. Appl. Phys. **89**, 6707 (2001); L.Q. English, M. Sato, A.J. Sievers, Phys. Rev. B **67**, 024403 (2003)
6. M. Sato, B.E. Hubbard, A.J. Sievers, Rev. Mod. Phys. **78**, 137 (2006)
7. N. Boechler, G. Theocharis, S. Job, P.G. Kevrekidis, M.A. Porter, C. Daraio, Phys. Rev. Lett. **104**, 244302 (2010)
8. I. Daumont, T. Dauxois, M. Peyrard, Nonlinearity **10**, 617 (1997)
9. P. Maniadis, S. Flach, Europhys. Lett. **74**, 452 (2006)
10. T. Roessler, J.B. Page, Phys. Lett. A **204**, 418 (1995)

11. T. Roessler, J.B. Page, Physica B **219–220**, 387 (1996)
12. A.J. Sievers, S. Takeno, Phys. Rev. Lett. **61**, 970 (1988)
13. J.B. Page, Phys. Rev. B **41**, 7835 (1990)
14. T. Kapitula, P.G. Kevrekidis, Nonlinearity **14**, 533 (2001)
15. L. Hadžievski, A. Maluckov, M. Stepić, D. Kip, Phys. Rev. Lett. **93**, 033901 (2004)
16. T.R.O. Melvin, A.R. Champneys, P.G. Kevrekidis, J. Cuevas, Phys. Rev. Lett. **97**, 124101 (2006)
17. J.L. Marín, S. Aubry, Nonlinearity **9**, 1501 (1996)
18. A.C. Scott, Am. J. Phys. **37**, 52 (1969)
19. K. Nakajima, T. Yamashita, Y. Onodera, J. Appl. Phys. **45**, 3141 (1974)
20. B. Denardo, B. Galvin, A. Greenfield, A. Larraza, S. Putterman, W. Wright, Phys. Rev. Lett. **68**, 1730 (1992).
21. W. Chen, Phys. Rev. B **49**, 15063 (1994)
22. S.Y. Lou, G. Huang, Mod. Phys. Lett. B **9**, 1231 (1995)
23. S.D. Santos, V. Sanchez-Morcillo, N. Jimenez, A.P. Abellard, A. Bouakaz, AIP Conf. Proc. **1433**, 311 (2012)
24. L.Q. English, F. Palmero, P. Candiani, J. Cuevas, R. Carretero-Gonzalez, P.G. Kevrekidis, A.J. Sievers, Phys. Rev. Lett. **108**, 084101 (2012)

Soliton Ratchets in sine-Gordon-Like Equations

Niurka R. Quintero

Abstract This chapter is an overview of the soliton ratchets (motion of solitons under zero-average external forces) in the driven and damped sine-Gordon equation. Soliton ratchets induced by the breaking of spatial, temporal and field symmetries are discussed in detail. Analytical methods, such as collective coordinates theory and symmetry analysis, are presented, together with their comparison with numerical simulations. Symmetry analysis, based on the time-shift invariance of the average velocity of solitons, explains in an unified way, intrinsic phenomena of the ratchets such as: the control of reversal currents by means of the amplitudes and phases of the drivers, the motion induced by damping, and the suppression of the current when certain symmetries of the forces hold. Finally, it is pointed out that some of the previous results require verification through numerical simulations and experiments.

Keywords Biharmonic force • Collective coordinates theory • Functional Taylor expansion • Gating ratchets • Long Josephson junctions • Ratchet transport • Soliton ratchets • Symmetric optical lattices • Time-reversal symmetry • Time-shift invariance • Time-shift symmetry

1 Introduction: From Particle-Ratchets to Soliton Ratchets

The transport of particles or solitons under zero-average forces (i.e., ratchet transport) has been extensively investigated in the last decade [1, 2]. This phenomenon has been predicted by means of the symmetries of the systems [1,3–5] and explained

N.R. Quintero (✉)
Departamento de Física Aplicada I, Universidad de Sevilla. Escuela Politécnica Superior, C/ Virgen de África, 7, 41011-Sevilla, Spain

Instituto de Matemáticas de la Universidad de Sevilla (IMUS). Edificio Celestino Mutis. Avda. Reina Mercedes s/n. 41012-Sevilla, Spain
e-mail: niurka@us.es

J. Cuevas-Maraver et al. (eds.), *The sine-Gordon Model and its Applications*, Nonlinear Systems and Complexity 10, DOI 10.1007/978-3-319-06722-3_6,

using specific theories for each particular physical system [1, 2]. Moreover, net motion has been observed in experiments and simulations in non-linear systems where spatio-temporal symmetries have been properly broken [1, 6–10]. As a first historical landmark, a spatially asymmetric ratchet model was proposed by Smoluchowski and Feynman [11]. Since the 1990s, different ratchet mechanisms have been widely used to elucidate the working principles of molecular motors [12, 13] and, recently, to design astonishing artificial molecular motors [14], to control the remarkable mobility of Leidenfrost drops [15, 16], to model the growth of pearls [17], etc. Furthermore, symmetry analysis suggests [1, 3–5] that ratchet transport can also appear if some temporal symmetries are broken by time-dependent forces [18–20]. This latter case has also been reported in experiments, where currents in semiconductors [21], unidirectional motion of fluxons in Josephson junctions [7, 9], and transport of cold atoms in optical lattices [22] were observed.

Theoretically, the ratchet effect was first studied in particle systems (see the review [1]). In many phenomena, such as scattering and diffusion, solitons also behave like particles [23]. This fact allows the number of degrees of freedom corresponding to the extended systems to be reduced to only a few, i.e., the dynamics of the soliton can be described in terms of so-called collective coordinates (CC), usually the position of the centre of mass and the width of the soliton. These two CC can be mapped into the position and the mass of a given particle. It is therefore natural to investigate the ratchet effect, which appears in particle systems, to the framework of solitons,[1] both topological [24–28] and non-topological[2] [29–32]. All these studies cover both the continuum and the discrete limits of solitonic equations. However, the soliton ratchet mechanisms have been explained only in the continuum limit by means of the collective coordinates theories [24, 25, 31], and questions related to the working principle of discrete soliton ratchets remain unsolved [27, 28].

The non-linear partial differential equations supporting soliton solutions lose their integrability properties when damping and external forces or fields are present. Damping makes the average velocity independent of the initial conditions and renders the ratchet velocity time-invariant (see the following section), whereas the periodic forces break the spatial, temporal, or field symmetries, and keep the systems out of equilibrium. It is well known that sine-Gordon solitons are very robust under these perturbations because they are topological [23]. This is the reason why, in this paradigmatic system, more than six realizations of solitons ratchets have been suggested [5, 24, 26, 33–35], including experiments in Josephson junctions [7, 8]. The aim of this work is to review the net transport of solitons induced by harmonic forces, i.e. the so-called soliton ratchets. To this end, by using symmetry properties of the ratchet velocity, certain common features appearing in soliton ratchets are explained. This analysis is complemented with collective coordinates theories.

---

[1] Along this work we will use solitons instead of solitary waves.

[2] Solitons can be classified into topological and non-topological according to the value of the topological charge.

This paper is organized as follows: In Sect. 2, the consequences of the time-shift invariance of the ratchet velocity are explored for four different classes of soliton ratchets. These include the cases of asymmetric potentials and symmetric temporal forces, and symmetric potentials with asymmetric temporal forces. All these cases are discussed in Sects. 3, 4, 5 and 6, where the main features of soliton ratchet dynamics are revisited. Finally, Sect. 7 summarizes the work carried out to date and examines the prospects for future work.

This review may serve as an introduction to the field. On the other hand, to the active researcher it offers a unifying point of view and the state of the art of Soliton Ratchets.

2 Implications of the Time-Invariance of the Soliton Ratchet Velocity

Assume that the position of the centre of mass of the soliton, $X(t)$, is a function of time that can be calculated when the system is driven by some periodic, time-dependent, external driving $\mathbf{f}(t)$. Therefore, the ratchet soliton current can be calculated as

$$v = \lim_{t\to\infty} \frac{1}{t} \int_0^t \dot{X}(\tau) d\tau = \lim_{t\to\infty} \frac{X(t) - X(0)}{t}. \tag{1}$$

From the mathematical point of view, this means that the current v is a functional[3] of the external driving $\mathbf{f}(t)$ (which we denote $v[\mathbf{f}(t)]$). If v is *sufficiently smooth*,[4] then it admits an nth-order functional Taylor expansion around $\mathbf{f} = \mathbf{0}$ [36].

For simplicity, throughout this work it is assumed that soliton ratchets are induced by two harmonic forces, i.e.,

$$\mathbf{f}(t) = \{\epsilon_1 \cos(q\omega t + \phi_1), \epsilon_2 \cos(p\omega t + \phi_2)\}. \tag{2}$$

In this case, the functional Taylor expansion reads

$$v[\mathbf{f}] = \sum_{n_1=0}^{\infty} \sum_{n_2=0}^{\infty} \langle c_{\mathbf{n}}(t_{11}, \ldots, t_{1n_1}, t_{21}, \ldots, t_{2n_2}) f_1(t_{11}) \cdots f_1(t_{1n_1}) f_2(t_{21}) \cdots f_2(t_{2n_2}) \rangle, \tag{3}$$

where $\mathbf{n} \equiv (n_1, n_2)$ and $\langle \Omega(t_1, \ldots, t_r) \rangle \equiv \frac{1}{T^r} \int_0^T dt_1 \cdots \int_0^T dt_r\, \Omega(t_1, \ldots, t_r)$. Kernels $c_{\mathbf{n}}(t_{11}, \ldots, t_{2n_2})$ are all real T-periodic functions of all their arguments, T being the period of $\mathbf{f}(t)$. Moreover, they can be assumed totally symmetric under

[3] A functional is a mapping from a set of functions to the real numbers.

[4] v must be n times Fréchet differentiable.

any exchange of their arguments. This formula reduces to the usual Taylor series only in a few special cases, for example when all forces are constant, or for certain special kernels (see [37]).

The time-shift invariance of v implies

$$v[\mathbf{f}(t+\tau)] = v[\mathbf{f}(t)] \tag{4}$$

for any $0 < \tau < T$. Thanks to this property, the experimental measurement of v can be repeated at any time, and, under the same conditions, the same results are obtained. Exceptions include Hamiltonian or non-dissipative systems [10], in which v also depends on the initial conditions and therefore the time-shift invariance (4) is lost. Time-shift invariance manifests itself also in the kernels $c_{\mathbf{n}}$.

Recently it has been shown that if a set of n-arbitrary harmonics, $\mathbf{f}(t)$, are used to generate the ratchet current, the fact that v is a time-invariant functional of $\mathbf{f}(t)$ is sufficient to determine the dependence of the ratchet current in all phases and all amplitudes of the harmonics, irrespective of the system under study, providing v is a *sufficiently smooth* functional [38]. For the specific case of the driving (2), the Theorem 1 in [38] reads

Theorem 1. *Let $v[f_1, f_2]$ be a time-shift invariant functional with Taylor series (3), where q and p are co-primes. Therefore,*

$$v[f_1, f_2] = \sum_{k=0}^{\infty} (\epsilon_1^p \epsilon_2^q)^k C_k(\epsilon_1, \epsilon_2) \cos\big(k\vartheta + \theta_k(\epsilon_1, \epsilon_2)\big), \tag{5}$$

with $\theta_0(\epsilon_1, \epsilon_2) = 0$, $\vartheta = p\phi_1 - q\phi_2$, and functions $C_k(\epsilon_1, \epsilon_2)$ and the phase lag $\theta_k(\epsilon_1, \epsilon_2)$ even in each ϵ_i, $i = 1, 2$.

What is shown in [38] is that v has a Fourier expansion as a function of certain combinations of the phases, and that the coefficients of this series depend on the amplitudes in a specific way. Notice that the integer number p (q), coming from the second (first) harmonic, appears in combination with the amplitude and the phase of the first (second) harmonic. This phenomenon is known in the literature of non-linear systems as *harmonic mixing*.

2.1 The Crucial Role of Symmetries

Regarding the transformations of $f_1(t)$ and $f_2(t)$ for which v remains invariant or for which its sign changes, further information on the shape of v can be obtained. Indeed, the time reversal symmetries $v[f_1(-t), f_2(-t)] = -v[f_1(t), f_2(t)]$ and $v[f_1(-t), f_2(-t)] = v[f_1(t), f_2(t)]$ reduce each phase lag in Eq. (5) to $\pm\pi/2$ and 0 or π, respectively. Moreover, v can be reversed using the force-reversal symmetries

$$v[-f_1(t), f_2(t)] = -v[f_1(t), f_2(t)], \tag{6}$$

$$v[-f_1(t), -f_2(t)] = -v[f_1(t), f_2(t)]. \tag{7}$$

In the former case, only the inversion of one harmonic inverts the current; in the latter, only the simultaneous inversion of the two harmonics can reverse the current. Furthermore, the symmetries (6)–(7) eliminate the even terms in the series (5), i.e. Eq. (5) becomes

$$v[f_1, f_2] = \sum_{k=1\,(k\ odd)}^{\infty} (\epsilon_1^p \epsilon_2^q)^k C_k(\epsilon_1, \epsilon_2) \cos\big(k\vartheta + \theta_k(\epsilon_1, \epsilon_2)\big), \tag{8}$$

and fix the values of p and q for which the current vanishes, regardless of the remaining parameters of the system. In fact, from (5) and (6) it follows that if p is an even number, then $v = 0$. This explains why the most popular choice for the generation of currents in the gating ratchets (see the next section) is $p = q = 1$. Similarly, from (5) and (7), $v = 0$ whenever $p + q$ is an even number. Therefore, the most suitable values for p and q are 2 and 1, respectively.

2.2 *Different Realizations of Soliton Ratchets in Damped and Driven sine-Gordon Equations*

Here we mainly focus on four different classes of soliton ratchets:

1. The first class involves the directed transport of solitons in the inhomogeneous damped sine-Gordon equation. The inhomogeneities are displaced such that the centre of the soliton travels along the asymmetric ratchet spatial potential. In this case, the aforementioned force-reversals do not hold and the soliton ratchet velocity is ruled by the full series (5). However, the soliton motion here can be induced solely by one harmonic, therefore we can set $\epsilon_2 = 0$. As a consequence, all terms in (5) vanish, except the constant term for $k = 0$, i.e.,

$$v = C_0(\epsilon_1) = \epsilon_1^2 Q(\epsilon_1), \tag{9}$$

   where Q is an even function of ϵ_1.
2. The second class is inspired from a ratchet potential for particles, and considers the asymmetric field potential in which the soliton lies. Moreover, the soliton is driven with one harmonic. Since the ratchet velocity satisfies the same symmetry properties of the previous issue, v is given by (9), where the coefficients of $Q(\epsilon_1)$ depend on the specific properties of the system.
3. The third class, which in analogy with particle ratchets can be referred to as *tilting ratchets*, are those for which the additive driving (usually the bi-harmonic

force) tilts the non-linear potential and is able to revert the soliton motion, i.e., the symmetry (7) holds. Therefore, for $p = 2$ and $q = 1$, and, without any loss of generality, setting $\phi_1 = 0$ and $\phi_2 = \phi$ for simplicity, the expression for the first two non-zero terms in the series (8) reads

$$v = v_1 \cos(\phi + \theta_1) + v_2 \cos(3\phi + \theta_2) + E_{15}(\epsilon_1, \epsilon_2), \tag{10}$$

where E_{15} contains terms of order 15 or higher in each ϵ; $v_1 = \epsilon_1^2 \epsilon_2 C_1(\epsilon_1, \epsilon_2)$; C_1 and θ_1 are even polynomials in ϵ_1 and ϵ_2 up to order 10; $v_2 = (\epsilon_1^2 \epsilon_2)^3 C_2(\epsilon_1, \epsilon_2)$, and C_2 and θ_2 are even polynomials in ϵ_1 and ϵ_2 up to order 4.

4. The last class, called *gating ratchets*, are those for which the first temporal harmonic is added and the second is used to modulate the non-linear potential. Here, the symmetry (6) signifies that the soliton velocity can be achieved for $p = q = 1$. Moreover, without lost of generality, $\phi_2 = 0$ and $\phi_1 = \phi$ are set. All these restrictions in (8) imply

$$v = v_1 \cos(\phi + \theta_1) + v_2 \cos(3\phi + \theta_2) + E_{10}(\epsilon_1, \epsilon_2), \tag{11}$$

where $v_1 = \epsilon_1 \epsilon_2 C_1(\epsilon_1, \epsilon_2)$; C_1 and θ_1, are even polynomials in ϵ_1 and ϵ_2 up to order 6; $v_2 = (\epsilon_1 \epsilon_2)^3 C_2(\epsilon_1, \epsilon_2)$; and C_2 and θ_2 are linear in ϵ_1^2 and ϵ_2^2.

Although less explored, combinations of these classes are also possible [39]. No matter which symmetry is broken in each particular class, the theory developed in Sect. 2.1 can be applied whenever v is a time-shift invariant smooth functional.

3 Soliton Ratchets Induced by Inhomogeneous External Forces

The generation of directed transport of sine-Gordon solitons with a lattice of point-like inhomogeneities can be achieved using

$$u_{tt} - u_{xx} + \sin u = R(x, t; u, u_t), \tag{12}$$

with

$$R(x, t; u, u_t) = -V(x) \sin u - \beta u_t + f(t), \tag{13}$$

where the periodic function $V(x)$ is used to modulate the sine-Gordon potential, β is the damping coefficient, and $f(t) = \epsilon \cos(\omega t + \phi)$ is an external force whose parameters ϵ, ω and ϕ represent the amplitude, frequency and phase of the periodic force, respectively. In particular, this system exhibits a ratchet-like phenomenon when $V(x)$ is represented by an asymmetric array of three delta functions (inhomogeneities), i.e., the unit cell configuration, of length L, is defined by three inhomogeneities, the first one located at the beginning of the cell, the second one at distance a from the first, and the third one at distance b from the second inhomogeneity,

$$V(x) = V_0 \sum_n [\delta(x - x_1 - nL) + \delta(x - x_2 - nL) \ + \delta(x - x_3 - nL)]$$

where $L = a+b+c$; $a = x_2-x_1, b = x_3-x_2$, and $c = L+x_1-x_3$, with $x_1 < x_2 < x_3$. Moreover, $a, b, c \sim l_0$ (static soliton width in the absence of inhomogeneities); $a, b < c$ with $a \neq b$. Specifically, in the framework of long Josephson junctions, if $V_0 > 0$ then the point-like inhomogeneities represent microshorts [40, 41]. The authors of [26], inspired by biological polymers, such as DNA, where the existence of three bases per codon seems to be the ideal configuration for the occurrence of net transport [42], chose three inhomogeneities per cell. It is also shown that the inclusion of more inhomogeneities per unit cell diminishes the efficiency of the transport [43]. However, the case in which only two inhomogeneities can be used to generate, enhance, and in general control the transport, has yet to be explored.

3.1 Collective Coordinates Theory

For the pure sine-Gordon equation, i.e., Eq. (12) with $R = 0$, the exact soliton solution is represented by

$$u(x,t) = 4\,\mathrm{atan}\left(\exp\left[\frac{x - X(t)}{l(t)}\right]\right) \tag{14}$$

where $X(t) = X(0) + Vt$ is the soliton position. The width, $l(t) = l_0\sqrt{1 - V^2}$, as well as the velocity, V, of the soliton are constants. Moreover, the soliton momentum,

$$P = -\int_{-\infty}^{+\infty} dx u_x u_t, \tag{15}$$

and also the soliton energy,

$$E = \int_{-\infty}^{+\infty} dx \left(\frac{1}{2}u_t^2 + \frac{1}{2}u_x^2 + 1 - \cos u\right), \tag{16}$$

are conserved.

In the presence of perturbations, such as those collected from Eq. (12) in $R(x, t, u, u_t)$ shape, velocity, energy and momentum are, in general, no longer fixed or constant. Rather than trying to solve Eq. (12), an *ansatz* for the soliton profile is chosen. Regarding the two collective variables, the most popular choice for the sine-Gordon equation is the so-called Rice *ansatz* [44,45]. This can also be represented by Eq. (14), but now $l(t)$ and $X(t)$ are unknown functions to be determined. By inserting the Rice *ansatz* (14) into (15) and (16), we obtain

$$P = \frac{M_0 l_0}{l(t)} \dot{X}, \tag{17}$$

and

$$E = \frac{1}{2}\frac{l_0}{l} M_0 \dot{X}^2 + \frac{1}{2}\frac{l_0}{l} \alpha M_0 \dot{l}^2 + \frac{1}{2} M_0 \left(\frac{l_0}{l} + \frac{l}{l_0} \right), \tag{18}$$

respectively, where $M_0 = 8$, $l_0 = 1$, $\alpha = \pi^2/12$, $q = 2\pi$. Expression (17) can be interpreted as the momentum of a particle of variable mass $M(t) = M_0 l_0 / l(t)$ and velocity $\dot{X}$.

3.2 Centre of Mass of the Soliton in the Ratchet Potential

There are several methods for calculating the evolution equations satisfied by $X(t)$ and $l(t)$. Here we focus on the time variation of the energy and the momentum [46]. If Eq. (12) with (13) is taken into account, then the time variations of the momentum and the energy read

$$\frac{dP}{dt} = - \int_{-\infty}^{+\infty} dx R u_x, \tag{19}$$

and

$$\frac{dE}{dt} = \int_{-\infty}^{+\infty} dx R u_t, \tag{20}$$

respectively. By inserting the *ansatz* in (19) and (20), and taking into account (17) and (18), the equations of motion for $X(t)$ and $l(t)$ read

$$\frac{dX}{dt} = \frac{P l}{M_0 l_0}, \tag{21}$$

$$\frac{dP}{dt} + \beta P = -\frac{\partial U}{\partial X} - q f(t), \tag{22}$$

$$\dot{l}^2 - 2 l \ddot{l} - 2 \beta l \dot{l} = \Omega^2 l^2 \left[1 + \frac{P^2}{M_0^2} \right] - \frac{1}{\alpha} + \frac{2 l^2}{\alpha M_0 l_0} \frac{\partial U}{\partial l}, \tag{23}$$

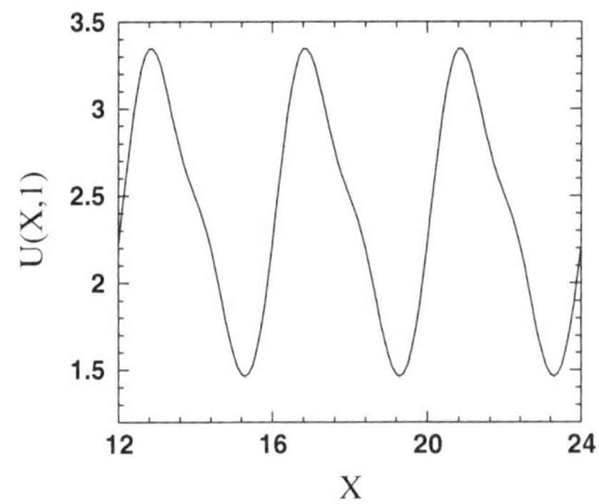

Fig. 1 Asymmetric effective potential $U(X, 1)$, Eq. (24) with $l = 1$, vs. X. Parameters: $n \in [-15, 15]$, $V_0 = 0.8$, $L = 4$, $x_1 = 0.5$, $x_2 = 1$ and $x_3 = 2.3$

where

$$U(X,l) = 2V_0 \sum_n \sum_{i=1}^{3} \operatorname{sech}^2\left(\frac{X - x_i - nL}{l}\right) \tag{24}$$

is the effective potential function and $\Omega = 1/(\sqrt{\alpha} l_0) \approx 1.1$. It can be shown [45] that when there is no perturbation in the system, $l(t)$ oscillates with this frequency if we start from any $l(0) \neq l_0$ or $\dot{l}(0) \neq 0$. Therefore, Ω is the characteristic frequency of Eq. (23). Although this means that the soliton width oscillates under small perturbations either on the initial conditions or directly on the sine-Gordon equation, this frequency is not related to the existence of an internal mode. Indeed, linear stability analysis of the perturbed sine-Gordon equation around the static soliton solution shows that the spectrum only contains one translational mode with zero frequency and the phonons with frequencies $\omega_{ph} = \sqrt{1 + k^2}$ (for all values of k), i.e. there is not any internal mode, since the sine-Gordon equation is an integrable system [47]. Moreover, [48] shows that some phonon modes are quasi-localized and therefore can affect the soliton width. In this sense the Rice *ansatz* considers in an effective way the action of the phonons on the soliton width.

Equations (21) and (22) represent the Newtonian equation for a damped and driven particle which moves along the asymmetric potential $U(X, l)$ (see Fig. 1). A remark on the collective coordinates theory is called for here: one may mistakenly gain the impression that we have proved that the centre and the width of the soliton exactly obey the classic equations of motion (21)–(23). However, the wavefunction (14) that we have used is only an *ansatz*, not an exact solution of Eq. (12). Notice that even the *ansatz* (14) fails to fulfil the boundary conditions corresponding to Eq. (12).

The simulation results of (21)–(24) [26, 43] show that the sine-Gordon soliton behaves very much like a point particle ratchet (see Fig. 5 in [6] and [12, 49]). Indeed, the existence of windows of motion separated by gaps where the motion is oscillatory with zero average velocity has been observed, and these windows were similar to those reported in particle ratchets [50]. Mathematically this could mean that the functional v is not smooth, and therefore the results of Theorem 1 cannot be applied. Curiously enough, a similar issue appears in discrete soliton ratchets [27, 28, 51]. However, as it is shown in Fig. 7 of [43], the smoothness of v can be achieved via thermal fluctuations. Further work on the regularization of the functional average velocity via thermal noise should be performed in the future.

4 Ratchet Transport in Asymmetric and Driven sine-Gordon Equation

Thermal fluctuations were first used in soliton ratchets for another purpose. Based on the overdamped ratchet dynamics of a particle in the asymmetric potential, and driven by a correlated noise in [52], the directed transport of the overdamped asymmetric sine-Gordon equation in the presence of noise was studied. For the overdamped system, a soliton ratchet was reduced to the usual single-particle ratchet simply by considering the ordinary differential equation describing the motion of the soliton centre of mass. This single-particle picture however, is adequate only in the overdamped case, i.e. when the radiative field would be strongly suppressed by the damping in the system. On the contrary, for underdamped or moderately damped systems, one could expect an interaction between the soliton centre and its width (internal degree of freedom) to play an important role in the generation of the ratchet dynamics.

In order to investigate the mechanism of soliton ratchets, a damped and asymmetric sine-Gordon model was introduced in [5]. As for the usual single-particle ratchet, a periodic forcing and noise were also considered. This leads to the following perturbed asymmetric double sine-Gordon equation

$$u_{tt} - u_{xx} + \sin u + \lambda \cos u = -\beta u_t + \epsilon \cos(\omega t + \phi) + \eta(x,t), \qquad (25)$$

where β denotes the damping constant, $\eta(x,t)$ is white noise with autocorrelation $\langle \eta(x,t)\eta(x',t')\rangle = D\delta(x-x')\delta(t-t')$, λ is the asymmetry parameter, and ϵ, ω, and ϕ represent the amplitude, frequency and phase of the driver, respectively. For $\lambda = 0$, we recover the sine-Gordon system, and no ratchet effect is expected unless spatial or temporal symmetries are broken. For simplicity, it is assumed that $\lambda \in [-1, 1]$ to prevent relative minima from appearing in the potential.

Several remarks are called for: (a) Here the soliton, rather than its centre, lies in the asymmetric potential, therefore the soliton is an asymmetric coherent excitation with an internal degree of freedom. This asymmetry is crucial for the induction of the ratchet transport, since $\lambda = 0$ implies $v = 0$. This is the reason why the size of this object cannot be neglected. The two-CC theory shows that the coupling between the soliton centre and its width explains the mechanism underlying this kind of soliton ratchets. (b) Once again, the damping assures the time-invariance of the ratchet velocity, v. This implies that v does not depend on the phase ϕ. Moreover, the damping can be used to control the direction of the motion. (c) The noise here

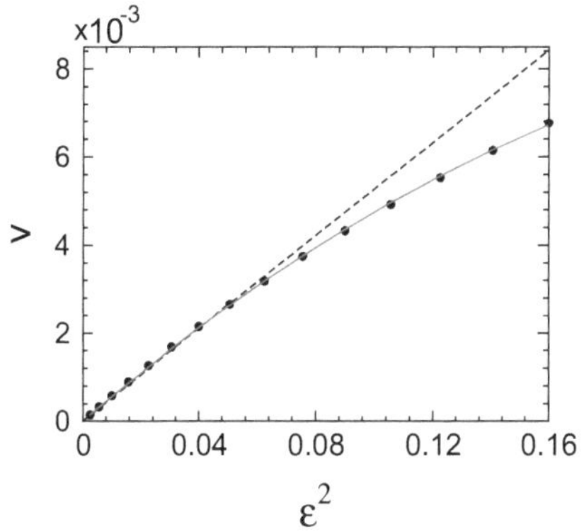

Fig. 2 v versus ϵ^2. *Circles*: v computed from simulations of Eq. (25). *Black dashed line*: fitted curve $v = 0.05264\epsilon^2$ (with values of v obtained when $\epsilon < 0.25$) with correlation coefficient $r = 0.9992$. *Red solid curve*: fitted curve $v = 0.05652\epsilon^2 - 0.09082\epsilon^4$ with correlation coefficient $r = 0.99993$. Parameters: $\lambda = -0.5$, $\beta = 0.4$, $\omega = 0.4$, $\eta = 0$ and $\phi = 0$

is not necessary for the regularization of the functional velocity and was considered for the sole purpose of verifying whether the ratchet soliton was robust under the effect of thermal fluctuations. Any $\epsilon > 0$ would induce the mobility of the soliton, i.e. there are no intervals of ϵ for which $v = 0$. (d) Due to the temporal symmetries, the dependence of the ratchet velocity as a function of ϵ is ruled out by Eq. (9). This is verified here by means of numerical simulations of Eq. (25) without noise (see Fig. 2). Indeed, for small values of ϵ ($\epsilon \leq 0.25$), v is proportional to ϵ^2. For $\epsilon > 0.25$ ($\epsilon^2 > 0.0625$), v is no longer a linear function in ϵ^2, i.e., $Q(\epsilon)$ in Eq. (9) is a polynomial $a_0 + b_0\epsilon^2$ rather than a constant.

4.1 Point Particle with Time-Dependent Mass Representation

The collective coordinates theory suggested in Sects. 3.1 and 3.2 can be used for the deterministic case, i.e. in Eq. (25) we set $\eta = 0$. The term with λ on the left-hand-side of Eq. (25) causes the asymmetry of the potential. In the search for the travelling wave solution of Eq. (25) with $\beta = 0$ and $\epsilon = 0$, the soliton solution can be found exactly. The terms on the right-hand-side of Eq. (25) added to the asymmetric double sine-Gordon equation can be considered as a small perturbation acting on the system. In this case, the momentum is defined by (15) and the energy reads

$$E = \int_{-\infty}^{+\infty} dx \left(\frac{1}{2}u_t^2 + \frac{1}{2}u_x^2 + C - \cos u + \frac{\lambda}{2}\sin(2u) \right), \tag{26}$$

($C = \cos u_0 - (\lambda/2)\cos(2u_0)$, $u_0 = \arcsin[(1 - A)/(4\lambda)]$, $A = \sqrt{1 + 8\lambda^2}$). Therefore, P and E depend on time and fulfill Eqs. (19) and (20) with

$R = -\beta u_t + \epsilon \cos(\omega t + \phi)$. The *ansatz* corresponding to the soliton solution for the perturbed asymmetric double sine-Gordon now reads

$$u(x,t) = u_0 + 2\tan^{-1}\left\{\frac{\text{sign}(\lambda)\, A\, B}{A - 1 - D\sinh\left[\frac{x - X(t)}{l(t)}\right]}\right\}, \tag{27}$$

where $B = \sqrt{2(4\lambda^2 - 1 + A)}$ and $D = 2\lambda\sqrt{2(1+A)}$. In the absence of perturbation, the soliton moves with constant velocity V and constant width $l(t) = l_0\sqrt{1 - V^2}$ ($l_0 = 2\sqrt{|\lambda|/[AB]}$), and hence $X(t) = Vt$. Notice that the *ansatz* (27) is consistent with the linear stability analysis performed in [24], where the existence of an internal mode frequency Ω_I below the phonon band is shown (this mode disappears in the sine-Gordon limit $\lambda = 0$). From the spatial profile of the corresponding localized eigenfunction, one can observe that the internal mode is linked to vibrations of the soliton width.

By substituting Eq. (27) into Eqs. (15), (16), (19), (20), and after straightforward calculations, it can be deduced that $l(t)$ satisfies Eq. (23) with $M_0 = I_4^2 I_1 / l_0$, $l_0 = 2\sqrt{\frac{|\lambda|}{A \cdot B}}$ $\alpha = (I_3 I_1 - I_2^2)/I_1^2$, where I_1, I_2, I_3 and I_4 are functions of the asymmetry parameter λ (see [24]). Moreover, the position fulfils

$$\frac{dX}{dt} = \frac{P(t)l(t)}{M_0 l_0} - \frac{I_2(\lambda)}{I_1(\lambda)}\dot{l}, \tag{28}$$

and since

$$U(l) = I(\lambda) f(t) l(t),$$

the momentum satisfies

$$\frac{dP}{dt} = -\beta P - q f(t). \tag{29}$$

Notice that from (28) one can observe that not only $P(t)$ does depend on the velocity, but also on the time derivative of the soliton width. Although Eq. (29) has been obtained within the framework of the collective coordinates theory, this is an exact result [53]. From any periodic function $f(t)$, the time average of P is zero in spite that the soliton is moving with an average velocity. The linear potential, in which the object with size $l(t)$ or mass $M(t) = l_0 M_0 / l(t)$ lies, is modulated by $f(t)$, and does depend on l rather than X. The present mechanism of soliton ratchets is slightly different to that considered in Sect. 3.1. Indeed, in the present case, the symmetric external force acts *directly* on the translational mode and on the soliton width, so that these two degrees of freedom resonate, thereby enabling the net motion. Although the equation for the momentum (29) is linear and can be solved exactly, it is difficult to find an exact solution for $X(t)$ and $l(t)$. However, by

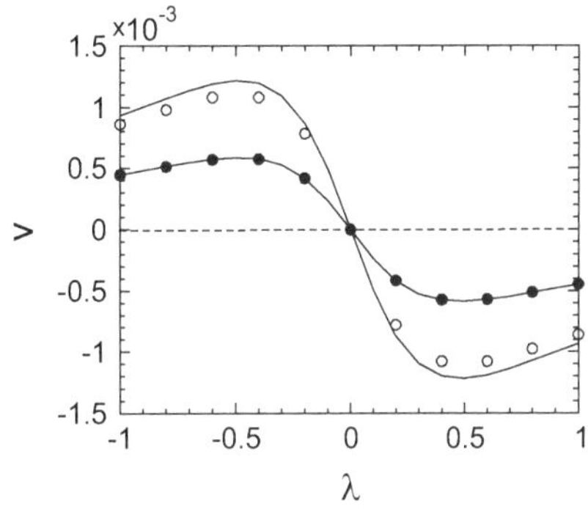

Fig. 3 v versus λ from simulations of Eq. (25), for two values of the damping coefficient: $\beta = 0.2$ (*open circles*) and $\beta = 0.5$ (*circles*). The *solid lines* represent the results from Eq. (30). *Dashed line*: $v = 0$. Other parameters: $\epsilon = 0.1$, $\omega = 0.1$ and $\phi = 0$

using a power series in ϵ for the soliton width $l(t)$, approximate solutions for $l(t)$ and the velocity of the centre of mass $\dot{X}(t)$ can be found. By taking the average value of this velocity over one period ($T = 2\pi/\omega$), $v \equiv \langle dX/dt \rangle = (1/T)\int_0^T (dX/dt)dt$, we finally obtain [54]

$$v = \frac{-\epsilon^2 q G(\lambda) \Omega^2 \beta}{2M_0(\beta^2 + \omega^2)[(\Omega^2 - \omega^2)^2 + \beta^2\omega^2]}, \tag{30}$$

where $G(\lambda)$ is given by Eq. (12) in [54]. As expected from the symmetry analysis, this expression shows that: (a) v does not depend on the phase of the driving; (b) for small ϵ, $v \sim \epsilon^2$ (see Fig. 2).

4.2 *Current Reversals and Resonance Phenomena*

The ratchet velocity exhibits a resonance as a function of the damping coefficient, the asymmetry parameter and the frequency of the driving, i.e. there are optimal values of these parameters that maximize the transport. Indeed, from (30), the optimal value of the damping to achieve the maximal v can be obtained analytically as a function of ω and λ (see Fig. 4 of [24]). Moreover, for small ϵ, it is straightforward to verify that the direction of the motion is determined only by $G(\lambda)$, i.e. by the parameter λ which controls the asymmetry of the potential, see Fig. 3. However, for a non-small amplitude, the approximate expression (30) is no longer valid, and interestingly the direction of the motion can be tuned via damping. From symmetry considerations, v satisfies Eq. (9) and therefore, in the next order correction $v = \epsilon^2(a_0 + b_0\epsilon^2)$, with a_0 and b_0 being two coefficients that depend on the damping and on the remaining parameters of the system, except the amplitude and the phase of the driving. This means that current reversals as a function of β and ω should be observed. This is confirmed by the numerical simulations in Fig. 5 in [54].

Here we verify another interesting prediction from (9), namely that current reversals can be induced by tuning the amplitude of the force. This phenomenon is shown in Fig. 4. Interestingly, from the fitted function $v = a_0\epsilon^2 + b_0\epsilon^4$, the

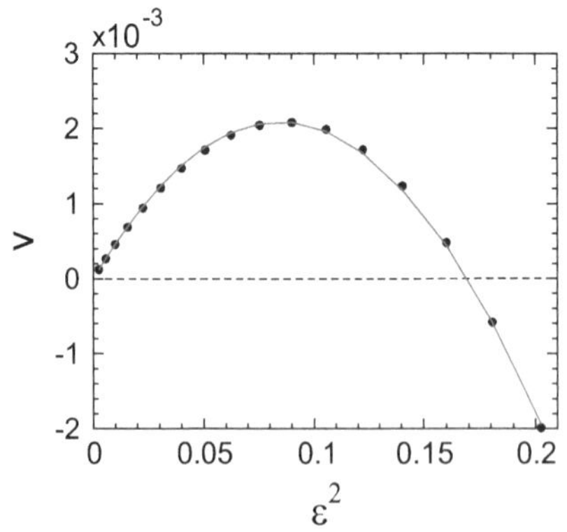

Fig. 4 v versus ϵ^2. *Circles*: v computed from simulations of Eq. (25). *Red solid line*: fitted curve $v = 0.04924\epsilon^2 - 0.2903\epsilon^4$ with correlation coefficient $r = 0.9996$. *Dashed line*: $v = 0$. Parameters: $\lambda = -0.5$, $\beta = 0.2$, $\omega = 0.4$ and $\phi = 0$

suppression of the current occurs at $\epsilon = \epsilon_0 = 0.41$. This implies that if the amplitude is set close to ϵ_0, and the damping, frequency or asymmetry parameter λ are changed to values similar to those used in Fig. 4, then current reversal can be observed. This explains the findings in [24, 54], where an unexpected current reversal was reported for low values of damping.

Comparison of the results shown in Figs. 2 and 4 leads to a striking conclusion on the small and not so small amplitude regimes. In the former case, Eq. (9) is reduced to $v = a_0\epsilon^2$; in the latter, $v = a_0\epsilon^2 + b_0\epsilon^4 + O(\epsilon^6)$. Notice that for the set of parameters of Fig. 4, the deviation from linear behaviour in ϵ^2 occurs earlier than for the frequency and damping chosen in Fig. 2. This means that the validity of these two regimes not only depends on the amplitude of the driver, but also on the remaining parameters of each particular system.

5 Ratchets in the Damped sine-Gordon Equation Driven by a Bi-harmonic Force

In this section, a bi-harmonic temporal force given by

$$f(t) = \epsilon_1 \cos(\omega t) + \epsilon_2 \cos(2\omega t + \phi), \tag{31}$$

is used to induce the soliton ratchet. This assures that, for non-zero ϵ_1 and ϵ_2, $f(t) \neq -f(t + T/2)$, $T = 2\pi/\omega$. Otherwise, a temporal translation will change the sign of f and, in accordance with the symmetry (7), the sign of the velocity will also change. Since v fulfills this time-shift invariance, v must be zero. This means that the soliton transport in a homogeneous system with symmetric field potential is impossible with only one harmonic for any value of the damping (see [55] and references therein).

Net transport in damped homogeneous extended fields driven by the bi-harmonic force has been studied extensively, first in the sine-Gordon equation [5, 7, 25, 56], and later on in the ϕ^4 and double-sine-Gordon systems [57, 58]. Here we focus on the case of the damped and driven sine-Gordon equation

$$u_{tt} - u_{xx} + \sin u = -\beta u_t + f(t), \tag{32}$$

which represents a model for long Josephson junctions, where the ratchet transport has also been experimentally observed [7]. The ratchet velocity is related to the measurement of the directed voltage induced by a bi-harmonic current. In the next section, we discuss some predictions from the symmetry analysis [38] summarized in Sect. 2 that have yet to be tested, either by simulations or by experiments.

5.1 Symmetries Shape the Current

The damped and driven sine-Gordon equation (32) is an example of the third class released in Sect. 2.2. The centre of mass of the soliton can be calculated by

$$X(t) = \frac{1}{q}\int_{-\infty}^{+\infty} dx\, x\, u_x, \tag{33}$$

where q is the topological charge. Therefore, the average of dX/dt, i.e. the ratchet velocity v, can be computed. Changing f by $-f$ in (32) is equivalent to the replacement of u with $-u$. This change reverses the sign of $X(t)$ in (33), hence the sign of the average velocity. This implies that v satisfies (7), where f_1 and f_2 are, respectively, the first and second harmonics in (31). This symmetry condition entails that v is given by Eq. (10). This expression enables three different regimes to be characterized:

1. In the small amplitude regime, only the first non-zero term is retained in (10), and v reads

$$v = C_0 \epsilon_1^2 \epsilon_2 \cos(\phi + \theta_0), \tag{34}$$

where C_0 and θ_0 depend on the specific parameters of the system, but do not depend on the amplitudes of the bi-harmonic force. The ratchet velocity is a sinusoidal function and current reversals appear by changing the phase ϕ. For a giving set of parameters (hence a given value of θ_0) we fix $\phi = -\theta_0 + \pi/2$ (hence $v = 0$), upon changing the damping, the frequency of the driver and other parameters we can vary θ_0 and therefore induce a current in either direction. Observing a sinusoidal behaviour of v as a function of ϕ does not imply that v is ruled by (34). Indeed, it is necessary to check that θ_0 does not depend on the amplitudes, and moreover that $v \sim \epsilon_1^2 \epsilon_2$.

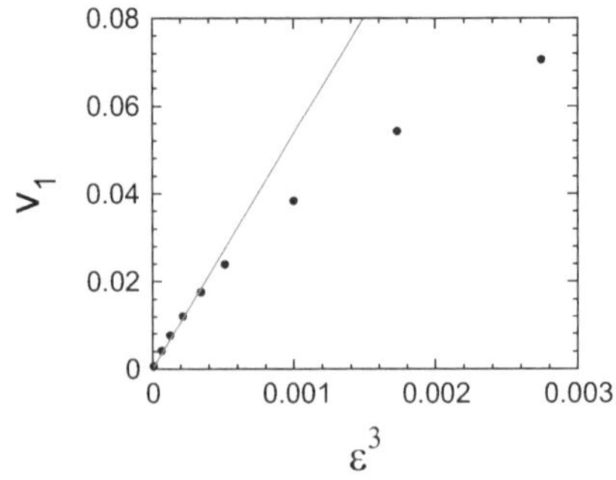

Fig. 5 v_1 versus ϵ^3. *Circles*: results computed from simulations of Eq. (32) with (31). *Red line* is the function $v_1 = 53.62\epsilon^3$, which was obtained from the fitting of simulation results when $\epsilon \leq 0.07$. Parameters: $\beta = 0.05$, $\omega = 0.1$

2. In the intermediate amplitude regime, the sinusoidal behaviour persists, however v is given by the full first term in (10)

$$v = C_1(\epsilon_1, \epsilon_2)\epsilon_1^2\epsilon_2 \cos(\phi + \theta_1(\epsilon_1, \epsilon_2)) = v_1(\epsilon_1, \epsilon_2)\cos(\phi + \theta_1(\epsilon_1, \epsilon_2)), \quad (35)$$

where C_1 and θ_1 are even polynomials in ϵ_1, ϵ_2 up to order 4. Figure 6 of [57] shows not only a sinusoidal behaviour of v as a function of the phase, but also good agreement between the ratchet velocity predicted by the collective coordinates theory and the simulations of (32) for $\epsilon_1 = \epsilon_2 = 0.02$, $\omega = 0.1$, $\beta = 0.05$. In the same figure, simulations for larger amplitudes, $\epsilon_1 = \epsilon_2 = 0.2$, are also reported, however the agreement between simulation and theory is only qualitative. Indeed, there is a factor 5 between the two velocities. This poor agreement when the amplitudes are not so small is due to the fact that v is ruled by Eq. (35) rather than Eq. (34). Here new simulations on Eq. (32) are provided to confirm this fact. Figure 5 shows v_1 as a function of ϵ^3, where $\epsilon = \epsilon_1 = \epsilon_2$. In order to obtain this result, the same procedure as that of the experiments reported in [59] is followed. First we compute the average velocity as a function of the phase ϕ. This allows us to check that v has sinusoidal behaviour, so that the results of the simulations can be fitted by Eq. (35), where v_1 and θ_1 are the coefficients of the fitting. This procedure is repeated for several values of $\epsilon \in [0.02, 0.14]$. For the chosen values of damping and frequency of the force of Fig. 5, the expression (34) is valid only for $\epsilon \leq 0.07$. For $0.07 < \epsilon < 0.14$, a departure from this regime is observed and the results are in perfect agreement with Eq. (35). In this case, the results from the simulations fit to $v = \epsilon^3(66.27 - 3455.46\epsilon^2 + 70775.1\epsilon^4)$ with a correlation coefficient 0.99995.
3. In the large amplitude regime, non-sinusoidal behaviour is expected. This anharmonicity has been observed in experiments shaking drops with two harmonics [60] and in simulations of classical particles in a one-dimensional driven superlattice [61], however it still has to be confirmed by experiments or simulations related to soliton ratchets. This issue is not trivial and merits further investigation. Indeed, solitons and antisolitons (i.e., antikinks) can appear in this regime instead of a single soliton. In order to prevent this from happening, it may be necessary to increase, not only the amplitudes of the bi-harmonic force, but also the damping of the system.

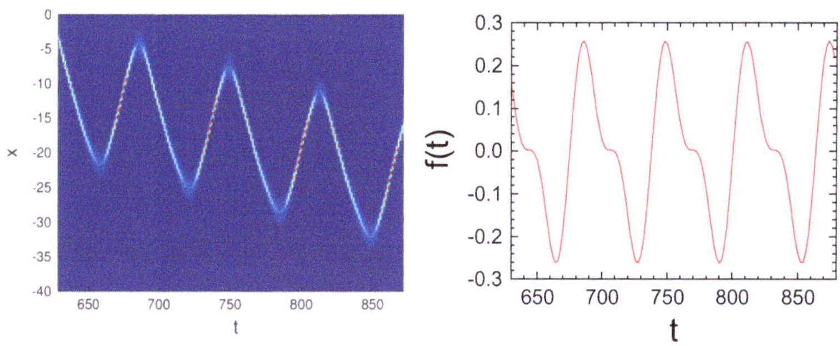

Fig. 6 *Left panel*: contour plot of the soliton obtained from simulations of Eq. (32). *Right panel*: the bi-harmonic force (31) versus time. Parameters: $\epsilon_1 = 0.2$, $\epsilon_2 = 0.1$, $\beta = 0.05$, $\omega = 0.1$ and $\phi = 1.6$. Initial condition: soliton solution of the sine-Gordon equation at rest and located at $X(0) = 20$

5.2 *Dynamics of the Driven and Damped Relativistic Particle*

The profile of the soliton solution of Eq. (32) is symmetric; it comes from a homogeneous extended system and therefore no asymmetry is expected in the spatial potential in which the centre of mass of the soliton lies. In Fig. 6, a contour plot of the soliton profile, solution of Eqs. (32) and (31), is represented and displays a ratchet dynamic. In the left panel, the net motion of the soliton to the left can be observed. In the right panel, the evolution of the bi-harmonic force used to generate this motion is represented. The breaking of the time-shift symmetry of the bi-harmonic force makes the existence of the soliton ratchet in Eq. (32) with (31) possible. The symmetry analysis of the previous section can be complemented with the collective coordinates theory, developed in detail in Sect. 3. Therefore, the equations of motion satisfied by the centre of the soliton, $X(t)$, and its width, $l(t)$, are given by the Eqs. (21)–(23), with $U = 0$ and $f(t)$ the bi-harmonic force (31). In order to shed light on the working principle of this kind of soliton ratchets, it is convenient to write Eqs. (21) and (22) as

$$\ddot{X} + \beta\dot{X} \quad \frac{\dot{X}\dot{l}}{l} \quad - -\frac{ql(t)f(t)}{M_0 l_0} \tag{36}$$

Cast in this form, it is easy to realize: first, that the centre of mass of the soliton is damped as the soliton itself, and second, that although the time average of the bi-harmonic force acting on the soliton is zero, the centre of the soliton is driven by an effective force, $-ql(t)f(t)/(M_0 l_0)$, and its average is not always zero (for a detailed discussion see [57]). An approximate expression for the soliton ratchet velocity can also be obtained (see Eq. (7) in [25]). It can be rewritten as Eq. (34). Moreover, the dependence of the coefficients C_0 and θ_0 on the frequency and damping can be obtained explicitly.

Inspired in the exact soliton solution (14) of the sine-Gordon equation without perturbations, the *ansatz* with one collective coordinates

$$u(x,t) = 4\,\mathrm{atan}\left(\exp\left[\frac{x - X(t)}{l_0\sqrt{1-\dot{X}^2}}\right]\right), \tag{37}$$

maps the dynamic of the soliton to the dynamic of a relativistic particle [5]. In this case, by inserting (37) in (17), we obtain the momentum

$$P(t) = \frac{M_0 \dot{X}}{\sqrt{1-\dot{X}^2}}. \tag{38}$$

This is a momentum corresponding to a relativistic particle of mass M_0 with a Lorentz factor $1/\sqrt{1-\dot{X}^2}$. As in the case of two collective variables, $P(t)$ satisfies the exact linear equation (22) with $u = 0$. Once $P(t)$ is obtained explicitly, the average velocity $v = \langle \dot{X} \rangle$ can be calculated from (38). The equation for the soliton position now reads

$$\ddot{X} + \beta \dot{X}(1-\dot{X}^2) = -\frac{q f(t)}{M_0 l_0}(1-\dot{X}^2)^{\frac{3}{2}}. \tag{39}$$

This differs slightly from (36), however the same mechanism underlying this kind of soliton ratchets can be identified, since the effective force is proportional to $f(t)\dot{X}^2$ and this function has, in general, non-zero average. In some cases, the use of the two collective coordinates *ansatz* instead of only one improves the approximate results obtained in the damped non-linear Klein–Gordon equations driven by a bi-harmonic force. For certain parameters, $\omega \ll 1$, $\epsilon_1 \ll 1$, $\epsilon_2 \ll 1$, and small damping, both *ansätze* provide the same average velocity [58]. The results obtained from the *ansatz* (37) imply at least two advantages: (a) the calculations performed in order to obtain the equations of motion are straightforward; and (b) the velocity of the relativistic particle, and therefore the average velocity, can be obtained analytically [37]. In fact, further order corrections are easily calculated, because

$$v = \left\langle \frac{dX}{dt} \right\rangle = \sum_{k=0}^{\infty} \left(-\frac{1}{2}\right)^k \frac{(2k-1)!!}{k!\,M_0^{2k+1}} \langle P(t)^{2k+1} \rangle. \tag{40}$$

The ratchet transport of the damped relativistic particle, driven by a bi-harmonic force, is an ideal system to study specific properties of the ratchet effect. This is a non-linear system that can be exactly solved. Therefore, it can be used as a test or as a counter-example of conjectures or general theories on the ratchets induced by bi-harmonic forces (or by other temporal forces) [37, 62].

Ratchets induced by bi-harmonic forces (31) are also characterized by current reversals and resonance phenomena upon changing the frequency, the phase and the amplitude of the driver, or the damping coefficient. Interestingly, the damping can

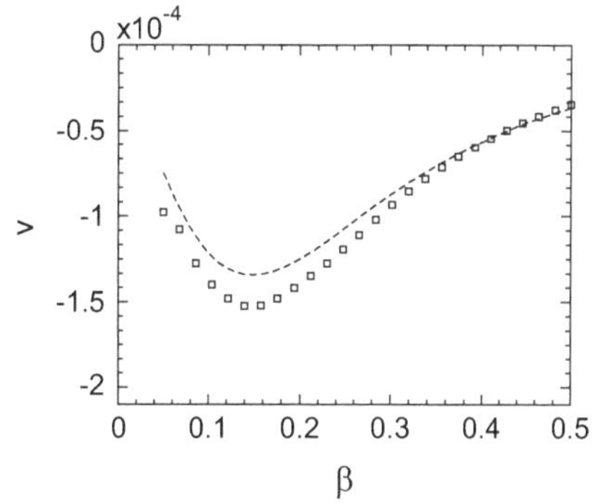

Fig. 7 Average velocity v as a function of the dissipation β. Squares are obtained from simulations of Eq. (32) with (31). *Dashed line* represents v calculated from Eq. (7) in [25]. Parameters: $\epsilon_1 = 0.04$, $\epsilon_2 = 0.026$, $\omega = 0.25$ and $\phi = \pi$

induce net motion. Indeed, for a small amplitude regime, the average velocity is ruled by the leading term in (40)

$$v = B\epsilon_1^2\epsilon_2 \cos(\phi + \theta_0), \tag{41}$$

with

$$B = \frac{3}{8M_0^3(\mu^2 + \omega^2)\sqrt{\mu^2 + 4\omega^2}}, \qquad \theta_0 = \tan^{-1}\left(\frac{2\omega^3}{\mu(\mu^2 + 3\omega^2)}\right), \qquad \mu = \frac{\beta}{M_0}. \tag{42}$$

Notice that in the undamped limit, $\beta \to 0$ (equivalently $\mu \to 0$), the parameters (42) become $B = 3/(16M_0^3\omega^3)$, and $\theta_0 = \pi/2$; whereas in the overdamped limit, $M_0 \to 0$ (and therefore $\mu \to \infty$ with $M_0\mu = \beta$), we obtain $B = 3/(8\beta^3)$, and $\theta_0 = 0$, and hence both limits agree with the predictions of [62]. Fixing $\phi = 0$ (or $\phi = \pi$), and starting experiments or simulations in the underdamped limit v vanishes. Indeed, in the limit $\beta \to 0$, the system (39) with $\phi = 0$ is invariant under $\mathcal{R} : t \mapsto -t$, and v changes its sign. Accordingly, $v = 0$ in this limit if $\phi = 0$. This effect is shown in Fig. 7, where v approaches zero as β goes towards zero. By increasing β, the absolute value of the average velocity is increased up to a maximum. Further increments in β reduce the mobility of the soliton. Although this phenomenon is well predicted by using a one collective coordinates *ansatz*, the agreement between simulations and Eq. (41) is poor (not shown in Fig. 7). However, the results obtained with a two collective coordinates *ansatz* agree well with the simulations (see Fig. 7).

6 Gating Ratchet in sine-Gordon Equation

In gating or pulsating ratchets, two harmonics are also used to generate the ratchet current. However, these harmonics act independently on the system and can have the same frequency. One harmonic is added directly to the system, whereas the other modulates the potential, where the particle lies. In the case of soliton ratchets, this harmonic modulates the non-linear sine-Gordon potential. Accordingly, in order to investigate the main features of this kind of ratchet, we analyze the following perturbed sine-Gordon equation (see [33]):

$$u_{tt} - u_{xx} + [1 + \epsilon_2 \cos(p\omega t + \phi_2)] \sin u = -\beta u_t + \epsilon_1 \cos(q\omega t + \phi_1). \quad (43)$$

The action of *parametric and additive* harmonics was first analyzed in [63] and later on in [64]. In the former work, drift motion of kinks in the ϕ^4 model was induced by the mixing of an additive and a parametric driver with the same frequency, where it was found that the average velocity approaches its maximum value when the frequencies of the drivers approach the internal mode frequency of ϕ^4. In the latter work, the motion of overdamped solitons in a deformable sine-Gordon symmetric potential was studied. For one soliton in a chain the ratchet phenomenon has been explained using a point-like particle representation with an oscillating mass.

Following the ideas of the previous section, it can be shown that v satisfies the force-reversal symmetry (6). Owing to this symmetry, if p is even and no restriction on q is imposed, the net motion is forbidden. Therefore, for $p = q = 1$, the average velocity is given by Eq. (11) [38]. This expression shows that the same three regimes explained in detail in Sect. 5.1, when the ratchet effect is induced by a bi-harmonic force, also appear in gating ratchets, although the powers of ϵ_1 and ϵ_2 are different from those of the expressions (34) and (35). Another interesting choice is $p = 1$ and $q = 2$. In this case, the small amplitude regime is fully determined by Eq. (34). This result does not imply that the average velocities in the two ratchet systems are the same. Notice that the constants C_0 and θ_0 do depend on all the system parameters, except the amplitudes of the harmonics.

6.1 *Current Reversals upon Changing the Phases and the Amplitudes of the Harmonics*

By increasing the amplitudes, v departs from the small amplitude regime. Since the coefficients v_1 and θ_1 in Eq. (35) do depend on the amplitudes, the current can be reversed not only by changing the phase, but also by tuning the amplitudes of the harmonics. As we have pointed out, this phenomenon was first reported in experiments in [59], where the average velocity of cold atoms in a symmetric optical potential was measured as a function of the phase of the bi-harmonic force. Despite the sinusoidal behaviour of v as a function of the phase, v_1 and θ_1 remain constant only for small amplitudes. This effect has yet to be verified by experiment or simulation in the case of gating ratchets.

6.2 Dynamics of Fluxons in Josephson Junctions and Cold Atoms in Optical Lattices

The control of the average velocity via damping, by frequency and phases was extensively investigated in [33] by means of simulations of Eq. (43) and the collective coordinates approach with the Rice *ansatz* (14). The soliton dynamic exhibits the same features as in the case of a ratchet induced by a bi-harmonic force. Although the additive and parametric forces act, respectively, on the centre and on the width of the soliton, as shown by the collective coordinates equations,

$$\ddot{X} + \beta \dot{X} - \frac{\dot{X}\dot{l}}{l} = -\frac{q l(t)\epsilon_1 \cos(q\omega t + \phi_1)}{M_0 l_0}, \quad P = \frac{M_0 l_0}{l(t)}\dot{X}, \tag{44}$$

$$\dot{l}^2 - 2\beta l \dot{l} - 2l\ddot{l} = \Omega^2 l^2 \left(\frac{P^2}{M_0^2} + 1\right) - \frac{1}{\alpha} + \Omega^2 l^2 \epsilon_2 \cos(p\omega t + \phi_2), \tag{45}$$

the same ratchet mechanism is identified. The non-zero effective force acting on the centre of mass of the soliton induces its net motion. In the case of the bi-harmonic force, $l(t)$ oscillates because its motion is non-linearly coupled with the motion of the centre of the soliton, that is driven directly by the bi-harmonic force. This coupling persists in Eq. (45) and the width of the soliton is parametrically driven with frequency $2q\omega$ through the term P^2. However, since $P(t)$ oscillates with frequency $q\omega$ after a transient time, this contribution to the spectrum of $l(t)$ does not cause any resonance. On the other hand, Eq. (45) shows that $l(t)$ satisfies the equation for a parametrically driven non-linear oscillator, and therefore the higher harmonics of $p\omega$ and $q\omega$ are also present in the spectrum of $l(t)$. Analogously, the higher harmonics of $q\omega$ do not provide any resonance with $P(t)$, and the net transport is expected only for $q = p$, $q = 2p$, $q = 3p$, etc. Moreover, a certain relationship between the phases of $P(t)$ and $l(t)$ should hold in order to guarantee non-zero average velocity.

The sine-Gordon equation serves as a model for the propagation of fluxons along Josephson junctions [40]. Soliton ratchets induced by a bi-harmonic force were reported experimentally in Josephson junctions [7]. A high mobility of the fluxon in long Josephson junctions was reported in experiments in [8], where fluxons, driven by an *ac* force, move in the asymmetric potential. This potential was created by means of a current injection (or non-uniform magnetic field). The motion of fluxons was also observed in a Josephson-junction parallel array, where they behave similar to a single particle in a periodic pinning asymmetric potential [6]. In this case, the system is intrinsically discrete and is modelled by a discrete version of the sine-Gordon equation, i.e. the Frenkel–Kontorova model. This shows that Josephson junctions provide a real physical system, where the soliton ratchets can be induced in different ways. However, to date, gating ratchets have been observed only in optical lattices, where average velocity of cold atoms has been measured, and there is no experimental verification on the gating soliton ratchets in Josephson junctions.

Symmetry analysis on gating ratchets predicts the same phenomenology in all cases. The fact that the same behaviour in these systems is ruled by the same symmetries needs to be confirmed experimentally.

7 Concluding Remarks

This work reviews the net transport of solitons induced by means of zero average forces (soliton ratchets) in the damped and driven sine-Gordon equation, which was originally put forth in [5, 24]. It is shown that the net motion of the soliton is generated by means of an inhomogeneous or asymmetric field potential [24, 26], via a bi-harmonic additive force [5, 25], a gating mechanism [33], etc. In all these cases, harmonic temporal functions are employed to induce the net transport. Using the time-shift invariance of the smooth functional velocity, it has recently been shown that the shape of the velocity has a specific dependence on all amplitudes and phases of harmonics [38]. This result is independent of the system and therefore can be tested in all realizations of soliton ratchets, including experiments in Josephson junctions [7]. This general theory provides not only a way for the prediction or suppression of the ratchet phenomenon, but also sheds light on the main features of soliton ratchets, such as the sinusoidal (non-sinusoidal) dependence of the velocity as a function of the phases of the harmonics for small (large) amplitudes, the current reversal by tuning the phases, amplitudes and frequencies of the forces, and the motion generated and enhanced by the damping. Many of these features have been extensively investigated by means of simulations and collective coordinates theory in the past [5, 24–26]. Certain other features are shown in this work by means of new simulations of the damped and driven sine-Gordon equation. However, there are some unsolved questions summarized in the current review which, in turn, suggest further works on simulations on the damped and driven sine-Gordon equation and on experiments on Josephson junctions. In addition, experimental results reported within the framework of soliton ratchets are expected to occur in other systems [38]. Further work may include the implementation of gating soliton ratchets, the observation of the net motion of fluxons by using a multifrequency force, the regularization of the functional ratchet velocity by means of the thermal fluctuations, the development of the collective coordinates theory for discrete soliton ratchets and for stochastic soliton ratchets, and the generalization of other kinds of ratchets reported in particle systems to the soliton ratchets.

Acknowledgements We thank José Cuesta, Franz G. Mertens, Renato Álvarez and Jesús Cuevas for the critical reading of this work. We acknowledge financial support through grants FIS2011-24540, from the Ministerio de Economía y Competitividad (Spain), grants FQM207, P09-FQM-4643, from Junta de Andalucía (Spain), and a grant from the Humboldt Foundation through a Research Fellowship for Experienced Researchers SPA 1146358 STP.

References

1. P. Reimann, Phys. Rep. **361**, 57 (2002)
2. P. Hänggi, F. Marchesoni, Rev. Mod. Phys. **81**, 387 (2009)
3. S. Flach, O. Yevtushenko, Y. Zolotaryuk, Phys. Rev. Lett. **84**, 2358 (2000)
4. P. Reimann, Phys. Rev. Lett. **86**, 4992 (2001)
5. M. Salerno, Y. Zolotaryuk, Phys. Rev. E **65**, 056603 (2002)
6. F. Falo, P.J. Martínez, J.J. Mazo, T.P. Orlando, K. Segall, E. Trías, Appl. Phys. A **75**, 263 (2002)
7. A.V. Ustinov, C. Coqui, A. Kemp, Y. Zolotaryuk, M. Salerno, Phys. Rev. Lett. **93**, 087001 (2004)
8. M. Beck, E. Goldobin, M. Neuhaus, M. Siegel, R. Kleiner, D. Koelle, Phys. Rev. Lett. **95**, 090603 (2005)
9. S. Ooi, S. Savel'ev, M.B. Gaifullin, T. Mochiku, K. Hirata, F. Nori, Phys. Rev. Lett. **99**, 207003 (2007)
10. T. Salger, S. Kling, T. Hecking, C. Geckeler, L. Morales-Molina, M. Weitz, Science **326**, 2141 (2009)
11. R.P. Feynman, R.B. Leighton, M. Sands, *The Feynman Lectures on Physics*, vol. 1 (Addison-Wesley, Reading, 1963), pp. 46.1–46.9
12. M.O. Magnasco, Phys. Rev. Lett. **71**, 1477 (1993)
13. P. Hänggi, F. Marchesoni, F. Nori, Ann. Phys. (Leipzig) **14**, 51 (2005)
14. E.R. Kay, D.A. Leigh, F. Zerbetto, Angew. Chem. Int. **46**, 72 (2007)
15. H. Linke, B.J. Alemen, L.D. Melling, M.J. Taormina, M.J. Francis, C.C. Dow-Hygelund, V. Narayanan, R.P. Taylor, A. Stout, Phys. Rev. Lett. **96**, 154502 (2006)
16. G. Dupeux, M. Le Merrer, Ch. Clanet, D. Quéré, Phys. Rev. Lett. **107**, 114503 (2011)
17. J.H.E. Cartwright, A.G. Checa, M. Rousseau, Langmuir **29**, 8370 (2013)
18. A. Ajdari, D. Mukamel, L. Peliti, J. Prost, J. Phys. I France **4**, 1551 (1994)
19. D.R. Chialvo, M.M. Millonas, Phys. Rev. A **209**, 26 (1995)
20. D. Cole, S. Bending, S. Savel'ev, A. Grigorenko, T. Tamegai, F. Nori, Nat. Mater. **5**, 305 (2006)
21. W. Schneider, K. Seeger, Appl. Phys. Lett. **8**, 133 (1966)
22. R. Gommers, S. Bergamini, F. Renzoni, Phys. Rev. Lett. **95**, 073003 (2005)
23. A. Sánchez, A.R. Bishop, SIAM Rev. **40**, 579 (1998)
24. M. Salerno, N.R. Quintero, Phys. Rev. E. **65**, 0625602 (2002)
25. L. Morales-Molina, N.R. Quintero, F.G. Mertens, A. Sánchez, Phys. Rev. Lett. **91**, 234102 (2003)
26. L. Morales-Molina, F.G. Mertens, A. Sánchez, Eur. Phys. J. B **37**, 79 (2004)
27. Y. Zolotaryuk, M. Salerno, Phys. Rev. E. **73**, 066621 (2006)
28. J. Cuevas, B. Sánchez-Rey, M. Salerno, Phys. Rev. E. **82**, 016604 (2010)
29. A.V. Gorbach, S. Denisov, S. Flach, Optics Lett. **31**, 1702 (2006)
30. D. Poletti, E.A. Ostrovskaya, T.J. Alexander, B. Li, Yu.S. Kivshar, Physica D **238**, 1338 (2009)
31. F.G. Mertens, N.R. Quintero, A.R. Bishop, Phys. Rev. E. **81**, 016608 (2010)
32. F.G. Mertens, N.R. Quintero, I. Barashenkov, A.R. Bishop, Phys. Rev. E. **84**, 026614 (2011)
33. E. Zamora-Sillero, N.R. Quintero, F.G. Mertens, Phys. Rev. E **74**, 046607 (2006)
34. F.G. Mertens, L. Morales-Molina, A.R. Bishop, A. Sánchez,, P. Müller, Phys. Rev. E **74**, 066602 (2006)
35. V. Stehr, P. Müller, F.G. Mertens, A.R. Bishop, Phys. Rev. E **79**, 036601 (2009)
36. A. Wouk, *Course of Applied Functional Analysis* (Wiley, New York, 1979)
37. N.R. Quintero, J.A. Cuesta, R. Álvarez-Nodarse, J. Phys. A Math. Theor. **44**, 425205 (2011)
38. J.A. Cuesta, N.R. Quintero, R. Álvarez-Nodarse, Phys. Rev. X **3**, 041014 (2013)
39. L. Morales-Molina, F.G. Mertens, A. Sánchez, Phys. Rev. E **73**, 046605 (2006)
40. D.W. McLaughlin, A.C. Scott, Phys. Rev. A **18**, 1652 (1978)
41. Yu.S. Kivshar, B.A. Malomed, Rev. Mod. Phys. **61**, 763 (1989)
42. M. Aldana-González, G. Cocho, H. Larralde, G. Martínez-Mekler, J. Theor. Biol. **220**, 27 (2003)

43. L. Morales-Molina, F.G. Mertens, A. Sánchez, Phys. Rev. E **72**, 016612 (2005)
44. M. Salerno, A.C. Scott, Phys. Rev. B **26**, 2474 (1982)
45. M.J. Rice, Phys. Rev. B **28**, 3587 (1983)
46. N.R. Quintero, A. Sánchez, F.G. Mertens, Phys. Rev. E **62**, 5695 (2000)
47. D.W. McLaughlin, A.C. Scott, Phys. Rev. A **18**, 1652 (1978)
48. N.R. Quintero, P.G. Kevrekidis, Phys. Rev. E **64**, 056608 (2001)
49. T.E. Dialynas, K. Lindenberg, G.P. Tsironis, Phys. Rev. E **56**, 3976 (1997)
50. R. Bartussek, P. Hänggi, J.G. Kissner, Europhys. Lett. **28**, 459 (1994)
51. Y. Zolotaryuk, Phys. Rev. E. **86**, 026604 (2012)
52. F. Marchesoni, Phys. Rev. Lett. **77**, 2364 (1996)
53. N.R. Quintero, B. Sánchez-Rey, J. Casado-Pascual, Phys. Rev. E. **71**, 058601 (2005)
54. N.R. Quintero, B. Sánchez-Rey, M. Salerno, Phys. Rev. E. **72**, 016610 (2005)
55. N.R. Quintero, A. Sánchez, Eur. Phys. J. B **6**, 133 (1998)
56. S. Flach, Y. Zolotaryuk, A.E. Miroshnichenko, M.V. Fistul, Phys. Rev. Lett. **88**, 184101 (2002)
57. L. Morales-Molina, N.R. Quintero, F.G. Mertens, A. Sánchez, Chaos **16**, 013117 (2006)
58. N.R. Quintero, R. Álvarez-Nodarse, F.G. Mertens, Phys. Rev. E. **80** 016605 (2009)
59. D. Cubero, V. Lebedev, F. Renzoni, Phys. Rev. E **82**, 041116 (2010)
60. X. Noblin, R. Kofman, F. Celestini, Phys. Rev. Lett. **102**, 194504 (2009)
61. T. Wulf, C. Petri, B. Liebchen, P. Schmelcher, Phys. Rev. E **86**, 016201 (2012)
62. N.R. Quintero, J.A. Cuesta, R. Álvarez-Nodarse, Phys. Rev. E **81**, 030102 (2010)
63. A. Sukstanskii, K.I. Primak, Phys. Rev. Lett. **75**, 3029 (1995); Yu.S. Kivshar, A. Sánchez, Phys. Rev. Lett. **77**, 582 (1996)
64. G. Costantini, F. Marchesoni, M. Borromeo, Phys. Rev. E **65**, 051103 (2002)

The sine-Gordon Equation in Josephson-Junction Arrays

Juan J. Mazo and Alexey V. Ustinov

Abstract Superconducting Josephson-junctions are excellent experimental systems for the general study of nonlinear phenomena and nonlinear localised excitations. Specifically, a long Josephson-junction is described by the continuous sine-Gordon equation and a Josephson-junction parallel array by its discrete counterpart. This chapter constitute a revision of the physics of such superconducting systems in the light of the sine-Gordon equation.

Keywords Breather • Discrete breather • Flux quantum • Fluxon • Frenkel–Kontorova model • Josephson junction • Josephson junction array • Kink • Long Josephson junction • Quantum solitons • Qubits • Rotobreather • Soliton • Vortex

---

J.J. Mazo (✉)
Dpto. Física de la Materia Condensada and Instituto de Ciencia de Materiales de Aragón, CSIC-Universidad de Zaragoza, 50009 Zaragoza, Spain
e-mail: juanjo@unizar.es

A.V. Ustinov
Physikalisches Institut, Karlsruhe Institut fuer Technologie, Wolfgang-Gaede str 1, D-76131, Karlsruhe, Germany

National University of Science and Technology MISIS, Leninskiy prospekt 4, 119049 Moscow, Russian Federation

Russian Quantum Center, Novaya str., 100, BC "URAL", SKOLKOVO, Moscow region, 143025 Russian Federation
e-mail: ustinov@kit.edu

J. Cuevas-Maraver et al. (eds.), *The sine-Gordon Model and its Applications*, Nonlinear Systems and Complexity 10, DOI 10.1007/978-3-319-06722-3_7,

1 Introduction

The concept of coherent structures or coherent excitations has important consequences when applied to condensed matter systems. Spatially or temporally coherent structures appear in many nonlinear extended systems. Such structures usually can be characterized by marked particle-like properties. In the past few years, these notions have become fundamental for understanding many problems and their implications extend over different fields of the physics of continuous and discrete systems [1,2].

Josephson-junction (JJ) arrays are a very good example of a physical system where such type of structures appear. In particular, long Josephson-junctions (LJJs) are paradigmatic devices to study soliton dynamics. This article is a personal, partial and subjective introduction to the physics of this system and its discrete counterpart. They have been widely studied in the past. Authoritative reviews on the subject can be found in [3–8].

We will introduce first the main aspects of the physics of superconducting JJs to present simple model descriptions of long Josephson arrays. We will revise the most prominent fluxon dynamics results in man-made and natural systems. Then, we will move to the discrete version of the long junction: a set of parallelly connected small JJs. We will finish with results on quantum dynamics of solitons in long JJs, and a few words about future perspectives.

2 Superconducting Josephson Effect

2.1 Josephson Effect

In his pioneering work of 1962 Brian Josephson studied the tunnel of Cooper pairs between superconducting metals and predicted the celebrated Josephson effect [9]; that is, the presence of a supercurrent between two superconductors coupled by a weak link, an artifice known as a Josephson junction. Since then, many works have been done which study the behavior of single junctions, JJ arrays, and other more complex devices with Josephson elements, and their outstanding applications [5, 10–13].

2.2 Superconducting Tunnel Junctions

A Josephson tunnel junction is a solid state physics device which consist of two superconducting electrodes (usually Niobium or Aluminum) separated by a thin insulating barrier (usually an Aluminum oxide), see Fig. 1.

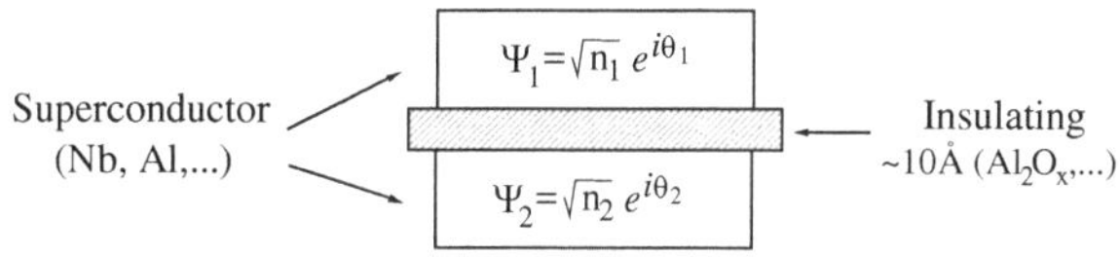

Fig. 1 Schematic diagram of a Josephson tunnel junction: two superconducting electrodes separated by a thin insulating barrier

The physics of the junction is given by the value of the gauge invariant phase difference between the superconducting electrodes $\varphi = \theta_1 - \theta_2 - \frac{2e}{\hbar}\int_1^2 \boldsymbol{A}(\boldsymbol{r},t)\,\boldsymbol{dl}$, with θ_i the phase of the macroscopic wave function in electrode i ($\Psi_i = \sqrt{n_i}e^{i\theta_i}$) and $\boldsymbol{A}$ the vector potential.

The basic equations for the Josephson effect are

$$I_s = I_c \sin\varphi \tag{1}$$

and

$$V = \frac{\hbar}{2e}\frac{d\varphi}{dt}. \tag{2}$$

They establish—*DC Josephson effect*—that at zero voltage (then φ is constant) it is possible to have a dc current across the junction. The maximum possible value of this current is I_c, the junction critical current. However—*AC Josephson effect*—in the presence of a constant voltage the junction responses with an ac current of frequency $2eV/\hbar$ (483.6 GHz for a typical value of 1 mV). Note that the $\hbar/2e$ quotient can be also written in terms of the flux quantum unit $\Phi_0 = h/2e$. Thus $\hbar/2e = \Phi_0/2\pi$.

The most important characteristic curve of the junction is the current-voltage ($I - V$) one, which measures the response, voltage, of the junction to an injected current. This can be modeled in its simplest version by the RCSJ (resistively and capacitively shunted junction) model [14, 15] (see Fig. 2). In this model the total current through the junction is the sum of three contributions: the Josephson supercurrent, a resistive normal current (tunneling of normal carriers from one electrode to the other) and a capacitive channel (associated with the junction capacitance); $I = I_J + I_R + I_C$ with $I_J - I_c \sin\varphi$, $I_R - V/R$ and $I_C = C dV/dt$. Then

$$I = C\dot{V} + \frac{1}{R}V + I_c \sin\varphi. \tag{3}$$

If we apply the second Josephson relation and normalize current with respect to the junction critical current, $i = I/I_c$, time with respect to the Josephson

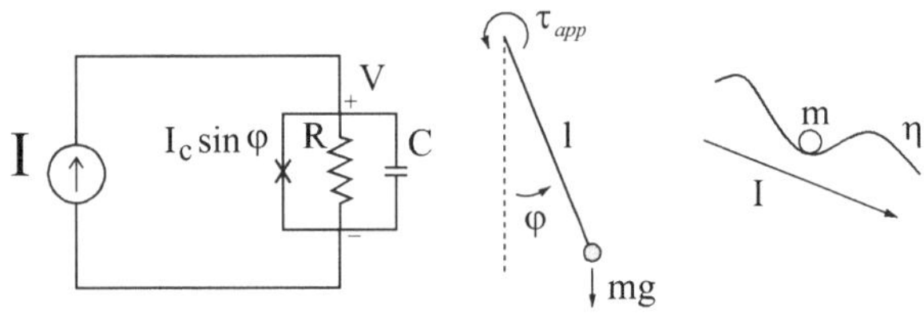

Fig. 2 RCSJ circuit model of the junction and two mechanical analogs: the forced and damped pendulum and the particle in the tilted washboard potential

plasma frequency $\omega_p = \sqrt{2\pi I_c/\Phi_0 C}$ and introduce the damping parameter $\Gamma = \sqrt{\Phi_0/2\pi I_c C R^2}$,[1] we obtain

$$i = \mathcal{N}(\varphi) = \ddot{\varphi} + \Gamma\dot{\varphi} + \sin\varphi. \tag{4}$$

This is the normalized equation for the dynamics of a single junction biased by an external current. This equation is identical to the equation for a forced and damped pendulum in a gravitational field or a particle in a tilted washboard potential $U(\varphi) = -E_J\cos\varphi - (\hbar I/2e)\varphi$ [mass $m = (\hbar/2e)^2 C$ and damping $\gamma = (\hbar/2e)^2(1/R)$], see Fig. 2. Both are simple mechanical analogs for the junction and illustrate that JJ devices are ideal experimental systems to probe basic nonlinear science results.

Thermal fluctuations can be included in the model by the addition of a noise current source $\tilde{I}(t)$ with $\langle\tilde{I}(t)\rangle = 0$ and $\langle\tilde{I}(t)\tilde{I}(t')\rangle = (2k_B T/R)\delta(t-t')$. In this case, the total current, in normalized units, is

$$i = \mathcal{N}(\varphi) = \ddot{\varphi} + \Gamma\dot{\varphi} + \sin\varphi + \tilde{\imath} \tag{5}$$

with $\langle\tilde{\imath}(\tau)\rangle = 0$ and $\langle\tilde{\imath}(\tau)\tilde{\imath}(\tau')\rangle = (2\Gamma k_B T/E_J)\delta(\tau-\tau')$.

3 Long Josephson Junctions

3.1 The Long JJ and the sine-Gordon Equation

A long Josephson junction (Fig. 3) is a junction which has one dimension (say x) long with respect to the so called Josephson penetration depth, λ_J [5], see below. Then, the phase difference is also a function of the spatial coordinate: $\varphi(x,t)$.

[1] The damping can be also defined in terms of the so-called quality factor $Q = 1/\Gamma$ or the Stewart-McCumber parameter $\beta_c = 1/\sqrt{\Gamma} = 2\pi I_c C R^2/\Phi_0$.

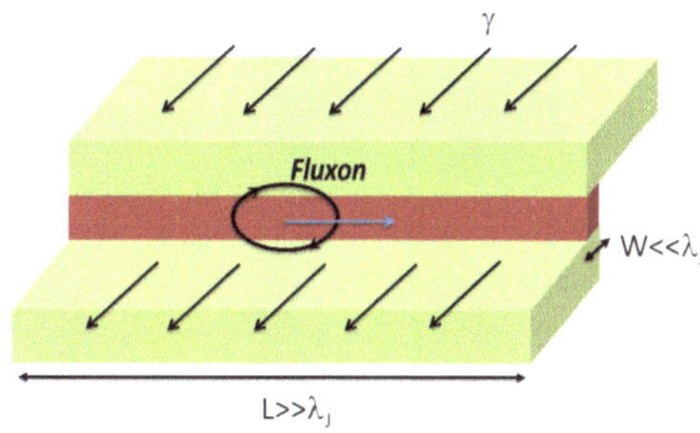

Fig. 3 Schematic diagram of a long JJ with a fluxon (also know as Josephson vortex) driven by an external current γ

Accordingly we introduce an spatial dependence in current and voltage along the junction, $V(x)$ and $I(x)$, which neglecting dissipative effects change as:

$$\frac{\partial V}{\partial x} = -L\frac{\partial I}{\partial t}$$
$$\frac{\partial I}{\partial x} = -C\frac{\partial V}{\partial t} - I_c \sin\varphi \tag{6}$$

Here, L and C are respectively the inductance and capacitance per unit of length of the junction.

Introducing the Josephson relations, we see that in its simplest approach the electrodynamics of the junction is described by the following sine-Gordon equation:

$$\lambda_J^2 \frac{\partial^2\varphi}{\partial x^2} - \frac{1}{\omega_p^2}\frac{\partial^2\varphi}{\partial t^2} = \sin\varphi. \tag{7}$$

Here $\lambda_J = \sqrt{\Phi_0/2\pi L I_c}$ is the *Josephson penetration depth* and $\omega_p = \sqrt{2\pi I_c/\Phi_0 C}$ the *Josephson plasma frequency* of the system, which give a measure of the natural length and time units. Measuring length in units of λ_J and time in units of ω_p^{-1} we get the dimensionless equation for the system,

$$\varphi_{xx} - \varphi_{tt} = \sin\varphi \tag{8}$$

When losses and bias are included the dynamics of the phase is described by the perturbed sine-Gordon equation [16]

$$\varphi_{xx} - \varphi_{tt} - \sin\varphi = \alpha\varphi_t - \beta\varphi_{xxt} - \gamma. \tag{9}$$

α measures the intensity of a dissipative term due to quasi-particle tunneling through the junction, β measures the surface resistance of the superconductors and γ the normalized bias current.

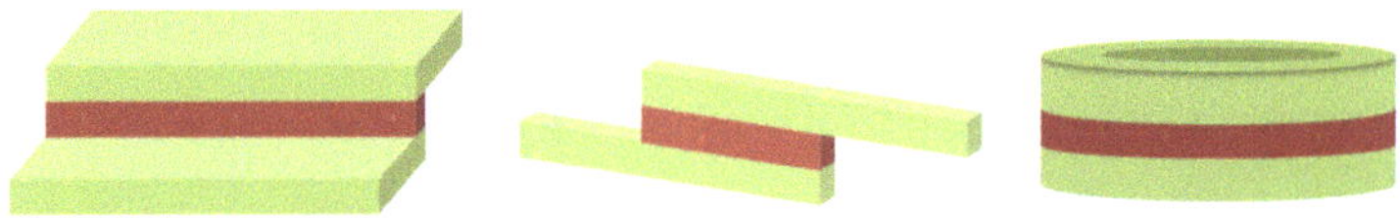

Fig. 4 Different geometries for a long JJ: overlap junction (*left*); in-line junction (*middle*); and annular junction (*right*)

3.2 Geometries for Long JJs

In any real system, the perturbed sine-Gordon equation (9) has to be solved together with appropriate boundary conditions. These conditions depend on the specific junction geometry. By *long* Josephson junction, we simply mean that the Josephson phase difference can vary along only one of its spatial dimensions, let's say the x-coordinate. We distinguish between such long junctions and short junctions (as presented in Sect. 2.2), for which the phase difference is (nearly) constant over the whole junction area.

Thus, for a long JJ we need: (1) a junction length $L \gg \lambda_J$ (φ changes along this direction); (2) a junction width $W \ll \lambda_J$ (φ nearly constant along this direction); and (3) the external magnetic field H applied perpendicular to the junction (a non-zero parallel component induces phase difference variations along the perpendicular direction).

Such quasi-one-dimensional Josephson junctions can be constructed in different ways, and junction geometry has a significant effect on soliton dynamics. The most extensively studied variations are the overlap geometry, the in-line geometry and the annular geometry, depicted in Fig. 4. They are studied by a perturbed sine-Gordon equation with appropriate boundary conditions.

In the more general case we deal with *large* (in both x and y dimensions) junctions, which are described by a partial differential equation (PDE) in two space dimensions, x and y, and time t: a 2d sine-Gordon equation.

3.3 Fluxons in Long JJs

The soliton solution of the sine-Gordon solution has a well defined physical meaning in the context of the long JJ systems, since it corresponds to a quantum of flux trapped in the junction. Thus, the soliton is usually referred to as *fluxon* or *Josephson vortex* in the context of long JJ.

Figure 5 schematically shows the fluxon in the array. It corresponds to a solution of the phase changing from 0 to 2π along the junction. The width scale for the change of the phase, the width of the fluxon, is given by the Josephson penetration depth λ_J. A spatial variation of the phase gives account of a magnetic flux in the system, as depicted in the figure, whose total integral is equal to one flux

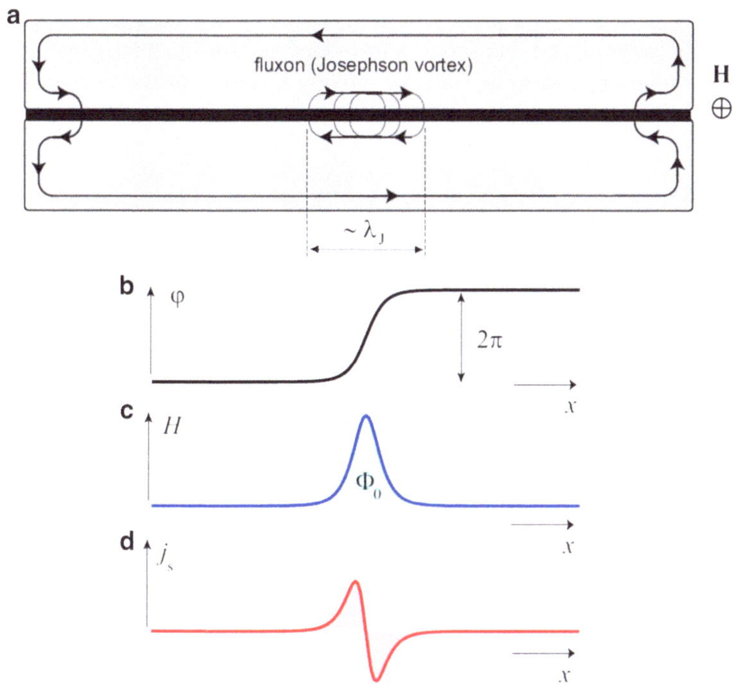

Fig. 5 (**a**) Schematic diagram of a long JJ with a fluxon in it and (**b**) phase φ, (**c**) field $H \sim \partial\varphi/\partial x$ and (**d**) density of superconducting current $j_s \sim \sin\varphi$ for a soliton in the junction. The fluxon extends over a physical distance of the order of λ_J

quantum Φ_0. In addition, from the Josephson relations we know that a non zero phase variation corresponds to a flowing supercurrent which, as shown in the figure, circulates around the fluxon core.

Similarly to single fluxons, the system also supports for multifluxon solutions formed by a set of fluxon, or antifluxons, or combinations of both.

3.4 Dynamical Solutions of Unperturbed Long JJs

In addition to static solutions ($\varphi = 0$ and fluxons), the unperturbed sine-Gordon equation, Eq. (8), also supports a family of dynamical solutions, mainly moving fluxons, small-amplitude waves and breathers.

Small-amplitude waves, $\varphi \ll 2\pi$, are obtained by taking the linear limit of the sinusoidal term in Eq. (8). The resulting equation has plane wave solutions

$$\varphi \sim e^{i(kx-\omega t)} \tag{10}$$

with $\omega^2 = 1 + k^2$ (we have previously normalized by ω_p and λ_J). This dispersion relation is linear at large k and shows a forbidden gap for small frequencies.

Fluxons solutions appear as static or moving solutions of the system. They are given by

$$\varphi = 4 \arctan \left[\exp \left(\pm \frac{x - vt - x_0}{\sqrt{1 - v^2}} \right) \right]. \tag{11}$$

The sign stands for fluxon or anti-fluxon solutions where the phase goes from 0 to $+2\pi$ or -2π respectively. Moving fluxons travel at constant velocity v without change in its shape.

Breather solutions are periodic oscillating localized solutions of the system without a topological charge. They can be depicted as a bound soliton-antisoliton pair and exist also as moving solutions along the junction.

4 Fluxon Dynamics in Long Josephson Junctions

4.1 Fluxon Dynamics

We have previously introduced the perturbed sine-Gordon equation, Eq. (9), describing a long quasi-one-dimensional Josephson junction and reviewed in Sect. 3.4 the most important exact solutions of the unperturbed sine-Gordon equation. Since the perturbation terms are usually small, one might think that the reviewed solutions would remain almost unchanged, or would acquire small corrections. To a large extent, this thinking is correct. Soliton and antisoliton solutions for the sine-Gordon equation are only slightly modified in a real junction. In contrast, breather solutions and isolated plasma wave packets do not survive dissipative perturbations, and decay with time. Though these solutions can be excited, e.g., by current pulses, they exist in real junctions only for very short times (in physical time units, several nanoseconds).

There exist several ways of addressing Eq. (9), for which there are not known analytical solutions. A first possibility is to perform a perturbative analysis, starting form the unperturbed sine-Gordon equation. The perturbation theory for solitons in Josephson junctions was put together in the classic paper by McLaughlin and Scott, published in 1978 [4]. Another perturbative analysis can be found in [17].

A standard tool of Josephson junction analysis is the energy balance approach. It allows for predicting the velocity v of a steadily moving soliton driven by a current γ balanced by the power loss due to the dissipative terms α and β,

$$\gamma = \frac{4}{\pi} \frac{v}{\sqrt{1 - v^2}} \left[\alpha + \frac{\beta}{3(1 - v^2)} \right] \tag{12}$$

The velocity predicted by Eq. (12) is in good agreement with direct numerical simulations of the perturbed sine-Gordon equation, Eq. (9). The soliton behaves in much the same way as does a relativistic particle having a normalized limiting velocity equal to unity. Similar expressions have been derived for the velocity of a soliton train propagating in the array.

In an experiment, α and β are usually fixed by the intrinsic properties of a given Josephson junction and by the temperature whereas γ is proportional to the bias current and, consequently, it can be easily varied in an experiment. Thus Eq. (12) determines the *current-voltage characteristics* of the system, the curve usually measured by experiments. Experimental curves can be almost perfectly fitted to the analytical power-balance equation (9).

There exist no infinite real junctions. Thus the dynamics of fluxons in long Josephson junctions depends importantly on the geometry and boundaries of the junction. The case of annular junctions is the simplest one, as traveling solitons (or antisolitons) will move interacting with the electromagnetic waves generated in its circular trajectory. For open junctions at zero magnetic field, the border behaves as a mirror and a soliton reaching the boundary is reflected as an antisoliton moving in the opposite direction. The soliton collision process with an open boundary can be viewed as a *virtual collision with an antisoliton*. There exists, however, a minimum soliton velocity below which a soliton and an antisoliton will not survive a collision. This minimum allows to define a minimum bias current for observing vortex reflection at the junction boundary. This simple picture is importantly modified by the presence of magnetic fields which affect the boundary conditions of the junction.

Characteristic curves of long Josephson-junctions typically show a series of branches, called *zero field steps*, associated to the dynamics of a number of solitons and antisolitons moving in the junction. First experimental indications of the magnetic flux motion in Josephson junctions date back to 1969 [18–20]. Soon after that, in 1972, Fulton and Dynes [21] performed a pioneering experiment with a long Josephson junction and clearly interpreted their data as due to the inertial motion of fluxons in the junction and named the resonant structures observed in the current-voltage characteristic of the junction as *zero-field steps*. Zero-field steps are related to the resonant oscillations of fluxons trapped in the long Josephson junction in the absence of an external field (see Fig. 6).

4.2 Annular Junctions and Fluxon Pinning Potential

Annular junctions are ideal systems for studying fluxon dynamics since they allow to observe undisturbed fluxon motion. Single sine-Gordon solitons trapped in an annular Josephson junction were observed for the first time by Davidson et al. [22, 23]. Particularly interesting, fluxon and antifluxons can be introduced at will in the systems, permitting to study fluxon dynamics in a very controlled way [24, 25]. In addition to the usual curves, experiments indicating the excitation

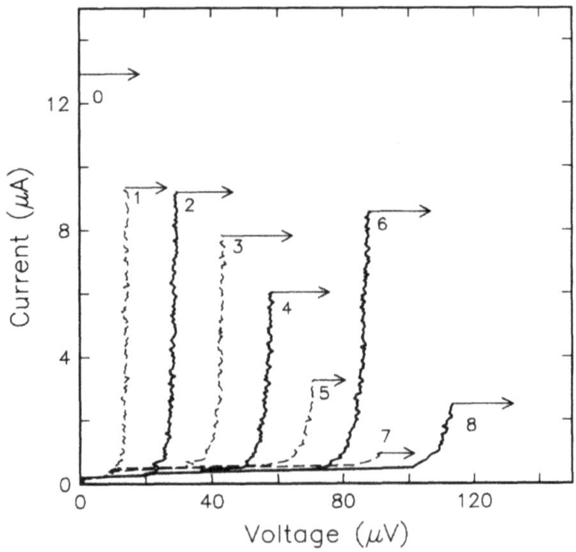

Fig. 6 Zero field steps appear in the current vs. voltage curve of a LJJ. Each number gives the total number of solitons and antisolitons moving in the junction. Reprinted figure with permission from A. Davidson, B. Dueholm, B. Kryger, and N. F. Pedersen, Physical Review Letters 55, 2059 (1985). Copyright 1985 by the American Physical Society

of electromagnetic modes by a vortex moving in an annular Josephson junction were reported [26]. Such modes were visible as resonances on the current-voltage characteristic of the junction.

In 1991 Grønbech-Jensen et al. [27] proposed a model for an annular Josephson junction in an external magnetic field. They showed that the field introduces an additional term in the perturbed sine-Gordon equation which describes the system. This extra term accounts for the coupling between the applied field and the flux density in the junction inducing a pinning potential for the fluxon. The static properties of the system were experimentally studied in [28] and the dynamical behavior of the fluxon in [29]. It was shown that, in the presence of a periodic pinning potential, the behavior of the fluxon in the LJJ is very similar to the response of a small junction; the fluxon can be seen as a single particle moving in a washboard potential. Thermal activation properties of the fluxon over the barrier were studied in [30]. There, a most interesting topic was discussed: the possibility of realizing fluxon potential engineering. In particular, a heart-shaped junction resulting in a double-well was shown. We will go discuss again this system in the last section when introducing quantum solitons in LJJs.

Shape engineering of closed LJJs allowed also for the proposal of a Josephson vortex ratchet [31]. In such system a Josephson vortex moves in a periodic asymmetric potential. Soon after, Carapella and Costabile [32] reported the observation of the ratchet effect for a fluxon trapped in an annular Josephson junction embedded in an inhomogeneous magnetic field. In [33] the ratchet effect is also experimentally studied, but in that case the asymmetric potential was caused by a current injector. Similar effects were studied later on in intrinsic BSCCO JJs [34].

Fig. 7 Example of series of stacked LJJs

5 Stacked Josephson Junctions

A new superconducting system that is described by special fluxon solutions is the long stacked Josephson junction sketched in Fig. 7 which consists of a series of alternating superconducting and isolating layers [8]. This type of systems can be studied in terms of N inductively coupled perturbed sine-Gordon equations [35] where the coupling is due to induced currents in superconducting regions shared by fluxons belonging to different layers.

Fluxons in coupled LJJs have been a subject of intensive theoretical and experimental investigation [8]. On the one side, man-made stacked LJJs with very few layers are interesting from a fundamental point of view [36–39]. In a magnetic field applied parallelly to the layers, fluxons are expected to penetrate into different Josephson junctions and may move coherently due to the interaction of their screening currents.

On the other hand, the discovery of the intrinsic Josephson effect in some high-temperature superconductors such as $Ba_2Sr_2CaCu_2O_{8+y}$ (BSCCO) convincingly showed that those materials are essentially natural superlattices of Josephson junctions formed on atomic structure scale [40–42]. The spatial period of such a superlattice is only 15 Å, so the Josephson junctions are extremely densely packed. The superconducting electrodes are formed by the copper oxide bilayers as thin as 3 Å and are separated by the non-superconducting BiO layers. Josephson superlattices with many active layers can naturally be expected to show very complex dynamics. Single crystals of BSCCO and some other materials show properties expected for multi-layer stacks of Josephson junctions [43–46]. Since it is rather difficult to measure the voltage on a single Josephson junction of atomic scale, natural stacks are usually measured in series.

In the last years most attention has been put in these systems triggered by the interest in generating and detecting electromagnetic fields in the elusive low terahertz range, where there is a severe lack of working devices. In particular, solid-state sources of terahertz (THz) radiation are being sought for sensing, imaging, and spectroscopy applications across the physical and biological sciences.

Millimeter-wave oscillators using intrinsic Josephson effect materials go up to sub-THz and THz frequencies and are promising devices for various sub-millimeter band applications [47–54].

6 Josephson Junction Parallel Arrays and the Discrete sine-Gordon Equation

6.1 Parallel Arrays

Systems with superconducting wires interrupted by *small* JJs are usually denoted as JJ arrays [55]. Fabrication techniques nowadays allow to make arrays of many sizes and geometries. Such systems have been extensively studied in the past from both basic and applied perspectives. For instance, series arrays have been designed for studying phase locking phenomena and building the voltage standard [56]. Superconducting loops interrupted by one or two junctions, known as SQUIDs (for *Superconducting Quantum Interference Devices*) [5, 11, 57–59], provide a sensitive measure of magnetic flux and are used nowadays as standard magnetic field detector in many laboratories. Ladder arrays were designed for studying the transition from one-dimensional to two-dimensional physics and allowed an experimental observation of discrete breathers [60, 61]. Two dimensional arrays are an ideal model system to study 2D phase transitions, frustration and disorder effects, vortex dynamics, phase synchronization and other non-linear dynamics results [62]. To finish, the JJ parallel arrays, which had been designed for studying fluxon transport devices, are also interesting from a fundamental point of view as they constitute an experimental realization of the discrete sine-Gordon equation [63–67].

A JJ parallel array is formed by a set of junctions connected in parallel by superconducting wires (as seen in Fig. 8). The equations for the array can be easily obtained when only mesh self-inductances are taking into account [55, 67]. In this case we get:

$$\ddot{\varphi}_j + \Gamma \dot{\varphi}_j + \sin \varphi_j = \lambda \left(\varphi_{j+1} - 2\varphi_j + \varphi_{j-1} \right) + i_{\text{ext}} \tag{13}$$

with $j = 1, \dots N$. Equation (13) has to be complemented by the appropriate boundary conditions. They are given by $\varphi_0 = \varphi_1 - 2\pi f_0$ and $\varphi_{N+1} = \varphi_N + 2\pi f_0$ for the case of open-ended arrays or $\varphi_{N+1} = \varphi_1 + 2\pi n_k$ and $\varphi_0 = \varphi_N - 2\pi n_k$ for ring arrays. In this last case the integer n_k gives the total number of kinks or fluxons trapped in the array.

It is easy to see that Eq. (13) constitutes the discrete version of the sine-Gordon equation. Let us denote $\varphi_j \equiv \varphi(x)$ and $\varphi_{j\pm1} \equiv \varphi(x \pm \delta)$. For large λ we can write $\varphi(x \pm \delta) \simeq \varphi(x) + \varphi_x \delta \pm \frac{1}{2} \varphi_{xx} \delta^2$ and then $\varphi_{j+1} - 2\varphi_j + \varphi_{j-1} \equiv \varphi_{xx} \delta^2$. Thus λ acts as a discreteness parameter: for large λ the physics of the array corresponds to that of the long JJ meanwhile for moderate and small λ values new discreteness effects appear in the dynamics of the system.

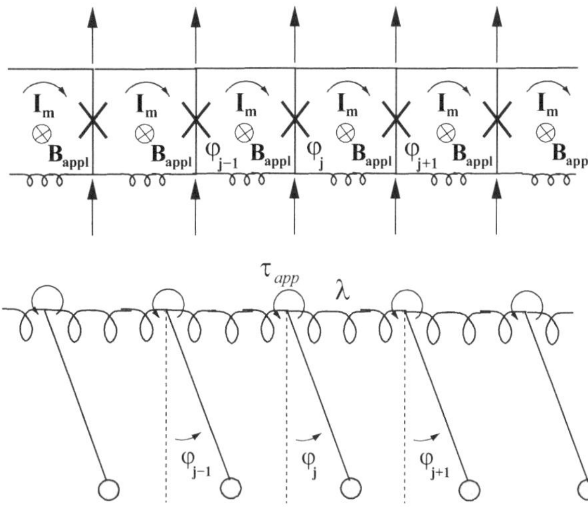

Fig. 8 JJ parallel array and its mechanical analog

Being the pendulum a mechanical analog of the single junction, the mechanical analog of the parallel array is a set of pendula connected by torsion springs (see Fig. 8), or a set of atoms in a periodic substrate potential interacting harmonically; a classical model in condensed matter physics coined under the name of Frenkel–Kontorova model [63–65].

6.2 *Kinks in Parallel Arrays*

A kink (anti-kink) corresponds to a solution where the phases go from 0 to $2\pi(-2\pi)$ along the array (see Fig. 9). They are the discrete counterpart of the continuous solitons found in the long Josephson-junction.

The existence of kinks does not depend crucially on the discreteness of the system and in many aspects the discrete solitons are similar to the continuous solitons found in the sine-Gordon equation. However, many other properties of the kinks do depend crucially on the discreteness of the array. The existence of a discrete lattice breaks the invariance under continuous translations of the solitons of the continuous model and discrete fluxons are pinned to the lattice: there exists a minimum energy barrier the kink needs to overcome in order to move through the lattice. This energy is the so-called Peierls–Nabarro (PN) barrier, E_{PN}, and decreases rapidly with λ, see [55].

We can go further and define not only a PN barrier but also a center of mass position for the kink ($X = C \pm \sum_j \varphi_j$) and a PN potential. Following this approach it is possible to identify the kink motion with the motion of a single particle over a

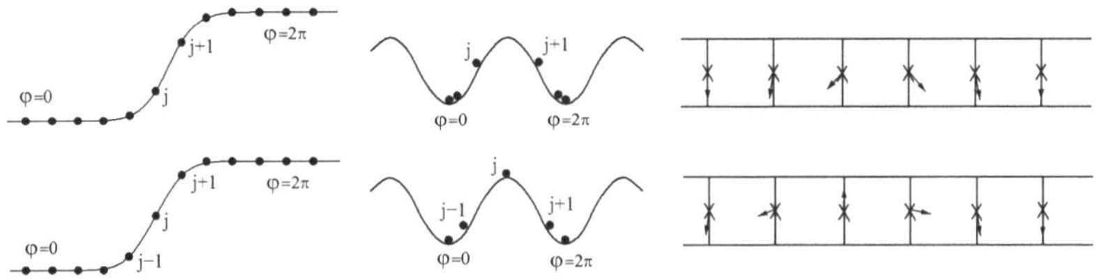

Fig. 9 Three different representations of the minimum energy configuration (*top figures*) and the saddle configuration, maximum of the PN potential (*bottom figures*), of a kink in the parallel array: phase representation (*left*), potential energy representation (*center*) and angular representation (*right*)

sinusoidal periodic potential. In the presence of external bias, an effective equation of motion for the kink is given by

$$m\ddot{X} + m\Gamma\dot{X} + \frac{E_{PN}}{2}\sin X = i. \tag{14}$$

The picture of the kink or fluxon as a single particle is particularly useful in arrays which are larger than the fluxon width and are driven by small currents (far from the whirling mode where all the junctions are in the voltage state and the fluxon delocalizes).

As for the continuous case, when the kink moves it radiates energy in form of small amplitude waves. This radiation is very strong for the case of underdamped arrays. There, phonons are easily excited by the kink in its wake and resonances between the kink velocity and these waves appear [66, 67]. The dispersion relation for such waves can be easily obtained as (for zero damping and current)

$$\omega_k^2 = 1 + 4\lambda\sin^2(k/2) \quad (-\pi < k \leq \pi). \tag{15}$$

Differently from the continuous case, this dispersion relation is characterized by a finite band with gap $\omega_{min} = \omega_0 = 1$ and maximum frequency $\omega_{max} = \omega_\pi = (1 + 4\lambda)^{1/2}$.

Fluxon ratchet potentials. We have considered arrays where all the junctions are identical and all the cells have the same size. However, it is possible to design different arrays which are adequate for studying new physical problems. One example is the use of Josephson parallel arrays to study the dynamics of kinks subjected to substrate ratchet potentials [68–70].

A ratchet potential is a periodic potential without inversion symmetry: $V(x) \neq V(-x)$; then it is easier to move a particle in one direction than in the other. A kink ratchet potential can be obtained with different suitable combinations of junctions and inductances. The simplest ones (see Fig. 10) are made alternating junctions of two critical currents and cells of two areas, and alternating junctions of three critical currents.

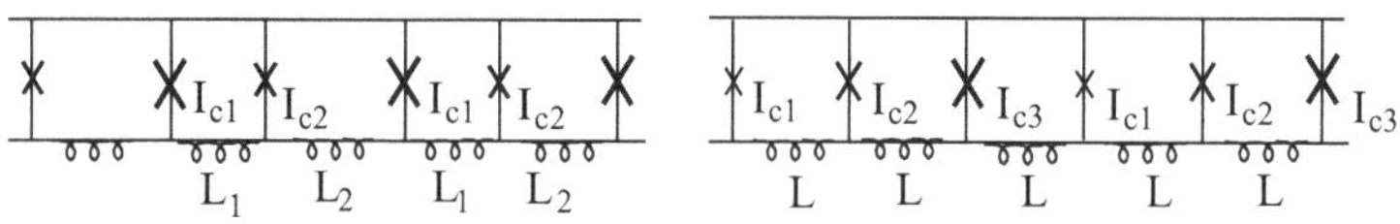

Fig. 10 Two different designs of parallel arrays where fluxons experiment a ratchet periodic potential

4π kinks. To finish, we mention that 4π kinks have been recently observed in the dynamics of Josephson parallel arrays [71]. A 4π kink carries two quanta of magnetic flux and results from the fusion of two standard 2π kinks.

6.3 Discrete Breathers in Parallel Arrays

Discrete breather [72–75] excitations (also called intrinsic localized modes) are solutions of the dynamics of nonlinear lattices for which the energy remains exponentially localized in a few sites of the array. As we have seen, Josephson arrays are ideal experimental discrete system to study nonlinear dynamics. Thus much effort in the last years has been devoted to the prediction and experimental observation of discrete breathers in Josephson arrays.

Josephson circuits are externally biased dissipative systems. Thus, the breather solutions that we are going to present are attractors of the dynamics of the array. The simplest Josephson array supporting breathers is the parallel array, which is described by a discrete sine-Gordon equation. This system supports only oscillobreather solutions where one junction describes a large amplitude oscillation meanwhile the others follow the external force and oscillate with a small amplitude. They appear in arrays biased by ac external currents. Figure 11 shows a picture of an oscillating discrete breather solution in a Josephson parallel array. Being the plasma frequency for a Josephson junction close to 100 GHz, it is not possible to follow the instantaneous dynamics of the phases of the system. The detection of oscillobreathers in parallel arrays is a formidable task which has not been made up to the date. Thus, the experimental effort and much of the theoretical studies of breather in Josephson arrays have been focused to study breathers (rotobreathers) in a variation of the parallel array: the so-called Josephson junction ladder [60, 61, 76, 77]. Different from oscillobreathers, in the rotobreather case one junction is set at the gap-voltage state (following the pendulum analog, it rotates at a given frequency) with the other junctions at zero dc voltage (oscillates with an exponentially decaying amplitude measured from the breather core).

Fluxons and breathers can coexist in ac-biased JJ parallel arrays. Their combined dynamics shows a rich phenomenology depending on the different parameter and conditions of the system. Figure 11 for instance, shows a situation were a fluxon is

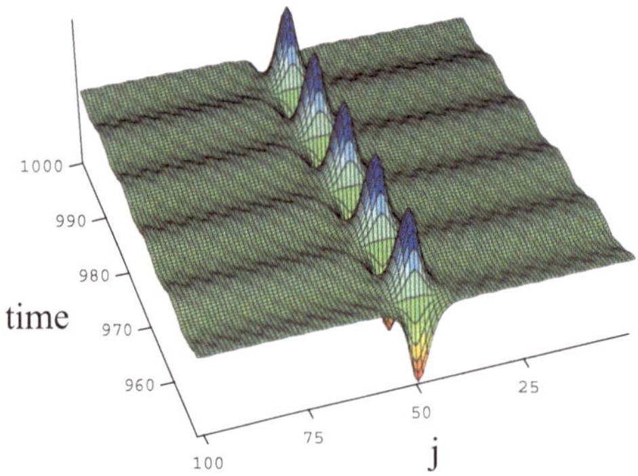

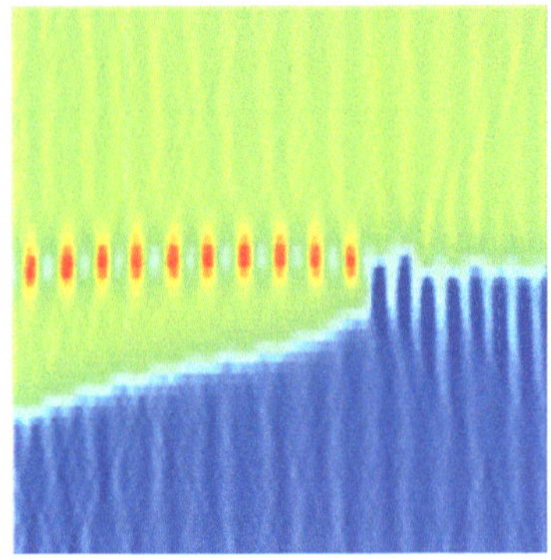

Fig. 11 *Left*: Numerical simulation of a breather solution in an ac biased Josephson parallel array. *Right*: Fluxon breather collision in an ac biased Josephson parallel array

attracted by a oscillobreather. After the interaction the discrete breather excitation is annihilated. Vortex-breather collisions in Josephson ladders were numerically studied in [78].

7 Quantum Solitons in LJJs

Being the Josephson effect a genuine quantum mechanical phenomena, in order to model and understand the behavior of the Josephson arrays previously considered, junctions have been treated as classical elements and Josephson arrays as classical circuits. This approximation, however, ceases to be valid for small enough junctions and at very low temperatures. In such cases, *secondary* quantum effects emerge in the system, and the phase of the junction requires to be treated as a true quantum operator. In such junctions, the phase difference across the junction and the charge on the junction electrode behave as quantum-mechanical conjugate variables. In the case of small junctions the Hamiltonian of the JJ can be written as the addition of Josephson and charging energies

$$H = -E_J \cos\varphi + \frac{Q^2}{2C} \tag{16}$$

where $Q \sim d\varphi/dt$ (that is, the charging energy is a kinetic energy term).

First experimental evidences of quantum mechanical behavior of small JJs were reported in the late 1980s and 1990s [79]. Recently Josephson arrays with small junctions are the base of many promising new quantum technologies [80] for on-chip quantum optics and quantum information processing (recent reviews on the subject are available in [81–83]).

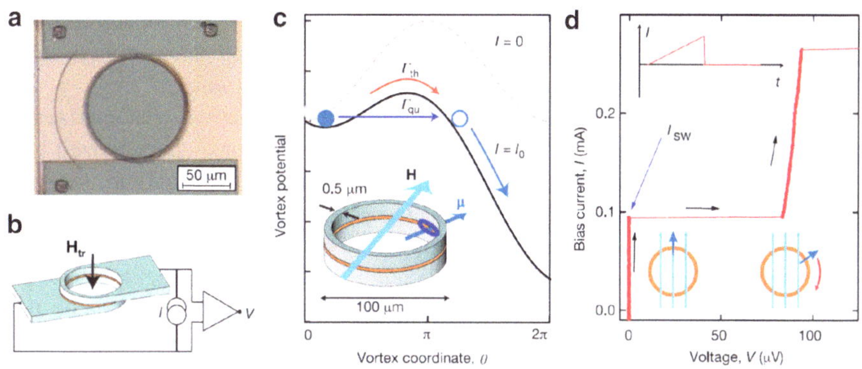

Fig. 12 *Left*: Circular array to explore vortex dynamics. *Middle*: A vortex can escape from a metastable potential due to thermal excitation or to quantum tunneling through a potential barrier. *Right*: Typical current-voltage curve. Switching is due to thermal excitation at high temperatures and to vortex tunneling below a certain temperature crossover [84]

We have shown above that annular junctions are ideal systems for studying Josephson vortices, which in turn behave in many respects as relativistic particles. Such particle should exhibit quantum mechanical properties if studied in suitable arrays and at low temperatures. This study was done by A. Wallraff et al. 10 years ago [84]. They proved the quantum dynamics of an individual Josephson vortex in a long Josephson junction. By measuring the statistics of the vortex escape from a controllable pinning potential, they demonstrated the existence of quantized levels of the vortex energy within the trapping potential well and quantum tunnelling of the vortex through the pinning barrier with a crossover from thermal activation to the macroscopic quantum tunneling observed at about 100 mK, see Fig. 12. In a different work, the thermal and the quantum dissociation of a single vortex-antivortex (VAV) pair in an annular Josephson junction was also experimentally observed [85].

The flexibility in the fabrication of Josephson arrays and the detection of the fluxon quantum tunneling allows to design qubit systems in LJJs. Heart-shaped junctions [30] or annular junction with a microshort [86] shows a double well potential structure for the fluxon; allowing for single vortex tunneling between both states and quantum superposition. The use of circuits based in LJJs for fluxon readout of superconducting qubit has been also recently achieved [87, 88].

8 Final Remarks

We have presented here a short introduction to the physics of long Josephson junctions. Measured for the first time more than 40 years ago, LJJs systems are still nowadays an active research field. Work today is mainly devoted to the applications of the arrays in two very different fields: developing terahertz technologies and quantum information theory. Substantial advances in both fields are rapidly occurring and foreseeably will occur in the next future.

This chapter has been devoted to *standard* Josephson junctions characterized by the Josephson relation given in Eq. (1). So-called π junctions, characterized by a different Josephson relation ($\varphi = \pi$ instead of 0 in its ground state) are today an important field of research. Then, long π Josephson junctions and combined 0-π junctions, where *fractional vortices* are found, are being proposed and studied with new interesting physics and potential applications in many diverse fields [89–95].

The discrete sine-Gordon model is at the basis for employing magnetic flux quanta in parallel arrays, often referred as Josephson Transmission Lines (JTLs), as basic bits for computation elements. A magnetic flux quantum, or fluxon (2.07×10^{-15} Wb), is alias of a Josephson vortex as it is formed by persistent current loops circulating around a certain point of the superconducting transmission line with embedded Josephson junctions. As discussed earlier in this review, Josephson fluxons behave as particle-like solitary waves—solitons. The idea of using Josephson flux quanta as elementary bits of information goes back to Fulton, Dynes and Anderson [96]. Since two decades ago, fluxons have been successfully employed for the implementation of Rapid Single Flux Quanta (RSFQ) logic in superconducting digital circuits [97–99] using shunted (overdamped) Josephson junctions. The fact that a single flux quantum can be manipulated, moved around by bias currents, erased by magnetic fields etc., stimulates very naturally the variety of ideas and conjectures related to elementary digital operations. More recent ideas involve underdamped dynamics of fluxons in discrete JTLs (alias of parallel arrays discussed earlier in this chapter), which have been proposed to be used for reversible computing [100]. This kind of computing dissipating less than the thermodynamic threshold $k_B T \ln 2$ energy per elementary operation can be performed in either classical or quantum regime [101]. Here fluxons are modified to create an internal degree of freedom with two stable states, which is possible by replacing the elementary Josephson junctions in the JTL by so-called nSQUID elements [102]. A recent historical review on fluxon logic circuits can be found in [103].

Acknowledgements JJM acknowledge financial support from Spanish MINECO through Project No. FIS2011-25167, cofinanced by FEDER funds.

References

1. A. Scott, *Nonlinear Science: Emergence & Dynamics of Coherent Structures* (Oxford University Press, Oxford, 1999)
2. T. Dauxois, M. Peyrard, *Physics of Solitons* (Cambridge University Press, Cambridge, 2006)
3. R.D. Parmentier, in *Solitons in Action*, ed. by K. Lonngren, A.C. Scott (Academic, New York, 1978), pp. 173–200
4. D.W. McLaughlin, A.C. Scott, Phys. Rev. A **18**, 1652 (1978)

5. A. Barone, G. Paternò, *Physics and Applications of the Josephson Effect* (Wiley, New York, 1982)
6. N.F. Pedersen, in *Solitons*, ed. by S.E. Trullinger, V.E. Zakharov, V.L. Pokrovsky (Elsevier, Amsterdam, 1986), pp. 469–502
7. R.D. Parmentier, in *The New Superconducting Electronics*, ed. by H. Weinstock, R.W. Ralston (Springer, Berlin, 1993), pp. 221–245; and arXiv:patt-sol/9303003
8. A.V. Ustinov, Physica D **123**, 315 (1998)
9. B.D. Josephson, Phys. Lett. **1**, 251 (1962)
10. J.R. Waldram, Rep. Prog. Phys. **39**, 751 (1976)
11. K.K. Likharev, *Dynamics of Josephson Junctions and Circuits* (Gordon and Breach Science, 1984)
12. T.P. Orlando, K.A. Delin, *Foundations of Applied Superconductivity* (Addison Wesley, Reading, 1991)
13. M. Tinkham, *Introduction to Superconductivity* (McGraw–Hill, New York, 1996)
14. W.C. Stewart, Appl. Phys. Lett. **12**, 277 (1968)
15. D.E. McCumber, J. Appl. Phys. **39**, 3113 (1968)
16. A.C. Scott, F.Y.F. Chu, S.A. Reible, J. Appl. Phys. **47**, 3272 (1976)
17. Y.S. Kivshar, B.A. Malomed, Rev. Mod. Phys. **61**, 800 (1989)
18. A.C. Scott, W.J. Johnson, Appl. Phys. Lett. **14**, 316 (1969)
19. J.T. Chen, T.F. Finnegan, D.N. Langenberg, in *Proceedings of the Internatioanl Conference on Superconductivity*, ed. by F. Chilton (Elsevier, Amsterdam, 1971), p. 413
20. A. Barone, J. Appl. Phys. **42**, 2747 (1971)
21. T.A. Fulton, R.C. Dynes, Solid State Commun. **12**, 57 (1973)
22. A. Davidson, B. Dueholm, B. Kryger, N.F. Pedersen, Phys. Rev. Lett. **55**, 2059 (1985)
23. A. Davidson, B. Dueholm, N.F. Pedersen, J. Appl. Phys. **60**, 1447 (1986)
24. A.V. Ustinov, T. Doderer, R. Huebener, Phys. Rev. Lett. **69**, 1815 (1992)
25. A.V. Ustinov, Appl. Phys. Lett. **80**, 3153 (2002)
26. A. Wallraff, A.V. Ustinov, V. Kurin, I. Shereshevsky, N. Vdovicheva, Phys. Rev. Lett., **84**, 151 (2000)
27. N. Grønbech-Jensen, P.S. Lomdahl, M.R. Samuelsen, Phys. Lett. A **154**, 14 (1991)
28. N. Martucciello, R. Monaco, Phys. Rev. B. **53**, 3471 (1996)
29. A.V. Ustinov, B. Malomed, N. Thyssen, Phys. Lett. A **233**, 239 (1997)
30. A. Wallraff, Y. Koval, M. Levitchev, M.V. Fistul, A.V. Ustinov, J. Low Temp. Phys **118**, 543 (2000)
31. E. Goldobin, A. Sterck, D. Koelle, Phys. Rev. E **63**, 031111 (2001)
32. G. Carapella, G. Costabile, Phys. Rev. Lett. **87**, 077002 (2001)
33. M. Beck, E. Goldobin, M. Neuhaus, M. Siegel, R. Kleiner, D. Koelle, Phys. Rev. Lett. **95**, 090603 (2005)
34. H.B. Wang, B.Y. Zhu, C. Gürlich, M. Ruoff, S. Kim, T. Hatano, B.R. Zhao, Z.X. Zhao, E. Goldobin, D. Koelle, R. Kleiner, Phys. Rev. B **80**, 224507 (2009)
35. S. Sakai, P. Bodin, N.F. Pedersen, J. Appl. Phys. **73**, 2411 (1993)
36. M.B. Mineev, G.S. Mkrtchjan, V.V. Schmidt, J. Low Temp. Phys., **45**, 497 (1981)
37. P.R. Auvil, J.B. Ketterson, J. Appl. Phys. **61** 1957 (1987)
38. A.V. Ustinov, H. Kohlstedt, M. Cirillo, N.F. Pedersen, G. Hallmanns, C. Heiden Phys. Rev. B **48**, 10614 (1993)
39. S. Sakai, A.V. Ustinov, H. Kohlstedt, A. Petraglia, N.F. Pedersen, Phys. Rev. B **50**, 12905 (1994)
40. R. Kleiner, F. Steinmeyer, P. Kunkel, P. Müller, Phys. Rev. Lett. **68**, 2394 (1992)
41. G. Oya, N. Aoyama, A. Irie, S. Kishida, H. Tokutaka, Jpn. J. Appl. Phys. **31**, L829 (1992)
42. R. Kleiner, P. Müller, Phys. Rev. B **49**, 1327 (1994)
43. J.U. Lee, J.E. Nordman, G. Hohenwarter, Appl. Phys. Lett. **67**, 1471 (1995)
44. A. Yurgens, D. Winkler, N. Zavaritsky, T. Claeson, Phys. Rev. B **53**, R8887 (1996)
45. Y.I. Latyshev, J.E. Nevelskaya, P. Monceau, Phys. Rev. Lett. **77**, 932 (1996)

46. G. Hechtfischer, R. Kleiner, K. Schlenga, W. Walkenhorst, P. Müller, H.L. Johnson, Phys. Rev. B **55**, 14638 (1997)
47. V.P. Koshelets, S.V. Shitov, L.V. Filippenko, A.M. Baryshev, W. Luinge, H. Golstein, H. van de Stadt, J.-R. Gao, T. de Graauw, IEEE Trans. Appl. Supercond. **7**, 3589 (1997)
48. M. Tachiki, M. Iizuka, K. Minami, S. Tejima, H. Nakamura, Phys Rev. B **71**, 134515 (2005)
49. S. Savel'ev, A.L. Rakhmanov, V.A. Yampol'skii, F. Nori, Nature Phys. **2**, 521 (2006)
50. L. Ozyuzer, A.E. Koshelev, C. Kurter, N. Gopalsami, Q. Li, M. Tachiki, K. Kadowaki, T. Yamamoto, H. Minami, H. Yamaguchi, T. Tachiki, K.E. Gray, W.-K. Kwok, U. Welp, Science **318**, 1291 (2007)
51. A. Dienst, M.C. Hoffmann, D. Fausti, J.C. Petersen, S. Pyon, T. Takayama, H. Takagi, A. Cavalleri, Nature Photon. **5**, 485 (2011)
52. D.Y. An, J. Yuan, N. Kinev, M.Y. Li, Y. Huang, M. Ji, H. Zhang, Z. L. Sun, L. Kang, B. B. Jin, J. Chen, J. Li, B. Gross, A. Ishii, K. Hirata, T. Hatano, V.P. Koshelets, D. Koelle, R. Kleiner, H.B. Wang, W.W. Xu, P.H. Wu, Appl. Phys. Lett. **102**, 092601 (2013)
53. A. Dienst, E. Casandruc, D. Fausti, L. Zhang, M. Eckstein, M. Hoffmann, V. Khanna, N. Dean, M. Gensch, S. Winnerl, W. Seidel, S. Pyon, T. Takayama, H. Takagi, A. Cavalleri, Nat. Mater. **12**, 535 (2013)
54. U. Welp, K. Kadowaki, R. Kleiner, Nat. Photon. **7**, 702 (2013)
55. J.J. Mazo, in *Energy Localisation and Transfer*, ed. by T. Dauxois, A. Litvak-Hinenzon, R.S. MacKay, A. Spanoudaki. Advanced Series in Nonlinear Dynamics, vol. 22 (World Scientific, River Edge, 2004), pp. 193–246
56. C.A. Hamilton, C.J. Burroughs, S.P. Benz, IEEE Trans. Appl. Supercond. **7**, 3756 (1997)
57. J. Clarke, Proc. IEEE **77**, 1208 (1989)
58. S.T. Ruggiero, D.A. Rudman (eds.), *Superconducting Devices* (Academic, Boston, 1990)
59. D. Koelle, R. Kleiner, F. Ludwig, E. Dantsker, J. Clarke, Rev. Mod. Phys. **71**, 631 (1999)
60. A.V. Ustinov, Chaos **13**, 716 (2003)
61. J.J. Mazo, T.P. Orlando, Chaos **13**, 733 (2003)
62. R.S. Newrock, C.J. Lobb, U. Geigenmüller, M. Octavio, Solid State Phys. **54**, 263 (2000)
63. L.M. Floría, J.J. Mazo, Adv. Phys. **45** 505 (1996)
64. O. Braun, Yu.S. Kivshar, Phys. Rep. **306**, 1 (1998)
65. O. Braun, Yu.S. Kivshar, *The Frenkel-Kontorova Model. Concepts, Methods and Applications* (Springer, Berlin, 2004)
66. A.V. Ustinov, M. Cirillo, B.A. Malomed, Phys. Rev. B **47**, 8357 (1993)
67. S. Watanabe, H.S.J. van der Zant, S.H. Strogatz, T.P. Orlando, Physica D **97**, 429 (1996)
68. F. Falo, P.J. Martínez, J.J. Mazo, S. Cilla, Europhys. Lett. **45**, 700 (1999)
69. E. Trías, J.J. Mazo, F. Falo, T.P. Orlando, Phys. Rev. E **61**, 2257 (2000)
70. F. Falo, P.J. Martínez, J.J. Mazo, T.P. Orlando, K. Segall, E. Trías, Appl. Phys. A **75**, 263 (2002)
71. J. Pfeiffer, M. Schuster, A.A. Abdulmalikov Jr., A.V. Ustinov, Phys. Rev. Lett. **96**, 034103 (2006)
72. A.J. Sievers, S. Takeno, Phys. Rev. Lett. **61**, 970 (1988)
73. S. Flach, C.R. Willis, Phys. Rep. **295**, 181 (1998)
74. Focus Issue on "Nonlinear localized modes: physics and applications". Chaos **13**(2) (2003)
75. S. Flach, A.V. Gorbach, Phys. Rep. **467**, 1 (2008)
76. E. Trías, J.J. Mazo, T.P. Orlando, Phys. Rev. Lett. **84**, 741 (2000)
77. P. Binder, D. Abraimov, A.V. Ustinov, S. Flach, Y. Zolotaryuk, Phys. Rev. Lett. **84**, 745 (2000)
78. E. Trías, J.J. Mazo, T.P. Orlando, Phys. Rev. B **65**, 054517 (2002)
79. J. Clarke, A.N. Cleland, M.H. Devoret, D. Esteve, J.M. Martinis, Science **239**, 992 (1988)
80. I. Georgescu, F. Nori, Phys. World **25**, 16 (2012)
81. J. Clarke, F.K. Wilhelm, Nature **453**, 1031 (2008)
82. J.Q. You, F. Nori, Nature **474**, 589 (2011)
83. M.H. Devoret, R.J. Schoelkopf, Science **339**, 1169 (2013)
84. A. Wallraff, A. Lukashenko, J. Lisenfeld, A. Kemp, M.V. Fistul, Y. Koval, A.V. Ustinov, Nature **425**, 155 (2003)

85. M.V. Fistul, A. Wallraff, Y. Koval, A. Lukashenko, B.A. Malomed, A.V. Ustinov, Phys. Rev. Lett. **91**, 257004 (2003)
86. A.N. Price, A. Kemp, D.R. Gulevich, F.V. Kusmartsev, A.V. Ustinov, Phys. Rev. B **81**, 014506 (2010)
87. K.G. Fedorov, A.V. Shcherbakova, M.J. Wolf, D. Beckmann, A.V. Ustinov. Phys. Rev. Lett. **112**, 160502 (2014)
88. K.G. Fedorov, A.V. Shcherbakova, R. Schäfer, A.V. Ustinov, Appl. Phys. Lett. **102**, 132602 (2013)
89. E. Goldobin, K. Vogel, O. Crasser, R. Walser, W. Schleich, D. Koelle, R. Kleiner, Phys. Rev. B **72**, 054527 (2005)
90. M. Weides, M. Kemmler, H. Kohlstedt, R. Waser, D. Koelle, R. Kleiner, E. Goldobin, Phys. Rev. Lett. **97**, 247001 (2006)
91. K. Cedergren, J.R. Kirtley, T. Bauch, G. Rotoli, A. Troeman, H. Hilgenkamp, F. Tafuri, F. Lombardi, Phys. Rev. Lett. **104**, 177003 (2010)
92. S. Scharinger, M. Turad, A. Stöhr, V. Leca, E. Goldobin, R.G. Mints, D. Koelle, R. Kleiner, Phys. Rev. B **86**, 144531 (2012)
93. H. Sickinger, A. Lipman, M. Weides, R.G. Mints, H. Kohlstedt, D. Koelle, R. Kleiner, E. Goldobin, Phys. Rev. Lett. **109**, 107002 (2012)
94. E. Goldobin, R. Kleiner, D. Koelle, Phys. Rev. B **87**, 224501 (2012)
95. D.M. Heim, K. Vogel, W.P. Schleich, D. Koelle, R. Kleiner, E. Goldobin, New J. Phys. **15**, 053020 (2013)
96. T. Fulton, R. Dynes, P.W. Anderson, Proc. IEEE **61**, 28 (1973)
97. K.K. Likharev, in *The New Superconducting Electronics*, ed. by H. Weinstock, R.W. Ralston (Springer, Berlin, 1993), pp. 423–452
98. K.K. Likharev, V.K. Semenov, A. Zorin, in *Superconducting Devices*, ed. by S.T. Ruggiero, D.A. Rudman (Academic, Boston, 1990), pp. 1–49
99. P. Bunyk, K.K. Likharev, D. Zinoviev, Int. J. High Speed Electron. **11**, 257 (2001)
100. V.K. Semenov, G.V. Danilov, D.V. Averin, IEEE Trans. Appl. Supercond. **13**, 938 (2003)
101. V.K. Semenov, G.V. Danilov, D.V. Averin, IEEE Trans. Appl. Supercond. **17**, 938 (2007)
102. J. Ren, V.K. Semenov, IEEE Trans. Appl. Supercond. **21**, 780 (2011)
103. K.K. Likharev, Physica C **482**, 6 (2012)

Some Selected Thoughts Old and New on Soliton-Black Hole Connections in 2d Dilaton Gravity

Floyd L. Williams

To the memory of John Langford

Abstract We discuss in an expository style some results,old and new,regarding connections between solitons and black holes in 2d dilaton gravity, for both Lorentzian and Euclidean metrics. Black hole vacua (the ground state or mass zero case), and non sine-Gordon solitons such as dissipatons (solutions of a resonance nonlinear Schrödinger equation) are also considered.

Keywords Black hole vacuum • Bright soliton • Dilaton gravity • Dini surface • Elliptic modulus • Ricci scalar curvature • Ricci tensor

1 Introduction: sine-Gordon- J-T Black Hole Connection, and Beyond

In 1997 J. Gegenberg and G. Kunstatter [1] found an intriguing connection between black holes in Jackiw–Teitelboim (J-T) dilaton gravity and solitons in sine-Gordon field theory. See also [2, 3]. For example, these authors demonstrated that using the one-soliton sine-Gordon solution u given by

F.L. Williams (✉)
Department of Mathematics and Statistics, University of Massachusetts Amherst, Amherst, MA 01003, USA
e-mail: williams@math.umass.edu

J. Cuevas-Maraver et al. (eds.), *The sine-Gordon Model and its Applications*, Nonlinear Systems and Complexity 10, DOI 10.1007/978-3-319-06722-3_8,

$$u(x,t) = 4 \arctan e^{\rho(x,t)}$$
$$\rho(x,t) \overset{def.}{=} \frac{m}{a}(x - vt)$$
$$a \overset{def.}{=} (1+v^2)^{1/2}: \quad u_{xx} + u_{tt} = m^2 \sin u, \tag{1}$$

one could construct (locally) a black hole metric

$$\begin{aligned} ds^2_{sol} &= \cos^2\left(\frac{u(x,t)}{2}\right) dx^2 - \sin^2\left(\frac{u(x,t)}{2}\right) dt^2 \\ &= \tanh \rho(x,t)\, dx^2 - \operatorname{sech}^2 \rho(x,t)\, dt^2. \end{aligned} \tag{2}$$

That is, ds^2_{sol} has a Killing vector field that is time-like in an asymptotic region and which vanishes at an interior point. This metric is also of Kruskal type, being free of coordinate singularities at the horizon. Moreover, by a sequence of (implicit) coordinate transformations it was shown that ds^2_{sol} transforms to the (Lorentzian) J-T black hole metric

$$ds^2_{bh} = -(m^2r^2 - v^2)\, dT^2 + (m^2r^2 - v^2)^{-1}\, dr^2, \tag{3}$$

which provides for a soliton-black hole connection. It was also established in [1], among other results, that by way of Bäcklund transformations, which relate different soliton solutions, a flat $SL(2,\mathbb{R})$ connection could be constructed from which the authors derived a gauge theory formulation of J-T dilaton gravity.

An *explicit* transformation Θ of variables $(x,t) \to (T,r)$ by which ds^2_{sol} in (2) goes to ds^2_{bh} in (3) was constructed by the author in [4]; also see [5,6]. Θ is given by

$$\begin{aligned} T &= -\frac{1}{2mv} \log\left[\frac{a \tanh \rho(x,t) + 1}{a \tanh \rho(x,t) - 1}\right] + \frac{x}{v} = -\frac{1}{mv} \coth^{-1}[a \tanh \rho(x,t)] + \frac{x}{v}, \\ r &= \frac{a}{m} \operatorname{sech} \rho(x,t). \end{aligned} \tag{4}$$

The squared soliton velocity v^2 corresponds to the black hole mass. The inverse transformation $\Psi: (T,r) \to (x,t)$, by which (conversely) ds^2_{bh} goes to ds^2_{sol}, is given by

$$\begin{aligned} x &= vT + \frac{1}{2m} \log\left[\frac{\sqrt{a^2 - m^2r^2} + 1}{\sqrt{a^2 - m^2r^2} - 1}\right] = vT + \frac{1}{m} \coth^{-1}[\sqrt{a^2 - m^2r^2}], \\ t &= \frac{x}{v} - \frac{a}{mv} \log\left[\frac{a + \sqrt{a^2 - m^2r^2}}{mr}\right]. \end{aligned} \tag{5}$$

One could replace $\rho(x,t)$ in (1) by $-\rho(x,t)$, which gives the *anti-kink* solution. The corresponding metric ds_{sol}^2 in (2) would be unchanged and the same transformation Θ would map it to ds_{bh}^2 in (3).

Gegenberg and Kunstatter in [2] extend the discussion of a one-soliton (or an anti-soliton) solution to n-soliton solutions of the Euclidean sine-Gordon equation and their relationship to Lorentzian black holes in the J-T theory. They indicate, in particular, how an n-soliton solution can be used to construct "sine-Gordon" coordinates for a black hole, and they find an explicit formula for its mass in terms of soliton parameters in the case $n = 2$. An explicit formula is also found in this case (using MAPLE) for the dilaton field. Similar results along these lines appear in the thesis of J. LeBlanc [7], with some focus on the stationary breather solution

$$u(x,t) = 4\arctan\left[\frac{\coth(w)\cos(m\sinh(w)t)}{\sinh(m\cosh(w)x)}\right]. \tag{6}$$

Regarding the issue of finding an *explicit* transformation Θ of variables $(x,t) \to (T,r)$ for more complicated sine-Gordon solutions u, extending results of the author, we mention a result of S. Beheshti [8]. For example, let u be the oscillating kink-antikink soliton

$$u(x,t) = 4\arctan\frac{v\sinh(amx)}{a\cos(vmt)},$$
$$a = (1+v^2)^{1/2}: \quad u_{xx} + u_{tt} = m^2\sin u, \tag{7}$$

and let ds_{sol}^2 be the corresponding soliton metric given by the first equation in (2):

$$ds_{sol}^2 = \cos^2\left(\frac{u(x,t)}{2}\right)dx^2 - \sin^2\left(\frac{u(x,t)}{2}\right)dt^2. \tag{8}$$

Then there exists an explicit transformation of variables Θ: $(x,t) \to (T,r)$ such that ds_{sol}^2 goes to the black hole metric given by

$$ds_{bh}^2 = -(m^2r^2 - M)\,dT^2 + \frac{dr^2}{m^2r^2 - M}, \quad M = 4A^2v^4m^2, \tag{9}$$

which compares with (3), for a real number A. Θ is given by

$$T = \frac{x}{2Amv^2} + \frac{1}{2Am^2v^2}\tanh^{-1}\left[\frac{a^2\cos^2(vmt) + (v^2+2)\sinh^2(amx)}{2a\cosh(amx)\sinh(amx)}\right]$$
$$r = \frac{4Av^2a\sin(vmt)\sinh(amx)}{a^2\cos^2(vmt) + v^2\sinh^2(amx)}. \tag{10}$$

In (9), we see (again) the dependence of the black hole mass M on the soliton parameters—v and m in this case. The inverse transformation Ψ of Θ, which takes ds_{bh}^2 in (9) to ds_{sol}^2 in (8) for u given by (7), is given by

$$\begin{aligned} x &= v\Psi_2(T,r) - \frac{a}{m}\log\frac{a\sqrt{M}+\sqrt{a^2M-(vmr)^2}}{vmr}, \\ t &= \Psi_2(T,r) \\ &= \frac{\sqrt{M}}{v} - \frac{1}{mv}\left[\tanh^{-1}\frac{\sqrt{a^2M-(vmr)^2}}{\sqrt{M}} - a\tanh^{-1}\frac{\sqrt{a^2M-(vmr)^2}}{a\sqrt{M}}\right]. \end{aligned} \tag{11}$$

As in [2], the dilaton field τ associated with the solution u given by (7) is also computed in [8]—also with the help of MAPLE. That is, the system

$$\begin{aligned} \tau_{xx} + \frac{1}{2}u_x\tau_x\tan\frac{u}{2} + \frac{1}{2}u_t\tau_t\cot\frac{u}{2} - m^2\tau\cos^2\frac{u}{2} &= 0 \\ \tau_{tt} - \frac{1}{2}u_x\tau_x\tan\frac{u}{2} - \frac{1}{2}u_t\tau_t\cot\frac{u}{2} + m^2\tau\sin^2\frac{u}{2} &= 0 \\ \tau_{xt} + \frac{1}{2}u_t\tau_x\tan\frac{u}{2} - \frac{1}{2}u_x\tau_t\cot\frac{u}{2} &= 0 \end{aligned} \tag{12}$$

which is the system

$$(\nabla_i\nabla_j - m^2 g_{ij})\tau = 0, \quad 1 \leq i,j \leq 2 \tag{13}$$

for the metric g given by (8), is solved. Namely, τ is given by

$$\tau(x,t) = \frac{4Av^2am\sin(vmt)\sinh(amx)}{a^2\cos^2(vmt) + v^2\sinh^2(amx)}, \tag{14}$$

which is mr for r in (10).

Further transformation results are presented in [8–10], including results for other dilaton models such as string inspired gravity, for example.

Another paper of interest, among many, is that of S. Vacaru [11] where, among other results, generalized sine-Gordon solutions (in two and three dimensions) are used to generate 4d Einstein gravity solutions—solutions where 4d energy-momentum values are related to soliton-dilaton parameters. A generalization of the Gegenberg-Kunstatter correspondence is found in the paper [12] of M. Cadoni, which involves a *two field* generalization of the sine-Gordon model. A concrete realization of the Cadoni correspondence is presented in [13], where the author with S. Beheshti construct (for static configurations) the desired, explicit transformations of variables Θ, Ψ analogous to (but simpler than) the Θ, Ψ in (4), (5), (10), (11) above.

In the present work, which is expository in nature, we consider further soliton-black hole connections, old and new, in 2d dilaton gravity. The dilaton models are more general than the J-T model, as alluded to in the earlier paragraph, and the discussion is not necessarily restricted to sine-Gordon solitons-nor to black holes of positive mass M. Section 5 contains thoughts on the vacuum case $M = 0$, and in Sect. 3 we consider connections that involve, for example, dissipative soliton solutions (= "bright" solitons) of a reaction diffusion system (RDS) derived from a resonant nonlinear Schrodinger equation. In particular we show that a black hole "dissipative elliptic soliton" cannot move faster than a prescribed value involving its elliptic modulus κ; see the inequality (47) that generalizes the known inequality in [14, 15], for example, established in case $\kappa = 1$.

In Sect. 4, we discuss a class of black hole solutions (of M. Cadoni and S. Mignemi [16–18]) of field equations that involve a *modulus* field—in addition to a usual dilaton field. This class contains the J-T solution of course and it also contains, for example, the E. Witten "cigar" solution [19]. We relate the solutions to static bright solitons. There is also a very brief discussion of Ricci solitons—in connection with Witten's cigar solution—which actually is predated by work of R. Hamilton [20].

Some geometry of the vacuum is discussed in Sect. 5, in particular in connection with a Dini surface corresponding to a sine-Gordon soliton. More specifically, we explain how to construct the Euclidean vacuum as a limit of the first fundamental form (suitably scaled) of a Dini surface. We do this by constructing an explicit transformation Ψ of variables—one analogous to the Ψ in (5); see (118). In this limit, the Dini surface "approaches" a tractroid.

Section 2 contains some notation and some basic definitions from Riemannian geometry needed for the later sections.

This work is dedicated to the memory of my friend Mr. John Langford whose humble life was a quest for knowledge, but knowledge coupled with wisdom and humanity—a life steered by peace and justice matters and by an insatiable love of music and mathematics.

2 Notation and Some Basic Definitions: Ricci Tensor, Ricci Scalar Curvature, Einstein Vacuum Equations, etc.

The pair (M, g) will denote a pseudo Riemannian manifold. Thus g is a pseudo Riemannian (or possibly a Riemannian) metric on a smooth manifold M. Smoothness will also be assumed for all associated data: functions f, Φ, real vector fields X, Y, Z, etc. Let ∇ denote the Levi-Civita affine connection on M, whose existence is provided for by the fundamental theorem of Riemannian geometry. That is, ∇ is the *unique* linear connection on M that is (i) compatible with g: $Zg(X, Y) = g(\nabla_Z X, Y) + g(X, \nabla_Z Y)$, and (ii) is torsion free: $\nabla_X Y - \nabla_Y X = [X, Y]$. ∇ is given/defined in fact by the *Koszul formula*:

$$2g(Z, \nabla_X Y) = Xg(Z, Y) + Yg(Z, X) + g(X, [Z, Y]) \\ + g(Y, [Z, X]) + g(Z, [X, Y]) - Zg(Y, X).$$

The *Riemann curvature* of ∇ is the function $\Omega(X, Y)$ that assigns to Z the vector field

$$\Omega(X, Y)Z \overset{def.}{=} \nabla_X \nabla_Y Z - \nabla_Y \nabla_X Z - \nabla_{[X,Y]} Z. \tag{15}$$

For the purpose of local computations, we choose a system $(U, \phi = (x_1, x_2, \ldots, x_m))$ of local coordinates on M, where $U \subset M$ is an open subset and $m = \dim M$. Since the vector fields $\left\{\frac{\partial}{\partial x_i}\right\}_{i=1}^m$ form a local frame on U, one can define Christoffel symbols Γ_{ij}^k and curvature components R_{ijk}^ℓ by

$$\nabla_{\frac{\partial}{\partial x_i}} \frac{\partial}{\partial x_j} = \sum_{k=1}^m \Gamma_{ij}^k \frac{\partial}{\partial x_k}$$

$$\Omega\left(\frac{\partial}{\partial x_i}, \frac{\partial}{\partial x_j}\right) \frac{\partial}{\partial x_k} = \sum_{\ell=1}^m R_{ijk}^\ell \frac{\partial}{\partial x_\ell}. \tag{16}$$

As usual let $g_{ij} \overset{def.}{=} g\left(\frac{\partial}{\partial x_i}, \frac{\partial}{\partial x_j}\right)$ and let $g^{ij} \overset{def.}{=}$ the ij^{th} entry of $[g_{ij}]^{-1}$. Then the following formulas hold.

$$\Gamma_{ij}^k = \frac{1}{2} \sum_{\ell=1}^m g^{\ell k} \left(\frac{\partial g_{i\ell}}{\partial x_j} + \frac{\partial g_{j\ell}}{\partial x_i} - \frac{\partial g_{ij}}{\partial x_\ell}\right)$$

$$R_{ijk}^\ell = \frac{\partial \Gamma_{jk}^\ell}{\partial x_i} - \frac{\partial \Gamma_{ik}^\ell}{\partial x_j} + \sum_{p=1}^m \left(\Gamma_{ip}^\ell \Gamma_{jk}^p - \Gamma_{jp}^\ell \Gamma_{ik}^p\right) = -R_{jik}^\ell. \tag{17}$$

The *Ricci tensor* components $R_{ij} = R_{ij}(g)$ and *Ricci scalar curvature* $R = R(g)$ are given by

$$R_{ij} = \sum_{\ell=1}^m R_{i\ell j}^\ell = R_{ji}, \quad R = \sum_{i,j=1}^m g^{ij} R_{ij}. \tag{18}$$

The symmetry of the Ricci tensor in (18) really does depend on the g-compatibility and torsion freeness of ∇. Moreover these two conditions (conditions (i), (ii) above) imply the following:

$$R_{ij} = \sum_{\ell,k=1}^m g^{\ell k} R_{i\ell jk}, \quad R_{ijk}^\ell = \sum_{r=1}^m g^{r\ell} R_{ijkr} \tag{19}$$

for the Riemann curvature tensor

$$R_{ijk\ell} \overset{def.}{=} g\left(\Omega\left(\frac{\partial}{\partial x_i}, \frac{\partial}{\partial x_j}\right)\frac{\partial}{\partial x_k}, \frac{\partial}{\partial x_\ell}\right) = \sum_{r=1}^{m} g_{r\ell} R^r_{ijk}, \tag{20}$$

which is subject to the Bianchi identities

$$R_{ijk\ell} + R_{jki\ell} + R_{kij\ell} = 0. \tag{21}$$

Our sign convention for the scalar curvature R in (18) generally differs from that in the physics literature where usually the scalar curvature there corresponds to $-R$.

Many gravitational and string field equations involve the *Hessian* $Hf = \nabla\nabla f$ and the gradient ∇f of a function f.

$$\begin{aligned}(Hf)(X,Y) &\overset{def.}{=} X(Yf) - (\nabla_{X^Y})f \\ &= X(Yf) - (\nabla_{Y^X} + [X,Y])f \\ &\overset{def.}{=} X(Yf) - (\nabla_{Y^X} + XY - YX)f \\ &= Y(Xf) - (\nabla_{Y^X})f = (Hf)(Y,X),\end{aligned} \tag{22}$$

where (again) we have used that ∇ is torsion free. From (16) the local expression

$$\nabla_i \nabla_j f \overset{def.}{=} (Hf)\left(\frac{\partial}{\partial x_i}, \frac{\partial}{\partial x_j}\right) = \frac{\partial^2 f}{\partial x_i \partial x_j} - \sum_{k=1}^{m} \Gamma^k_{ij} \frac{\partial f}{\partial x_k} \tag{23}$$

follows. The gradient of f is the vector field ∇f defined by $Yf = g(Y, \nabla f)$ for all vector fields Y. Locally

$$\nabla f = \sum_{\ell=1}^{m}\left(\sum_{k=1}^{m} g^{k\ell}\frac{\partial f}{\partial x_k}\right)\frac{\partial}{\partial x_\ell}. \tag{24}$$

We shall also need the Laplace operator $\Box$ of g:

$$\Box = \frac{1}{\sqrt{|\det g|}}\sum_{j=1}^{m}\frac{\partial}{\partial x_j}\left(\sum_{i=1}^{m} g^{ij}\sqrt{|\det g|}\frac{\partial}{\partial x_i}\right). \tag{25}$$

$$|\nabla f|^2 \overset{def.}{=} g(\nabla f, \nabla f) \tag{26}$$

denotes the square of the length of the gradient of f, which is also written as $(\nabla f)^2$ by some physicists. More generally

$$\nabla f \cdot \nabla \Phi \stackrel{def.}{=} g(\nabla f, \nabla \Phi) \tag{27}$$

for two functions f, Φ.

In the special case when $m = 2$, the following formulas can be established:

$$R_{11} = g^{22} R_{1212}, \quad R_{12} = -g^{12} R_{1212} \; (= R_{21}), \quad R_{22} = g^{11} R_{1212}, \tag{28}$$

and hence [by (18)]

$$R = \frac{2R_{1212}}{\det g}. \tag{29}$$

As a consequence of these formulas we see that since

$$g^{-1} \stackrel{def.}{=} \begin{bmatrix} g^{11} & g^{12} \\ g^{21} & g^{22} \end{bmatrix} = (\det g)^{-1} \begin{bmatrix} g_{22} & -g_{12} \\ -g_{21} & g_{11} \end{bmatrix} \tag{30}$$

we have

$$\begin{aligned} R_{11} - \frac{R}{2} g_{11} &= g^{22} R_{1212} - R_{1212} (\det g)^{-1} g_{11} \\ &= (\det g)^{-1} g_{11} R_{1212} - R_{1212} (\det g)^{-1} g_{11} = 0, \\ R_{12} - \frac{R}{2} g_{12} &= -g^{12} R_{1212} - R_{1212} (\det g)^{-1} g_{12} \\ &= (\det g)^{-1} g_{12} R_{1212} - R_{1212} (\det g)^{-1} g_{12} = 0, \\ R_{22} - \frac{R}{2} g_{22} &= g^{11} R_{1212} - R_{1212} (\det g)^{-1} g_{22} \\ &= (\det g)^{-1} g_{22} R_{1212} - R_{1212} (\det g)^{-1} g_{22} = 0. \end{aligned}$$

That is, the Einstein vacuum field equations

$$R_{ij} - \frac{R}{2} g_{ij} = 0 \tag{31}$$

are *always* satisfied/solved by (M, g) in case $\dim M = 2$.

3 Reaction Diffusion Systems, Resonant Nonlinear Schrödinger Equation, and Jackiw–Teitelboim Gravity

In Sect. 2 it was noted that the Einstein equations (31) were automatically satisfied for *any* metric g on a two-dimensional space-time M. Thus Einstein gravity in two dimensions is trivial. A non-trivial theory of gravity is provided by the

Jackiw–Teitelboim (J-T) *model* [21, 22]. Here, in addition to a metric g on M, the theory involves a *dilaton field* τ—a scalar field on M. The action integral is given by

$$S(g,\tau) = \text{constant} \int_M d^2x \sqrt{|\deg g|}\, \tau \left(R(g) - \frac{2}{\ell^2} \right) \tag{32}$$

for a *negative* cosmological constant $\Lambda = -\frac{1}{\ell^2}$, from which one derives the equations of motion

$$R(g) = \frac{2}{\ell^2}, \quad \nabla_i \nabla_j \tau = \frac{g_{ij}\tau}{\ell^2}. \tag{33}$$

The second set of equations in (33) gives the motion of the field τ, and the first equation shows that the scalar curvature $R = R(g)$ of g is always *constant.* Compare equations (18), (23).

It turns out, on the other hand, by some interesting work of L. Martina, O. Pashaev, G. Soliani, and J-H. Lee [14, 15, 23–25], for example, that one can use solutions of a particular two-component reaction diffusion (RD) system to construct metrics of constant scalar curvature. These solutions, moreover, are in correspondence with solutions $\Psi(x,t)$ of a *resonance* nonlinear Schrödinger (RNLS) equation

$$i\Psi_t + \Psi_{xx} + \gamma|\Psi|^2\Psi = \frac{2|\Psi|_{xx}}{|\Psi|}\Psi \tag{34}$$

(so called by these authors), where γ is a real constant, and where the $\Psi(x,t)$ in turn can be constructed to model a *Madelung fluid* [26]. Among these RD solutions are *dissipative solitons* (called dissipatons) that describe black holes in the J-T model.

We review some of these interesting results, and we present some new solutions of RD systems in terms of Weierstrass' elliptic phi-function.

Some typical RD systems in chemistry, physics, or biology, for example, assume the form

$$\begin{aligned} r_t &= d_r r_{xx} + F(r,s) \\ s_t &= d_s s_{xx} + G(r,s) \end{aligned} \tag{35}$$

where r, s are functions of two variables (say space and time variables x, t), d_r, d_s are diffusion constants for r, s, and where F, G are growth and interaction functions. r and s could be concentration functions for two chemicals, or prey and predator functions in a two species ecological model, for example. Our main (but not exclusive) interest is in the RD system

$$\begin{aligned} r_t - r_{xx} + B(rs)r &= 0 \\ s_t + s_{xx} - B(rs)s &= 0 \end{aligned} \tag{36}$$

with $d_r = 1, d_s = -1, F(x, y) = -Bx^2y, G(x, y) = Bxy^2$, for a real constant B.

If one takes $\Psi = e^{R-iS}$ for real-valued functions R, S of x, t, then equation (34) is equivalent to the system of equations

$$\begin{aligned} R_t - S_{xx} - 2R_xS_x &\overset{(i)}{=} 0 \\ S_t - R_{xx} - R_x^2 - S_x^2 + \gamma e^{2R} &\overset{(ii)}{=} 0 \end{aligned} \tag{37}$$

for a Madelung fluid; again γ is assumed to be a real scalar: $\gamma \in \mathbb{R}$. Moreover, define

$$r = e^{R+S}, \quad s = -e^{R-S}. \tag{38}$$

Then by adding equation (i), (ii) in (37) and by subtracting equation (ii) from equation (i), respectively, one sees that (respectively)

$$\begin{aligned} r_t - r_{xx} - \lambda(rs)r &= 0 \\ s_t + s_{xx} + \lambda(rs)s &= 0, \end{aligned} \tag{39}$$

which is the RD system (36) with $B = -\lambda$. Here $r > 0$, $s < 0$ with $-rs \overset{def.}{=} e^{2R} > 0$, $-r/s = e^{2S} > 0$. Conversely, if $r > 0$ and $s < 0$ are given solutions of the RD system (39), define

$$R = \frac{1}{2}\log(-rs), \quad S = \frac{1}{2}\log(-r/s). \tag{40}$$

Then it can be seen that R, S satisfy the system (37) for a Madelung fluid and hence

$$\Psi \overset{def.}{=} e^{R-iS} \overset{def.}{=} (-rs)^{1/2}(-r/s)^{-i/2} \tag{41}$$

is a solution of the RNLS equation (34).

Now here is a key observation of this section: Suppose r, s are solutions of the RD system (36). Form the metric $g = [g_{ij}]$ where

$$\begin{aligned} &g_{11} \overset{def.}{=} -r_xs_x, \quad g_{12} \overset{def.}{=} \frac{1}{2}(sr_x - rs_x), \\ &g_{22} \overset{def.}{=} rs\text{:} \quad ds^2 = g_{11}\,dt^2 + 2g_{12}\,dt\,dx + g_{22}\,dx^2. \end{aligned} \tag{42}$$

Then g has *constant scalar curvature* given by

$$R = R(g) = 4B. \tag{43}$$

This is discussed in the earlier references [14,15,23–25], where a gauge formulation of the J-T model and Cartan's zweibein formalism are employed. In these references (and in others) the scalar curvature there is the negative of our $R(g)$. Formula (43) could also be established in a more pedestrian way. One could use a Maple program (tensor), for convenience, to verify it. In what follows, some concrete solutions of (36) (old and new) are presented. By considering elliptic functions, for example, one can obtain a large class of solutions. In particular, we find some new solutions of the RNLS equation (34).

For real constants b, a, v (v a velocity parameter), $b, a \neq 0$, define

$$\begin{aligned} r(x,t) &= ba\,\mathrm{dn}(a(x-vt),\kappa)e^{\left[\frac{v^2}{4}+a^2(2-\kappa^2)\right]t-\frac{vx}{2}} \\ s(x,t) &= -ba\,\mathrm{dn}(a(x-vt),\kappa)e^{-\left[\frac{v^2}{4}+a^2(2-\kappa^2)\right]t+\frac{vx}{2}}, \end{aligned} \tag{44}$$

where dn is a Jacobi elliptic function with elliptic modulus κ. For $0 \leq \kappa \leq 1$, dn and the other Jacobi elliptic functions sn, cn, sd, cd, sc are real-valued for real arguments; in particular r, s are real-valued functions of x, t. They are solutions of the RD system (36) for the choice there of $B = 2/b^2$, as argued in [24]. By formula (43) the corresponding metric g in (42) has constant scalar curvature $R(g) = 8/b^2$. The formulas that follow explicate g, where we have used the identities $\mathrm{cn}^2 x + \mathrm{sn}^2 x = 1, \mathrm{dn}^2 x + \kappa^2 \mathrm{sn}^2 x = 1$ [27,28] in the computations: At a point (x,t)

$$\begin{aligned} g_{11} &= b^2a^2\left[a^2\kappa^4\,\mathrm{sn}^2(a(x-vt),\kappa)\,\mathrm{cn}^2(a(x-vt),\kappa) - \frac{v^2}{4}\,\mathrm{dn}^2(a(x-vt),\kappa)\right], \\ g_{12} &= \frac{b^2a^2v}{2}\,\mathrm{dn}^2(a(x-vt),\kappa), \\ g_{22} &= -b^2a^2\,\mathrm{dn}^2(a(x-vt),\kappa). \end{aligned} \tag{45}$$

From the first formula, the points that correspond to an event horizon (i.e. $g_{11}(x,t) = 0$) are subject to the condition (with the continuing assumption $0 \leq \kappa \leq 1$)

$$\frac{\kappa^2|\,\mathrm{sn}(a(x-vt),\kappa)\,\mathrm{cn}(a(x-vt),\kappa)|}{\mathrm{dn}(a(x-vt),\kappa) > 0} = \frac{1}{2}\left|\frac{v}{a}\right|. \tag{46}$$

Again, $\mathrm{dn}^2(x) = 1-\kappa^2\,\mathrm{sn}^2 x = \mathrm{cn}^2 x+\mathrm{sn}^2 x-\kappa^2\,\mathrm{sn}^2 x \doteq \mathrm{cn}^2 x+(1-\kappa^2)\,\mathrm{sn}^2 x \geq \mathrm{cn}^2 x$, so that Eq. (46) and the inequality $|\,\mathrm{sn}\,x| \leq 1$ impose the requirement

$$|v| \leq 2|a|\kappa^2. \tag{47}$$

The meaning of this inequality is that we have generalized the statement that a black hole dissipative soliton (a dissipaton) *cannot* move faster than the prescribed value $2|a|$. Namely in [14, 15], for example, (47) is established in the special case when the elliptic modulus $\kappa = 1$, in which case $\mathrm{dn}(x, \kappa) = \mathrm{sech}\, x$ and thus r, s in (44) reduce to

$$r(x,t) = ba\ \mathrm{sech}(a(x - vt))e^{\left[\frac{v^2}{4}+a^2\right]t - \frac{vx}{2}}$$
$$s(x,t) = -ba\ \mathrm{sech}(a(x - vt))e^{-\left[\frac{v^2}{4}+a^2\right]t + \frac{vx}{2}}, \tag{48}$$

which are the dissipatons denoted by q^+, q^- respectively in [15] for the choice $b = 1$, and by e^+, e^- respectively in [14] for the choice $b = (-8/\Lambda)^{1/2}$, for a cosmological constant $\Lambda < 0$. In these two references the notation k is used instead of the above notation a. We have based our derivation of (47) of course on the concrete formula for g_{11} in (45), which does not appear in the literature.

r and s in (48) are also referred to as *bright solitons*—or at least analogues of bright solitons. For example, instead of the RNLS equation (34), consider the nonlinear Schrodinger equation

$$i\Psi_t + \frac{1}{2}\Psi_{xx} + \gamma|\Psi|^2\Psi = 0. \tag{49}$$

For $\gamma > 0$ this equation has the bright soliton solution

$$\Psi_B(x,t) = ba\ \mathrm{sech}(a(x - vt))e^{i\left[\frac{a^2-v^2}{2}t + vx\right]} \tag{50}$$

for $b = \lambda^{-1/2}$, which one compares with (48). Similar to r, s in (48) one has the *dark soliton* solution [25] r, s given by

$$r(x,t) = ba \tanh(a(x - vt + c))e^{\left[\frac{v^2}{4}-2a^2\right]t - \frac{vx}{2}}$$
$$s(x,t) = -ba \tanh(a(x - vt + c))e^{-\left[\frac{v^2}{4}-2a^2\right]t + \frac{vx}{2}} \tag{51}$$

of the RD system

$$r_t - r_{xx} + \frac{2}{b^2}(rs)r = 0, \quad s_t + s_{xx} - \frac{2}{b^2}(rs)s = 0, \tag{52}$$

which is the system (36) for $B = 2/b^2$, and which therefore leads to a metric (42) with constant scalar curvature $R = 8/b^2$, by (43).

By definition (44) and Eq. (41) we see that

$$\Psi(x,t) = |ba|e^{-i\left(\left[\frac{v^2}{4}+a^2(2-\kappa^2)\right]t - \frac{vx}{2}\right)}\, \mathrm{dn}(a(x - vt), \kappa) \tag{53}$$

is a solution of the RNLS equation (34), for the choice $\gamma = -2/b^2$. In particular when $\kappa = 1$, Ψ in (53) is a one-soliton solution.

In the next example, a new class of solutions of a RD system is presented. Using the Weierstrass phi function $\wp = \wp(; g_2, g_3)$ with *invariants* g_2, g_3 [28], we solve the system

$$r_t - r_{xx} + \frac{r^2 s}{2b^2} = \frac{b^4 C^2}{4rs^2}$$
$$s_t + s_{xx} - \frac{rs^2}{2b^2} = -\frac{b^4 C^2}{4r^2 s}, \tag{54}$$

where $b \neq 0$ and C are constants, which is an example of (35), and which reduces to the system (36) when $C = 0$. Namely, given constants A, B, C define

$$g_2 = \frac{A^2}{3} + \frac{B}{2}$$
$$g_3 = -\frac{A^3}{27} - \frac{AB}{12} - \frac{C^2}{16}$$
$$\Delta = g_2^3 - 27g_3^2 = \frac{B^3}{8} + \frac{A^2 B^2}{16} - \frac{A^3 C^2}{8} - \frac{9}{32}ABC^2 - \frac{27}{256}C^4. \tag{55}$$

With v a velocity parameter as before and c any constant, we define

$$r(x,t) = be^{c - \frac{vx}{2} + \left(\frac{v^2}{4} + \frac{A}{2}\right)t}\left[-4\wp(x - vt; g_2, g_3) + \frac{2A}{3}\right]^{1/2}$$
$$s(x,t) = -be^{-c + \frac{vx}{2} - \left(\frac{v^2}{4} + \frac{A}{2}\right)t}\left[-4\wp(x - vt; g_2, g_3) + \frac{2A}{3}\right]^{1/2}. \tag{56}$$

Then it is possible to show that r, s is a solution of the system (54) and that in particular for the choice $C = 0$, the metric g in (42) has constant scalar curvature given by $R(g) = 2/b^2$, in accordance with formula (43). Also for $C = 0$, g is given explicitly by

$$g_{11} = -b^2\left[4\wp(x - vt; g_2, g_3)^2 + \left(\frac{2A}{3} - v^2\right)\wp(x - vt; g_2, g_3) - \frac{B}{2} + \frac{v^2 A}{6} - \frac{2}{9}A^2\right]$$
$$g_{12} = -2b^2 v\wp(x - vt; g_2, g_3) + b^2 v\frac{A}{3}$$
$$g_{22} = 4b^2\wp(x - vt; g_2, g_3) - \frac{2}{3}b^2 A. \tag{57}$$

For $C = 0$ (which we continue to assume) the equation $4z^3 - g_2 z - g_3 = 0$ has roots

$$e_1 = -\frac{A}{12} + \frac{\sqrt{A^2+2B}}{4}, \quad e_2 = \frac{A}{6}, \quad e_3 = -\frac{A}{12} - \frac{\sqrt{A^2+2B}}{4}. \tag{58}$$

Given $A > 0, B \geq 0$, for example, we express the solution in (56) in terms of the Jacobi elliptic functions sn, cn—the advantage of which is to obtain an expression in terms of the elementary function hyperbolic sine in the special case when the elliptic modulus $\kappa = 1$. First assume $A > 0, B > 0$. Then in (55), (58), $\Delta = B^3/8 + A^2B^2/16 > 0$, $e_1 > e_2 > e_3$, and we can apply the formula

$$\wp(z; g_2, g_3) = e_1 + (e_1 - e_3)\frac{\mathrm{cn}^2(w;\kappa)}{\mathrm{sn}^2(w;\kappa)} \tag{59}$$

for

$$w \overset{def.}{=} \sqrt{e_1 - e_3}\, z, \quad \kappa = \left(\frac{e_2 - e_3}{e_1 - e_3}\right)^{1/2}. \tag{60}$$

Using (58), (59), (60), we have a solution of the RD system

$$r_t - r_{xx} + \frac{r^2 s}{2b^2} = 0, \quad s_t + s_{xx} - \frac{rs^2}{2b^2} = 0 \tag{61}$$

given by

$$r(x,t) = be^{c - \frac{vx}{2} + \left(\frac{v^2}{4} + \frac{A}{2}\right)t}\left[A - \sqrt{A^2+2B} - 2\sqrt{A^2+2B}\,\frac{\mathrm{cn}^2(w;\kappa)}{\mathrm{sn}^2(w;\kappa)}\right]^{1/2}$$

$$s(x,t) = -be^{-c + \frac{vx}{2} - \left(\frac{v^2}{4} + \frac{A}{2}\right)t}\left[A - \sqrt{A^2+2B} - 2\sqrt{A^2+2B}\,\frac{\mathrm{cn}^2(w;\kappa)}{\mathrm{sn}^2(w,\kappa)}\right]^{1/2} \tag{62}$$

for

$$w \overset{def.}{=} \frac{(A^2+2B)^{1/4}(x - vt)}{\sqrt{2}}, \quad \kappa \overset{def.}{=} \left[\frac{A + \sqrt{A^2+2B}}{2\sqrt{A^2+2B}}\right]^{1/2}, \quad A, B > 0. \tag{63}$$

Formula (59) [and hence formulas (62)] also holds in the special case $B = 0$ in which case one has various simplifications: In (63), $w = \sqrt{\frac{A}{2}}(x - vt)$ and $\kappa = 1$. Also $\mathrm{cn}(z;1) = \mathrm{sech}\, z, \mathrm{sn}(z;1) = \tanh z \Rightarrow \mathrm{cn}(z;1)/\mathrm{sn}(z/1) = 1/\sinh z$, so that (62) reduces to

$$r(x,t) = b\sqrt{2A}\, i e^{c - \frac{vx}{2} + \left(\frac{v^2}{4} + \frac{A}{2}\right)t}\left[\sinh\sqrt{\frac{A}{2}}\,(x - vt)\right]^{-1}$$

$$s(x,t) = -b\sqrt{2A}\, i e^{-c+\frac{vx}{2}-\left(\frac{v^2}{4}+\frac{A}{2}\right)t} \left[\sinh\sqrt{\frac{A}{2}}\,(x-vt)\right]^{-1}. \tag{64}$$

By replacing b by bi in (64) with c, b real, we obtain *real* solutions

$$r(x,t) = -b\sqrt{2A}\, e^{c-\frac{vx}{2}+\left(\frac{v^2}{4}+\frac{A}{2}\right)t} \left[\sinh\sqrt{\frac{A}{2}}\,(x-vt)\right]^{-1}$$

$$s(x,t) = b\sqrt{2A}\, e^{-c+\frac{vx}{2}-\left(\frac{v^2}{4}+\frac{A}{2}\right)t} \left[\sinh\sqrt{\frac{A}{2}}\,(x-vt)\right]^{-1}, \tag{65}$$

not of the system (61) but of the system

$$r_t - r_{xx} - \frac{r^2 s}{2b^2} = 0, \quad s_t + s_{xx} + \frac{rs^2}{2b^2} = 0. \tag{66}$$

By (41) and (56) we obtain the following solution Ψ of the RNLS equation (34), for $\lambda = -1/2b^2$:

$$\Psi(x,t) = |b| \left[-4\wp(x-vt; g_2, g_3) + \frac{2A}{3}\right]^{1/2} e^{-i\left[c-\frac{vx}{2}+\left(\frac{v^2}{4}+\frac{A}{2}\right)t\right]} \tag{67}$$

with $g_2 = A^3/3 + B/2, g_3 = -A^3/27 - AB/12$ [by (55)]. We have noted that $\wp$ can also be expressed the terms of the Jacobi elliptic functions cn, sn, so that by (41) and (62) we can also write

$$\Psi(x,t) = |b| \left[A - \sqrt{A^2+2B} - 2\sqrt{A^2+2B}\,\frac{\mathrm{cn}^2(w,\kappa)}{\mathrm{sn}^2(w,\kappa)}\right]^{1/2} e^{-i\left[c-\frac{vx}{2}+\left(\frac{v^2}{4}+\frac{A}{2}\right)t\right]} \tag{68}$$

for w, κ given by (63), $A > 0, B \geq 0$ since we pointed out that (59) [and hence (62)] also holds for $B = 0$.

4 Dilatonic Black Holes with a Modulus Field, Ricci Solitons, and Static Bright Solitons

Because of connections to string theory and for various other reasons, the study of dilaton gravity theories in two dimensions continues to flourish. Such studies serve, in part, as a laboratory by which insight into certain four (or higher)-dimensional, less tractable problems can be gained. These theories contain many exact two-dimensional black hole solutions whose analysis aids in the comprehension of semi-classical effects, for example, of four-dimensional black holes in Einstein's general relativity.

We consider, in particular, a class of black hole solutions (of M. Cadoni and S. Mignemi [16–18]) where in addition to a dilaton field Φ, the two-dimensional action involves a *modulus* field Ψ—a field that arises from the compactification of heterotic string models. The class contains, in particular, the Jackiw–Teitelboim (J-T) solution and E. Witten's black hole (which is also referred to as R. Hamilton's "cigar" soliton). The action is given by

$$2\pi S(g,\Phi,\Psi) = \int d^2x\sqrt{|\det g|}\,e^{-2\Phi}\left[-R(g) + 4|\nabla\Phi|^2 - \frac{2}{3}|\nabla\Psi|^2 + \lambda_1^2 - \lambda_2^2 e^{2(\Phi-\frac{q}{3}\Psi)}\right] \tag{69}$$

for real scalars $\lambda_1, \lambda_2 \in \mathbb{R} - \{0\}$ and a coupling constant $q > 0$. By our sign convention the scalar curvature $R = R(g)$ of g [see (18)] is the negative of that in [17].

The equations of motion for the action $S(g,\Phi,\Psi)$ given in (69) are

$$\begin{aligned}
&\Box\Psi - 2(\nabla\Phi)\cdot(\nabla\Psi) + \frac{q}{2}\lambda_2^2 e^{2(\Phi-\frac{q}{3}\Psi)} = 0,\\
&4\,\Box\Phi - 4|\nabla\Phi|^2 - \frac{2}{3}|\nabla\Psi|^2 - R + \lambda_1^2 = 0,\\
&2\nabla_i\nabla_j\Phi - \frac{2}{3}\nabla_i\Psi\nabla_j\Psi = g_{ij}\left[2\,\Box\Phi - 2|\nabla\Phi|^2 - \frac{1}{3}|\nabla\Psi|^2 + \frac{\lambda_1^2}{2} - \frac{\lambda_2^2}{2}e^{2(\Phi-\frac{q}{3}\Psi)}\right]
\end{aligned} \tag{70}$$

for $1 \le i, j \le 2$, $\nabla_i\Psi = \frac{\partial\Psi}{\partial x_i}$, where $\Box$ is the Laplacian of g, $\nabla\phi$ is the gradient of a function ϕ, etc. See (23), (24), (25), (26), (27). *Exact* solutions (black holes) of these equations have been found [17] (for any q) under the assumption that Ψ is suitably related to Φ [in which case the exponential in (70) is actually a constant]. They will have the general (Lorentzian) form

$$ds^2 = -f(r)\,dt^2 + \frac{dr^2}{f(r)} \tag{71}$$

for coordinates $(t,r), r > 0$, where

$$\begin{aligned}
f(r) &= b^2r^2 - a^2r^{\frac{2k}{k+1}}, \quad -1 < k < 1, \quad a,b\in\mathbb{R}, \quad b \neq 0\\
k &= \frac{3-2q^2}{3+2q^2}; \quad q^2 = \frac{3(1-k)}{2(k+1)}.
\end{aligned} \tag{72}$$

The scalar curvature of the metric in (71) is given [at (t,r)] by

$$R = f''(r) = 2\left[b^2 + a^2\frac{k(1-k)}{(k+1)^2}r^{-\frac{2}{k+1}}\right]. \tag{73}$$

One takes

$$\begin{aligned}\Phi(t,r) &= \phi(r) = \frac{1}{2}\left(\frac{k-1}{k+1}\right)\log r + \text{any constant},\\ \Psi(t,r) &= \psi(r)\end{aligned} \tag{74}$$

where Φ and Ψ are related by

$$\psi(r) = \frac{3}{q}\phi(r) + \frac{3}{2q}\log\frac{q^2}{3}: \quad e^{2(\Phi-\frac{q}{3}\Psi)} = \frac{3}{q^2}. \tag{75}$$

Moreover for the choices

$$\lambda_1^2 = \left(\frac{3+k}{1-k}\right)\lambda_2^2, \quad b = \frac{(k+1)\lambda_2}{\sqrt{2(1-k)}}: \quad f(r) = \frac{(k+1)^2\lambda_2^2}{2(1-k)}r^2 - a^2 r^{\frac{2k}{k+1}} \tag{76}$$

one has that g defined in (71) solves the field equations (70). Also with q, k, Φ, Ψ, and λ_1, λ_2 related by (72), (74), (76) the action $S(g, \Phi, \Psi)$ in (69) reduces to

$$S = \frac{1}{2\pi}\int dx^2\sqrt{|\det g|}\,e^{-2\Phi}\left[-R + \frac{8k}{k-1}|\nabla\Phi|^2 + \lambda_2^2\right]. \tag{77}$$

Next, consider the change of variables $(t,r) \to (T,\sigma) \to (\tau,\sigma)$ where

$$\begin{aligned}T &= bt, \quad r = \cosh^{k+1}\left(\frac{b\sigma}{k+1}\right) = \cosh^{k+1}\left(\frac{\lambda_2\sigma}{\sqrt{2(1-k)}}\right),\\ \tau &= \frac{\sqrt{2(1-k)}}{\lambda_2}T,\end{aligned} \tag{78}$$

by (76). Also choose $a = b$. Then ds^2 in (71) goes to

$$\begin{aligned}ds^2 &= -\cosh^{2k}\left(\frac{\lambda_2\sigma}{\sqrt{2(1-k)}}\right)\sinh^2\left(\frac{\lambda_2\sigma}{\sqrt{2(1-k)}}\right)dT^2 + d\sigma^2\\ &= -\left(\frac{\lambda_2}{\sqrt{2(1-k)}}\right)^2\cosh^{2k}\left(\frac{\lambda_2\sigma}{\sqrt{2(1-k)}}\right)\sinh^2\left(\frac{\lambda_2\sigma}{\sqrt{2(1-k)}}\right)d\tau^2 + d\sigma^2.\end{aligned} \tag{79}$$

Here ds^2 is defined for $k = -1$:

$$ds^2 = -\tanh^2\left(\frac{\lambda_2\sigma}{2}\right) dT^2 + d\sigma^2$$
$$= -\left(\frac{\lambda_2}{2}\right)^2 \tanh^2\left(\frac{\lambda_2\sigma}{2}\right) d\tau^2 + d\sigma^2. \tag{80}$$

In the coordinates (T, σ) the dilaton and modulus fields Φ, Ψ have the expressions

$$\Phi(T,\sigma) = \left(\frac{k-1}{2}\right) \log\cosh\left(\frac{\lambda_2\sigma}{\sqrt{2(1-k)}}\right) + \delta$$
$$\Psi(T,\sigma) = \frac{3}{q}\Phi(T,\sigma) + \frac{3}{2q}\log\frac{q^2}{3}:$$
$$q^2 = \frac{3(1-k)}{2(1+k)}, \quad e^{2(\Phi(T,\sigma)-\delta)} = \cosh^{k-1}\left(\frac{\lambda_2\sigma}{\sqrt{2(1-k)}}\right), \tag{81}$$

by (72), (74), for a constant δ.

We also consider *Euclidean* black holes. Here ds^2 in (71) is replaced by

$$ds^2 = f(r)\, dt^2 + \frac{dr^2}{f(r)}$$
$$= \cosh^{2k}\left(\frac{\lambda_2\sigma}{\sqrt{2(1-k)}}\right)\sinh^2\left(\frac{\lambda_2\sigma}{\sqrt{2(1-k)}}\right) dT^2 + d\sigma^2$$
$$= \left(\frac{\lambda_2}{\sqrt{2(1-k)}}\right)^2 \cosh^{2k}\left(\frac{\lambda_2\sigma}{\sqrt{2(1-k)}}\right)\sinh^2\left(\frac{\lambda_2\sigma}{\sqrt{2(1-k)}}\right) d\tau^2 + d\sigma^2$$

for $k = -1$

$$= \left(\frac{\lambda_2}{2}\right)^2 \tanh^2\left(\frac{\lambda_2\sigma}{2}\right) d\tau^2 + d\sigma^2, \tag{82}$$

by the previous changes of variables, where in the latter three equations here we again assume that $a = b$.

A further change of variables is as follows: $(\tau, \sigma) \to (\theta, r) \to (x, y)$ for

$$\theta = \left(\frac{\lambda_2}{2}\right)^2 \tau, \quad r = \sinh\frac{\lambda_2\sigma}{2}, \qquad x = r\cos\theta, \quad y = r\sin\theta. \tag{83}$$

That is,

$$\tanh^2\left(\frac{\lambda_2\sigma}{2}\right) = \frac{\sinh^2\frac{\lambda_2\sigma}{2}}{1+\sinh^2\frac{\lambda_2\sigma}{2}} = \frac{r^2}{1+r^2},$$
$$\left(\frac{\lambda_2}{2}\right)^2 d\sigma^2 = \left(\frac{d}{dr}\sinh^{-1} r\right)^2 dr^2 = \frac{dr^2}{1+r^2},$$
$$(d\tau)^2 = \left(\frac{\lambda_2}{2}\right)^{-4} d\theta^2 \tag{84}$$

which means that ds^2 in (82) goes to

$$ds^2 = \left(\frac{\lambda_2}{2}\right)^{-2}\left(\frac{r^2\,d\theta^2 + dr^2}{1+r^2}\right) = \left(\frac{\lambda_2}{2}\right)^{-2}\left(\frac{dx^2+dy^2}{1+x^2+y^2}\right). \tag{85}$$

That is for $k = -1$, the Euclidean metric in (82) is transformed to a multiple $4/\lambda_2^2$ of Witten's black hole metric

$$ds_H^2 \overset{def.}{=} \frac{dx^2+dy^2}{1+x^2+y^2} \tag{86}$$

on $\mathbb{R}^2$. The subscript "H" here is in recognition of the fact that ds_H^2 is also known as Hamilton's *cigar soliton*, which occurs in discussions on the *Ricci flow*—a notion introduced in the seminal 1982 paper [20] of R. Hamilton, which therefore predates the 1991 paper [19] of E. Witten. The idea here is to evolve an initial metric into one that is more canonical—to do so in a way that hopefully improves one's understanding of the topology of the underlying Riemannian manifold. Hamilton's innovative ideas have lead to (among other important results) G. Perelman's proof of W. Thurston's *geometrization conjecture* for compact three-manifolds—a conjecture of which the Poincaré conjecture is a special case. The Ricci flow equation (or parabolic Einstein equation) is given by

$$\dot{g}_{ij}(t) = \frac{\partial}{\partial t} g_{ij}(t) = 2R_{ij}(t). \tag{87}$$

For example, one can evolve the soliton metric in (86) by

$$ds_H^2(t) \overset{def.}{=} \frac{dx^2+dy^2}{e^{4t}+x^2+y^2}, \tag{88}$$

where t varies over $\mathbb{R}$. The non-vanishing Ricci tensor components are given by

$$R_{11}(t) = R_{22}(t) = \frac{-2e^{4t}}{(e^{4t}+x^2+y^2)^2}. \tag{89}$$

Since $g_{11}(t) = g_{22}(t) = (e^{4t} + x^2 + y^2)^{-1}$ (with $g_{12}(t) = g_{21}(t) = 0$), we see that the Ricci flow equation (87) is indeed satisfied in this case. Also $(\mathbb{R}^2, ds_H^2)$ is a *Ricci soliton*—meaning that there is a smooth function and a real scalar λ for which

$$R_{ij} + \nabla_i \nabla_j f = \lambda g_{ij}. \tag{90}$$

In fact we can take $\lambda = 0$ [so that $(\mathbb{R}^2, ds_H^2)$ is a *steady* soliton], and we can choose the *potential* function f to be given by

$$f(x, y) \stackrel{def.}{=} \log(1 + x^2 + y^2). \tag{91}$$

One can verify Eq. (90) for these choices of f, λ by using Eqs. (23), (89) [with $t = 0$ in (89)], and the formulas

$$\Gamma_{11}^1 = \frac{-x}{1 + x^2 + y^2} = \Gamma_{12}^2 = -\Gamma_{22}^1, \quad \Gamma_{12}^1 = \frac{-y}{1 + x^2 + y^2} = \Gamma_{22}^2 = -\Gamma_{11}^2 \tag{92}$$

for the non-vanishing Christoffel symbols. In particular $\nabla_1 \nabla_2 f = 0$.

A remarkable property of steady solitons is that they are extremals of Perelman's energy functional

$$F(g, f) \stackrel{def.}{=} \int d^m x \sqrt{\det g}\, e^{-f} \left[-R + |\nabla f|^2\right]. \tag{93}$$

This functional is in fact the (Euclidean) low energy supergravity action for gravitational and dilaton fields [29]. These various remarks [following equation (85)] illustrate the importance of the case $k = -1$, for example.

Note that for $k = 0$, $f(r) = b^2 r^2 - a^2$ in (71), (72) so that one obtains the J-T model with *constant* scalar curvature $R = 2b^2$ by (73)—which is consistent with (33).

We point out now that the metrics in (42) and (80) are actually related when the former is constructed from the bright soliton solution r, s in (48), in the *static* case $v = 0$, for a suitable choice there of the parameter a. Thus for $v = 0$, r, s assume the simpler form

$$r(x,t) = bae^{a^2 t} \operatorname{sech} ax, \quad s(x,t) = -bae^{-a^2 t} \operatorname{sech} ax. \tag{94}$$

The a, b here have nothing to do with the notation a, b in (72), for example. Again for $\kappa = 1$, [the assumption made which lead to (48)] $\operatorname{sn}(z; 1) = \tanh z$ and $\operatorname{cn}(z; 1) = \operatorname{dn}(z; 1) = \operatorname{sech} z$. The formulas in (45) therefore reduce simply to

$$g_{11} = b^2 a^4 \tanh^2 ax \operatorname{sech}^2 ax, \quad g_{12} = 0, \quad g_{22} = -b^2 a^2 \operatorname{sech}^2 ax. \tag{95}$$

That is, r, s in (94) induce the metric

$$ds_B^2 = -b^2 a^2 \operatorname{sech}^2 a\sigma \left[-a^2 \tanh^2 a\sigma \, d\tau^2 + d\sigma^2\right], \tag{96}$$

in (42), where by a switch of notation we have written τ for t and σ for x, and where the subscript "B" in (96) denotes "brightness". Thus we see that for the choice $a = \lambda_2/2$ the bright soliton metric ds_B^2 in (96) is the multiple $-\frac{b^2}{4}\lambda_2^2 \operatorname{sech}^2 \frac{\lambda_2}{2}\sigma$ of the (Lorentzian) Hamilton-Witten black hole metric ds^2 (for $k = -1$) in (80), which provides for another example of a soliton-black hole connection.

In (95), $-g_{22} > 0$ and we can consider

$$\frac{g_{11}}{(-g_{22})^{k+2}} = \frac{b^2 a^4 \sinh^2 ax}{(\cosh^2 ax)^2} \left(\frac{\cosh^2 ax}{b^2 a^2}\right)^{k+2} = \frac{(\sinh^2 ax)(\cosh^2 ax)^k}{(b^2)^{k+1}(a^2)^k}, \tag{97}$$

with $g_{11} = -r_x s_x, g_{22} = rs$ by definition (42), which means that we can write (97) as

$$-(\sinh^2 ax) \cosh^{2k} ax = (b^2)^{k+1}(a^2)^k \frac{r_x s_x}{(-rs)^{k+2}}. \tag{98}$$

Going back to (79), we see that for the class of black holes that have been indexed by k, $-1 \leq k < 1$, the corresponding metric (for the record) can be expressed in terms of the static bright solitons r, s in (94) if we make the choice there of $a = \lambda_2/\sqrt{2(1-k)}$:

$$ds^2 = (b^2)^{k+1} \left(\frac{\lambda_2^2}{2(1-k)}\right)^k \frac{r_x s_x}{(-rs)^{k+2}} dT^2 + dx^2. \tag{99}$$

Also since $(-rs)(x, t) = b^2 a^2 / \cosh^2 ax$ implies that

$$(-rs)^{(k-1)/2}(x, t) = \frac{|ba|^{k-1}}{\cosh^{k-1} ax}$$

we can express the dilaton field Φ (and hence the modulus field Ψ) in terms of r, s—again for the choice of $a = \lambda_2/\sqrt{2(1-k)}$ in (94):

$$e^{2(\Phi(T,x)-\delta)} = \frac{|ba|^{k-1}}{(-rs)^{\frac{k-1}{2}}}, \tag{100}$$

by (81).

5 The Ground State: Black Hole Vacua by Way of sine-Gordon Solitons

For the choice $a = 0$ in (72) one obtains by (71), (74) the *dilaton vacuum*

$$ds^2 = -b^2 r^2\, dt^2 + \frac{dr^2}{b^2 r^2},$$
$$e^{2[\Phi(t,r)-\Phi_0]} = (br)^{\frac{k-1}{k+1}} \tag{101}$$

with *constant* scalar curvature $R = 2b^2$ by (73). δ denotes the constant in (74), (81), (100). Then the second equation in (101) follows by choosing $\Phi_0 = \delta + \frac{1-k}{2(k+1)} \log b$. We begin by constructing here an explicit transformation by which the Euclidean version

$$ds_E^2 = b^2 r^2\, dt^2 + \frac{dr^2}{b^2 r^2} \tag{102}$$

of the metric ds^2 in (101) is realized as a limit—a *mass zero* limit. The transformation is, in a sense, a cousin to the (Lorentzian) one constructed by the author in the Eq. (5). For our transformation we in fact start with a one-soliton sine-Gordon solution to which a *Dini surface* (a pseudo-spherical helicoid) corresponds. This then provides for a soliton-black hole vacuum connection.

We consider the following parametrization $\Phi\colon \mathbb{R}^2 \to \mathbb{R}^3$,

$$\Phi(u, v) = \Big(f(u, v), g(u, v), h(u, v)\Big)$$

of a Dini surface (a helicoid generated by a tractrix):

$$f(u, v) = \frac{2\lambda}{1+\lambda^2}\, \text{sech}\Big(\lambda u + \frac{v}{\lambda}\Big) \cos(u - v)$$
$$g(u, v) = \frac{2\lambda}{1+\lambda^2}\, \text{sech}\Big(\lambda u + \frac{v}{\lambda}\Big) \sin(u - v)$$
$$h(u, v) = \frac{2\lambda}{1+\lambda^2}\Big[\lambda u + \frac{v}{\lambda} - \tanh\Big(\lambda u + \frac{v}{\lambda}\Big)\Big] + \Big(\frac{1-\lambda^2}{1+\lambda^2}\Big)(u - v), \tag{103}$$

for $\lambda \in \mathbb{R} - \{0\}$. The first fundamental form

$$ds_I^2 = E\, du^2 + 2F\, du\, dv + G\, dv^2 \tag{104}$$

of this surface is defined by the dot products

$$E \stackrel{def.}{=} \Phi_u \cdot \Phi_u, \quad F \stackrel{def.}{=} \Phi_u \cdot \Phi_v, \quad G \stackrel{def.}{=} \Phi_v \cdot \Phi_v. \tag{105}$$

Computing these, using (103), one finds that

$$E = G = 1, \quad F = \frac{\cosh^2\left(\lambda u + \frac{v}{\lambda}\right) - 2}{\cosh^2\left(\lambda u + \frac{v}{\lambda}\right)}. \tag{106}$$

On the other hand there is the one-soliton solution ϕ of the sine-Gordon equation

$$\phi_{uv} = \sin\phi: \quad \phi(u, v) = 4 \arctan e^{\lambda u + \frac{v}{\lambda}}. \tag{107}$$

By the double angle formula $2\cosh^2 x = \cosh 2x + 1$, it turns out that $\cos\phi(u, v)$ coincides exactly with F in (106), which means that the first fundamental form can be expressed as

$$ds_I^2 = du^2 + 2\cos\phi(u, v)\, du\, dv + dv^2. \tag{108}$$

ds_I^2 has scalar curvature $R = 2\phi_{uv}/\sin\phi$ which is therefore *constant* by (107): $R = 2$.

For our purpose it is convenient to consider a slight generalization $\tilde{\phi}$ of ϕ. For $b \in \mathbb{R} - \{0\}$

$$\tilde{\phi}(u, v) \stackrel{def.}{=} \phi(bu, bv) = 4 \arctan e^{b\left(\lambda u + \frac{v}{\lambda}\right)}. \tag{109}$$

Here

$$\tilde{\phi}_{uv} = b^2 \sin\tilde{\phi} \tag{110}$$

[by (107)] and in place of (108) we work with the metric

$$ds^2 = du^2 + 2\cos\tilde{\phi}(u, v)\, du\, dv + dv^2, \tag{111}$$

which has scalar curvature $R = 2b^2$. By the change of variables $2u = x + y$, $2v = x - y$, ds^2 in (111) goes to

$$ds^2 = \cos^2\frac{u}{2}(x, y)\, dx^2 + \sin^2\frac{u}{2}(x, y)\, dy^2 \tag{112}$$

with scalar curvature $R = 2b^2$ for

$$u(x, y) \stackrel{def.}{=} \tilde{\phi}\left(\frac{x + y}{2}, \frac{x - y}{2}\right) = 4 \arctan e^{\frac{b}{2}\left[\left(\lambda + \frac{1}{\lambda}\right)x + \left(\lambda - \frac{1}{\lambda}\right)y\right]}, \tag{113}$$

which satisfies the sine-Gordon equation

$$u_{xx} - u_{yy} = b^2 \sin u. \tag{114}$$

Note that if we define

$$\alpha = \frac{2\lambda}{1+\lambda^2}, \quad v = \frac{1-\lambda^2}{1+\lambda^2}, \quad \rho(x,y) = \frac{b}{\alpha}(x - vy), \tag{115}$$

thus α, v are the factors in (103), then given (113), (115) we can write

$$u = 4 \arctan e^{\rho}, \quad \alpha^2 = 1 - v^2. \tag{116}$$

Moreover, since $\sin(u/2) = \operatorname{sech} \rho, \cos(u/2) = -\tanh \rho$ [by (116)], Eq. (112) assumes the form

$$ds^2 = \tanh^2 \rho(x,y)\, dx^2 + \operatorname{sech}^2 \rho(x,y)\, dy^2. \tag{117}$$

To construct our transformation $\Psi = (\psi_1, \psi_2)$ we assume $v \neq 0$: $\lambda \neq \pm 1$. Then we take

$$\begin{aligned} x &= \psi_1(t,r) \stackrel{def.}{=} vt + \frac{1}{2b} \log \left[\frac{\sqrt{1+v^2-b^2r^2}+1}{\sqrt{1+v^2-b^2r^2}-1} \right] \\ y &= \psi_2(t,r) \stackrel{def.}{=} \frac{\psi_1(t,r)}{v} - \frac{\alpha}{bv} \log \left[\frac{\alpha + \sqrt{1+v^2-b^2r^2}}{\sqrt{b^2r^2-2v^2}} \right]. \end{aligned} \tag{118}$$

Note that

$$\begin{aligned} dx^2 &= \left(\frac{\partial \psi_1}{\partial t} dt + \frac{\partial \psi_1}{\partial r} dr \right)^2 \\ &= \left(\frac{\partial \psi_1}{\partial t} \right)^2 dt^2 + 2 \frac{\partial \psi_1}{\partial t} \frac{\partial \psi_1}{\partial r} dt\, dr + \left(\frac{\partial \psi_1}{\partial r} \right)^2 dr^2 \end{aligned}$$

and similarly

$$dy^2 = \left(\frac{\partial \psi_2}{\partial t} \right)^2 dt^2 + 2 \frac{\partial \psi_2}{\partial t} \frac{\partial \psi_2}{\partial r} dt\, dr + \left(\frac{\partial \psi_2}{\partial r} \right)^2 dr^2.$$

If we use $\tanh^2 t = 1 - \operatorname{sech}^2 t$, we see by (117) that

$$\begin{aligned} ds^2 = &\left\{ \left(\frac{\partial \psi_1}{\partial t} \right)^2 + (\operatorname{sech}^2 \rho \circ \Psi) \left[\left(\frac{\partial \psi_2}{\partial t} \right)^2 - \left(\frac{\partial \psi_1}{\partial t} \right)^2 \right] \right\} dt^2 \\ &+ 2 \left\{ \frac{\partial \psi_1}{\partial t} \frac{\partial \psi_1}{\partial r} + (\operatorname{sech}^2 \rho \circ \Psi) \left[\frac{\partial \psi_2}{\partial t} \frac{\partial \psi_2}{\partial r} - \frac{\partial \psi_1}{\partial t} \frac{\partial \psi_1}{\partial r} \right] \right\} dt\, dr \\ &+ \left\{ \left(\frac{\partial \psi_1}{\partial r} \right)^2 + (\operatorname{sech}^2 \rho \circ \Psi) \left[\left(\frac{\partial \psi_2}{\partial r} \right)^2 - \left(\frac{\partial \psi_1}{\partial r} \right)^2 \right] \right\} dr^2. \end{aligned} \tag{119}$$

Now

$$\frac{\partial\psi_1}{\partial t} = v, \quad \frac{\partial\psi_1}{\partial r} = \frac{br}{(v^2 - b^2r^2)\sqrt{1 + v^2 - b^2r^2}},$$
$$\frac{\partial\psi_2}{\partial t} = 1, \quad \frac{\partial\psi_2}{\partial r} = \frac{-bvr\sqrt{1 + v^2 - b^2r^2}}{(v^2 - b^2r^2)(b^2r^2 - 2v^2)} \tag{120}$$

where the relation $\alpha^2 = 1 - v^2$ [by (116)] is kept in mind. Also by (115), (118)

$$\rho(\Psi(t,r)) \overset{def.}{=} \frac{b}{\alpha}\,[\psi_1(t,r) - v\psi_2(t,r)] = \log\left[\frac{\alpha + \sqrt{1 + v^2 - b^2r^2}}{\sqrt{b^2r^2 - 2v^2}}\right], \tag{121}$$

and since $\operatorname{sech}(\log t) = 2t/(1+t^2)$, $\tanh(\log t) = (t^2-1)/(t^2+1)$, it follows from (121) that (again as $\alpha^2 = 1 - v^2$)

$$\operatorname{sech}\rho(\Psi(t,r)) = \frac{\sqrt{b^2r^2 - 2v^2}}{\alpha}, \quad \tanh\rho(\Psi(t,r)) = \frac{\sqrt{1 + v^2 - b^2r^2}}{\alpha}. \tag{122}$$

Given the formulas in (120), (122), and again the relation $\alpha^2 = 1 - v^2$, we can compute the coefficients of $dt^2, dt\,dr, dr^2$ in (119). The coefficient of dt^2 is $v^2 + (1 - v^2)(b^2r^2 - 2v^2)/\alpha^2 \overset{(i)}{=} b^2r^2 - v^2$. The coefficient of $dt\,dr$ is

$$2\left\{\frac{vbr}{(v^2 - b^2r^2)\sqrt{1 + v^2 - b^2r^2}} + \frac{b^2r^2 - 2v^2}{\alpha^2}\left[\frac{-bvr\sqrt{1 + v^2 - b^2r^2}}{(v^2 - b^2r^2)(b^2r^2 - 2v^2)} - \frac{vbr}{(v^2 - b^2r^2)\sqrt{1 + v^2 - b^2r^2}}\right]\right\}, \tag{123}$$

where the sum of the latter two terms here is

$$-\frac{brv[1 + v^2 - b^2r^2 + (b^2r^2 - 2v^2)]}{\alpha^2(v^2 - b^2r^2)\sqrt{1 + v^2 - b^2r^2}} = \frac{-bvr}{(v^2 - b^2r^2)\sqrt{1 + v^2 - b^2r^2}}, \tag{124}$$

which means that the coefficient of $dt\,dr$ [the expression in (123)] $\overset{(ii)}{=} 0$. The coefficient of dr^2 is

$$\frac{b^2r^2}{(v^2 - b^2r^2)^2(1 + v^2 - b^2r^2)} + \frac{b^2r^2 - 2v^2}{\alpha^2}\left[\frac{b^2v^2r^2(1 + v^2 - b^2r^2)}{(v^2 - b^2r^2)^2(b^2r^2 - 2v^2)^2} - \frac{b^2r^2}{(v^2 - b^2r^2)^2(1 + v^2 - b^2r^2)}\right], \tag{125}$$

where the sum of the first and third terms here is

$$\frac{\alpha^2 b^2 r^2}{\alpha^2(v^2-b^2r^2)^2(1+v^2-b^2r^2)} - \frac{(b^2r^2-2v^2)b^2r^2}{\alpha^2(v^2-b^2r^2)^2(1+v^2-b^2r^2)}$$
$$= \frac{b^2r^2[1-v^2-b^2r^2+2v^2]}{\alpha^2(v^2-b^2r^2)^2(1+v^2-b^2r^2)} = \frac{b^2r^2[1+v^2-b^2r^2]}{\alpha^2(v^2-b^2r^2)^2(1+v^2-b^2r^2)}, \tag{126}$$

which added to the second term in (125) gives

$$\begin{aligned}\frac{b^2r^2}{\alpha^2(v^2-b^2r^2)^2} &+ \frac{b^2v^2r^2(1+v^2-b^2r^2)}{\alpha^2(v^2-b^2r^2)^2(b^2r^2-2v^2)} \\ &= \frac{b^2r^2(b^2r^2-2v^2)+b^2v^2r^2(1+v^2-b^2r^2)}{\alpha^2(v^2-b^2r^2)^2(b^2r^2-2v^2)} \\ &= \frac{b^2r^2[b^2r^2-2v^2+v^2(1+v^2-b^2r^2)]}{\alpha^2(v^2-b^2r^2)^2(b^2r^2-2v^2)} \\ &= \frac{b^2r^2[-v^2(1-v^2)+b^2r^2(1-v^2)]}{\alpha^2(v^2-b^2r^2)^2(b^2r^2-2v^2)} \\ &= \frac{b^2r^2[-v^2+b^2r^2]}{(v^2-b^2r^2)^2(b^2r^2-2v^2)} \\ &\overset{(iii)}{=} \frac{b^2r^2}{(b^2r^2-v^2)(b^2r^2-2v^2)}\end{aligned} \tag{127}$$

for the coefficient of dr^2.

Thus in summary, we see by way of the explicit transformation Ψ of variables in (118) that [by (i), (ii), (iii)] the soliton metric $ds^2 = du^2 + 2\cos\tilde{\phi}(u,v)\,du\,dv + dv^2$ in (111) is transformed to the metric

$$ds^2 = (b^2r^2 - v^2)\,dt^2 + \frac{dr^2}{(b^2r^2-v^2)\left(1-\frac{2v^2}{b^2r^2}\right)}, \tag{128}$$

provided that $v \neq 0$: $\lambda \neq \pm 1$ in (103); see (115). From formula (128) we clearly see that as $v \to 0$ (say $\lambda \to 1$, so that the Dini surface in (103) reduces to a tractroid) $ds^2 \to b^2r^2\,dt^2 + (b^2r^2)^{-1}\,dr^2$, which is precisely the Euclidean vacuum ds_E^2 in (102).

Note of course that the soliton metric ds^2 in (111) is derived from the first fundamental form ds_I^2 of the Dini surface by way of the change of variables $(u,v) \to (u/b, v/b)$ in (108): $ds_I^2 \to b^2\,ds^2$.

From the note that followed equation (93), the J-T model with constant scalar curvature $R = 2b^2$ is obtained for $k = 0$: By (71)

$$ds^2 = -(b^2r^2 - a^2)\,dt^2 + (b^2r^2 - a^2)^{-1}\,dr^2. \tag{129}$$

Also the dilaton field and ADM mass have the expression

$$e^{-2\Phi} = e^{-2\Phi_0} br, \quad M = e^{-2\Phi_0} a^2 b/2, \tag{130}$$

respectively, by (101) and formula (4.10a) of [17]. Since ds_E^2 in (102) was obtained by taking $a = 0$ in (72) (regardless of the value of k) we see by the mass formula in (130) that ds_E^2 can be regarded as a mass zero limit. On the other hand we have also obtained ds_E^2 as the limit $v \to 0$ by way of formula (128). Comparing $b^2r^2 - v^2$ in (128) with $b^2r^2 - a^2$ in (129) we see that for v^2 corresponding to $a^2 = 2e^{2\Phi_0}M/b$ [again by (130)], we have consistency in thinking of ds_E^2 as a $M \to 0$ or as a $v \to 0$ limit. Here (again) $v \overset{def.}{=} (1 - \lambda^2)/(1 + \lambda^2)$ by (115). The metric ds^2 in (128) also has (for the record) constant scalar curvature $R = 2b^2$.

Besides the first fundamental form ds_I^2 in (108), there is the second fundamental form ds_{II}^2 given by

$$ds_{II}^2 = 2\sin\phi(u, v)\,du\,dv, \tag{131}$$

for ϕ in (107). ds_I^2 and ds_{II}^2 determine the Dini surface, up to a rigid motion by a theorem of P. Bonnet. Here

$$\sin\phi(u, v) = -\frac{2\sinh\left(\lambda u + \frac{v}{\lambda}\right)}{\cosh^2\left(\lambda u + \frac{v}{\lambda}\right)}. \tag{132}$$

Also, in addition to the one-soliton solution given by (107) of the sine-Gordon equation $\phi_{uv} = \sin\phi$, there is the well-known two-soliton solution ϕ given by

$$\phi(u, v) = 4\arctan\left[\left(\frac{\lambda_1 + \lambda_2}{\lambda_1 - \lambda_2}\right)\left(\frac{e^{\lambda_1 u + \frac{v}{\lambda_1}} - e^{\lambda_2 u + \frac{v}{\lambda_2}}}{1 + e^{(\lambda_1+\lambda_2)u + \left(\frac{1}{\lambda_1} + \frac{1}{\lambda_2}\right)v}}\right)\right] \tag{133}$$

for $\lambda_1, \lambda_2 \in \mathbb{R} - \{0\}, \lambda_1 \neq \lambda_2$.

If one takes $\lambda_1 = \sin\theta - i\cos\theta, \lambda_2 = \sin\theta + i\cos\theta, \theta \in \mathbb{R}$, then although λ_1, λ_2 are *not* real the corresponding bracket in (133) is real:

$$\phi(u, v) = 4\arctan\left[\left(\frac{\sin\theta}{\cos\theta}\right)\frac{\sin((u - v)\cos\theta)}{\cosh((u + v)\sin\theta)}\right], \tag{134}$$

which is a *breather* solution of the sine-Gordon equation. The solutions in (133), (134) give rise to *Kuen* and *breather* surfaces, respectively, with constant scalar curvature $R = -2$.

An extended discussion of the vacuum from an *automorphic* view point is presented in [30]. The misprint

$$“X_a \overset{def.}{=} \Gamma_a \setminus H^2 \overset{def.}{=} \left\{ \begin{bmatrix} 1 & nz \\ 0 & 1 \end{bmatrix} \middle| n \in \mathbb{Z} \right\} ”$$

of mine in definition (40) there should read

$$“X_a \overset{def.}{=} \Gamma_a \setminus H^2, \Gamma_a \overset{def.}{=} \left\{ \begin{bmatrix} 1 & nz \\ 0 & 1 \end{bmatrix} \middle| n \in \mathbb{Z} \right\} ”.$$

Also the “single zero” referred to at the end of Sect. 4 of [30] should be corrected (for the record) to read “single non-trivial zero” (at $s = 1/2$) since clearly the gamma function used to construct the zeta function in definition (72) there has trivial zeros at $s = -1/2, -3/2, -5/2, \ldots$.

References

1. J. Gegenberg, G. Kunstatter, Phys. Lett. B **413**, 274 (1997)
2. J. Gegenberg, G. Kunstatter, Phys. Rev. D **58**, 124010 (1998)
3. J. Gegenberg, G. Kunstatter, in *Solitons: Properties, Dynamics, Interactions, Applications*, ed. by R. Mackenzie, M. Paranjape, W. Zakrzewski (Springer, New York, 2000), pp. 99–106
4. F. Williams, in *Quantum Field Theory Under the Influence of External Conditions*, ed. by K. Milton (Rinton Press, New Jersey, 2004), pp. 370–372
5. F. Williams, in *Trends in Soliton Research*, ed. by L. Chen (Nova Science Publishers, New York, 2006), pp. 1–14
6. F. Williams, On solitons, nonlinear sigma-models, and two-dimensional gravity. Paper presented at the Fourth International Winter Conference on Mathematical Methods in Physics, Rio de Janeiro (2004). pos.sissa.it/archive/conferences/013/003/wc2004_003.pdf
7. J. LeBlanc, Sine-Gordon breather black holes. M.Sc. Thesis, The University of New Brunswick, 2001
8. S. Beheshti, Solutions to the dilaton field equations with application to the soliton-black hole correspondence in generalized J-T gravity. Ph.D. Thesis, University of Massachusetts, 2008
9. S. Beheshti, in *Mathematical Physics Research Developments*, ed. by M. Levy (Nova Science Publishers, New Jersey, 2008), pp. 379–417
10. S. Beheshti, in *A Window into Zeta and Modular Physics*, ed. by K. Kirsten, F. Williams. Mathemathical Sciences Research Institute Publications, vol. 57 (University of California/Cambridge University Press, Berkeley/Cambridge, 2010), pp. 295–306
11. S. Vacaru, J. High Energy Phys. **04**, 009 (2001)
12. M. Cardoni, Phys. Rev. D **58**, 104001 (1998)
13. S. Beheshti, F. Williams, J. Phys. A Math. Theor. **40**, 4017 (2007)
14. J.H. Lee, O. Pashaev, Mod. Phys. Lett. A **17**, 1601 (2002)
15. L. Martina, O. Pashaev, G. Soliani, Phys. Rev. D **58**, 084025 (1998)
16. M. Cadoni, S. Mignemi, Phys. Rev. D **48**, 5536 (1993)
17. M. Cadoni, S. Mignemi, Nucl. Phys. B **427**, 669 (1994)
18. M. Cadoni, S. Mignemi, Mod. Phys. Lett. A **10**, 367 (1995)
19. E. Witten, Phys. Rev. D **44**, 314 (1991)
20. R. Hamilton, J. Differ. Geom. **17**, 255 (1982)

21. R. Jackiw, in *Quantum Theory of Gravity*, ed. by S. Christensen (Adam Hilger Ltd., Bristol, 1984), pp. 403–420
22. C. Teitelboim, in *Quantum Theory of Gravity*, ed. by S. Christensen (Adam Hilger Ltd., Bristol, 1984), pp. 327–344
23. J.H. Lee, O. Pashaev, J. Nonlinear Math. Phys. **8**(suppl.), 230 (2001)
24. J.H. Lee, O. Pashaev, Theor. Math. Phys. **152**, 991 (2007)
25. L. Martina, O. Pashaev, G. Soliani, Class. Q. Gravity **14**, 3179 (1997)
26. E. Madelung, Z. Phys. **40**, 322 (1927)
27. P. Byrd, M. Friedman, *Handbook of Elliptic Integrals for Engineers and Physicists* (Springer, Berlin, 1954)
28. K. Chandrasekharan, *Elliptic Functions*, vol. 281 (Springer, Berlin, 1985)
29. J. Polchinski, *String Theory. Vol. 2: Superstring Theory and Beyond* (Cambridge University Press, Cambridge, 1998)
30. F. Williams, J. Phys. A Math. Theor. **45**, 374008 (2012)

Dressing with Control: Using Integrability to Generate Desired Solutions to Einstein's Equations

Shabnam Beheshti and Shadi Tahvildar-Zadeh

Abstract Motivated by integrability of the sine-Gordon equation, we investigate a technique for constructing desired solutions to Einstein's equations by combining a dressing technique with a control-theory approach. After reviewing classical integrability, we recall two well-known Killing field reductions of Einstein's equations, unify them using a harmonic map formulation, and state two results on the integrability of the equations and solvability of the dressing system. The resulting algorithm is then combined with an asymptotic analysis to produce constraints on the degrees of freedom arising in the solution-generation mechanism. The approach is carried out explicitly for the Einstein vacuum equations. Applications of the technique to other geometric field theories are also discussed.

Keywords Chiral field • Control theory • Dressing matrix • Einstein equations • Grassmann manifold • Harmonic map • Integrability • Kerr spacetime • Killing reduction • Ernst formulation • Lax system • Symmetric space • Zero-curvature condition

1 Introduction

Can solutions to the Einstein Equations be "made to order," i.e., is it possible to construct spacetimes with a prescribed set of asymptotic "observables" (such as total mass and angular momentum) and a prescribed causal and/or singular structure (e.g. number of components of event horizon, or number of ring singularities, etc.),

S. Beheshti (✉) • S. Tahvildar-Zadeh
Department of Mathematics, Rutgers, The State University of New Jersey, Piscataway, NJ 08854, USA
e-mail: beheshti@math.rutgers.edu; shadi@math.rutgers.edu

J. Cuevas-Maraver et al. (eds.), *The sine-Gordon Model and its Applications*, Nonlinear Systems and Complexity 10, DOI 10.1007/978-3-319-06722-3_9,

given a specified set of initial parameters? We provide a first step in answering this question by investigating a solution-generation mechanism for integrable *harmonic maps*, known as the *vesture*, or *dressing* method.

The seeds of our discussion, however, are first sowed in sine-Gordon theory. The special class of nonlinear equations to which the sine-Gordon equation belongs is that of classically integrable partial differential equations (PDEs). Even though nonlinear, these PDEs have the property that new solutions can be formed from old ones, in what amounts to a nonlinear generalization of the superposition principle. We exploit this very feature in order to study the Einstein equations. As a consequence, the main thread of this chapter can be summarized in the following diagram:

$$\text{sine-Gordon} \longrightarrow \text{chiral fields} \longrightarrow \text{Einstein's equations} \longrightarrow \text{harmonic maps.}$$

We shall first briefly introduce classically integrable PDEs. Then, building on the techniques developed for the sine-Gordon equation, we introduce Einstein's gravitational equations and put them in a similar context as sine-Gordon, by briefly reviewing both the Zakharov-Belinski and Ernst formulations of the stationary axisymmetric Einstein vacuum equations. We will then realize this same system of PDEs as an axially symmetric harmonic map.

With background established, we place the gravitational field equations in the context of integrability by way of two theorems; informally, the results state that axially symmetric harmonic maps satisfy an integrable system of equations and that new axially symmetric harmonic maps with any number of prescribed singularities can be constructed from a given one. We implement the algorithm afforded by the proofs of these theorems explicitly in the case of the Einstein vacuum and Einstein-Maxwell equations. It is in the final section that we give a partial answer to the question posed at the beginning of this chapter, namely, by combining a novel asymptotic expansion with the calculated results of the vesture method, we show that, given any real numbers $M > 0$ and J, it is indeed possible to treat the arbitrary constants of the dressing procedure as control parameters and produce a 1-solitonic harmonic map that, when viewed as a stationary axisymmetric spacetime metric, has total ADM mass equal to M and total ADM angular momentum equal to J, and is otherwise free of unwanted pathologies (such as a non-zero NUT parameter), in other words, it is a member of the Kerr family of metrics. We conclude by proposing a strategy for generating solutions in more general cases and indicate possible future directions of study.

2 Integrability and Inverse Scattering

A nonlinear PDE (or system of PDEs) is said to be *classically integrable* if there exists an overdetermined linear system, called a *Lax system*, the compatibility condition of which is precisely the nonlinear PDE.[1] We describe this in further detail.

Start with a nonlinear evolution equation for $u(t,x)$. Associate to it a corresponding overdetermined system for an isospectral family of linear differential operators $L(t,u)$, with u-dependent coefficients, whose eigenfunctions satisfy an additional linear evolution equation

$$L\psi = \lambda\psi, \quad \psi_t = B\psi, \tag{1}$$

where $B = B(u, u_x, u_{xx}, \ldots)$ is another linear operator with coefficients depending on u and its derivatives. Isospectrality implies a compatibility condition between L and B, namely

$$L_t - [B, L] = 0. \tag{2}$$

Under the appropriate choices[2] of L, B, this equation agrees with the nonlinear PDE of interest, and the operators L and B are called a Lax Pair for the PDE [1].

Example 1. One of the first equations to be successfully studied in this setting is the Korteweg-de Vries or KdV equation

$$u_t - 6uu_x + u_{xxx} = 0, \tag{3}$$

first used in modeling shallow water wave dynamics. Defining the operators $L = -\frac{d^2}{dx^2} + u$ and $B = -4\frac{d^3}{dx^3} + 6u\frac{d}{dx} + 3u_x$, it is easy to verify that the compatibility condition (2) is satisfied if and only if u satisfies (3). Notice that in this case the first equation in (1) is simply a classical eigenvalue problem for a linear Schrödinger operator having u as potential.

As the above example demonstrates, the operator B depends on the unknown u, and may indeed be considerably more complicated than L, so solving the Lax system (1) may not be a straightforward task, except possibly at $t = 0$. However, the appearance in (1) of time-evolution for the eigenfunction ψ, provides a starting point for the Inverse Scattering Method (ISM), a nonlinear analogue of the Fourier Transform, to be applicable. The three steps of the ISM are depicted in Fig. 1

[1] The term *integrable* has been adopted in various contexts, and is ambiguous in characterizing the features of such an equation (e.g., existence of closed-form solutions by quadrature, possessing infinitely many conservation laws, exhibiting solitonic dynamics, etc.). We shall restrict our attention to those equations which are integrable in the Lax sense.

[2] It is a long-standing open problem to characterize the PDEs which admit a Lax formulation.

Fig. 1 Classical ISM

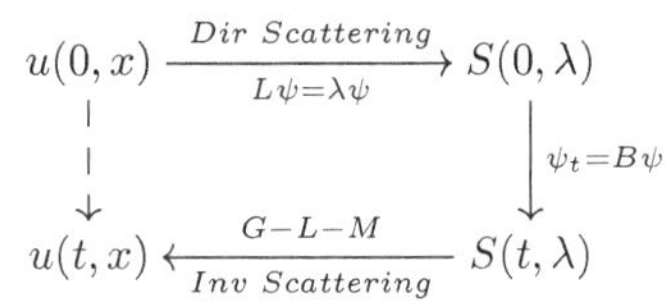

(excerpted from [2]), in which evolution of the scattering data determined by the eigenvalue problem is used to recover a potential $u(t,x)$ from the Cauchy data $u(0,x)$ of the nonlinear PDE of interest.

In the first step, direct scattering refers to finding a matrix $S(0,\lambda)$ relating asymptotic eigenstates at $x = \pm\infty$ by way of $\lim_{x\to\infty}\psi(0,x) = S(0,\lambda)\lim_{x\to-\infty}\psi(0,x)$. Note that S is a matrix since the asymptotic eigenspaces are multidimensional. This step requires identifying a Lax pair.[3] For integrable equations, isospectrality of the operator L implies that the time evolution of S is linear; consequently, in the second step, the scattering matrix $S(t,\lambda)$ can be found (rightmost arrow). Upon integration of the Gelfand-Levitan-Marchenko equations associated with the system, the third step, inverse scattering, recovers a time-evolved solution to the original nonlinear PDE of interest. Standard techniques are described in [3–5] and references therein.

For what follows, we will be careful to distinguish between integrability of a PDE (or system of PDEs) and existence of a solution-generating mechanism for that PDE, noting that many of the examples of this chapter will possess both features.

2.1 Classical Integrability, Extended: The sine-Gordon Equation

Expanding the scope of classical integrability to address second order evolution equations involves going from scalar linear operators to matrix operators, sometimes at the expense of introducing powers of λ, the spectral parameter.[4] In this case, matrix equations comparable to (1) are given by

$$\partial_x\psi = U\psi, \quad \partial_t\psi = V\psi, \tag{4}$$

where $\psi(t,x,\lambda) : \mathbb{R}^2 \times \mathbb{C} \to \mathbb{C}^{2\times 2}$ and U, V are 2×2 matrix functions depending on u, λ. The compatibility or zero-curvature condition corresponding to (2) is

$$U_t - V_x + [U, V] = 0. \tag{5}$$

[3] It also hinges on space being non-compact and one dimensional.

[4] Although one is no longer considering an eigenvalue problem in the classical sense, λ is often still called a spectral parameter.

We take a moment to justify use of the phrase "zero-curvature" by describing the geometry afforded by (4) and (5). Observe that (4) can be used to define a connection on a two-dimensional vector bundle over $\mathbb{R}^2$ by setting

$$\nabla_{\partial_x}\psi = (\partial_x - U)\psi, \quad \nabla_{\partial_t}\psi = (\partial_t - V)\psi, \tag{6}$$

and extending linearly to ∇_X, for a vector field X. In this manner, we may view the two equations in (6) as parallel transport of ψ in the x and t directions, respectively. It is then straightforward to calculate the curvature of this connection for the basis vector fields $X = \partial_x, Y = \partial_t$ to be

$$\nabla_X\nabla_Y\psi - \nabla_Y\nabla_X\psi - \nabla_{[X,Y]}\psi = (U_t - V_x + [U, V])\psi. \tag{7}$$

Thus, the connection has zero curvature if parallel transport of ψ along any path connecting two points agrees, i.e., precisely if compatibility condition (5) is satisfied.

Example 2. Matrix formulation of the ISM is used to address the integrability of the *sine-Gordon equation* (in null coordinates)

$$u_{\zeta\eta} = \sin u, \tag{8}$$

first used in the study of mechanical solitons and differential geometry of constant curvature surfaces.[5] Defining U, V as

$$U = -i\lambda\begin{pmatrix}1 & 0\\ 0 & -1\end{pmatrix} + \frac{1}{2}\begin{pmatrix}0 & u_\zeta\\ u_\zeta & 0\end{pmatrix} \quad \text{and} \quad V = \frac{i}{4\lambda}\begin{pmatrix}\cos u & \sin u\\ \sin u & -\cos u\end{pmatrix},$$

the Lax system analogous to (4) is given by $\partial_\zeta\psi = U\psi$, $\partial_\eta\psi = V\psi$ and the zero curvature condition $U_\eta - V_\zeta + [U, V] = 0$ ensures that $u(\zeta, \eta)$ evolves according to (8). A comparable initial value problem to the one depicted in Fig. 1 is explicitly solved for the sine-Gordon equation by the ISM in [6]; full details for the inverse scattering procedure may be found in [7, 8]. It is worth noting that various Lax formulations are possible, by choosing different U, V in (4) (see, for instance, [9]).

2.2 *From sine-Gordon to Chiral Fields*

Progression from matrix Lax pairs for a single nonlinear PDE to matrix Lax pairs for nonlinear *systems* of PDEs occurs naturally when studying simple geometric field theories, as the next example demonstrates. What is surprising is that from the

[5]Using the transformation $x = \zeta + \eta, t = \zeta - \eta$, one may easily recover the second recognizable form of the sine-Gordon equation, $u_{xx} - u_{tt} = \sin u$.

appropriate perspective, this matrix operator modification applies equally well to equations of gravitation and electrodynamics.

Example 3. Let U be an open subset of $\mathbb{R}^2$ with coordinates (ζ, η) and suppose $g : U \to G$ is a smooth function of (ζ, η) with values in a semisimple Lie group $G \subseteq GL_n(\mathbb{C})$. Denote $(g^{-1})_\eta$ by g_η^{-1} and consider the following action

$$S[g] = \int_U \frac{1}{2} \mathrm{Tr}\,\left(g_\zeta g_\eta^{-1}\right)\, d\zeta d\eta. \tag{9}$$

The Euler-Lagrange equations of this action are known as equations of the *principal chiral field* on the group G–a free field in two-dimensional spacetime with values in G [7, 10, 11]. The word *chiral* here refers to the fact that this Lagrangian is invariant separately under the left, and the right, action of the group G on itself. With a non-commutative group, this action can thus be used to model physical phenomena that appear to break the chiral symmetry, such as the interaction of mesons in the chiral limit [12]. In that case $G = SU(N)$ where N is the number of quark flavors.

By introducing the compact notation $A = -g_\zeta g^{-1}$, $B = g_\eta g^{-1}$, and observing that $g_\eta^{-1} = -g^{-1} g_\eta g^{-1}$, the Lagrangian density in (9) becomes $\frac{1}{2}\mathrm{Tr}AB$. Consequently, the field equations $(g_\zeta g^{-1})_\eta + (g_\eta g^{-1})_\zeta = 0$ are re-expressed as

$$\begin{aligned} A_\eta - B_\zeta &= 0 \\ A_\eta + B_\zeta + [A, B] &= 0, \end{aligned} \tag{10}$$

where the second equation is the compatibility condition arising from the definitions of A and B.

To solve these equations for g by means of inverse scattering, we cast (10) as the compatibility condition of a linear system of matrix equations of the form

$$\partial_\zeta \psi = U\psi, \quad \partial_\eta \psi = V\psi, \tag{11}$$

for ψ a vector function of ζ, η and U, V $n \times n$ matrix functions of ζ, η and a complex parameter $\lambda \in \mathbb{C}$. Equality of mixed partials in ψ imposes the compatibility condition just as in (5), replacing x and t by ζ and η, respectively, yielding

$$U_\eta - V_\zeta + [U, V] = 0. \tag{12}$$

There are several quantities to determine in (11), with varying degrees of freedom. The matrices U, V are to be determined in terms of ζ, η and λ, bearing in mind the compatibility conditions of both U, V and A, B; furthermore, given a solutions ψ of the linear system, the desired matrix g should be recoverable by fixing a value of the parameter λ, under the definitions of A and B. Zakharov and Shabat carried out a *vesture method* (later called the *dressing technique*) to solve this overdetermined system of conditions by assuming U, V have simple poles in λ:

$$U = U_0 + \sum_{n=1}^{N_1} \frac{U_n}{\lambda - \lambda_n}, \qquad V = V_0 + \sum_{n=1}^{N_2} \frac{V_n}{\lambda - \mu_n}. \tag{13}$$

Substituting these expressions into zero-curvature condition (12) and clearing denominators, one obtains a polynomial in λ of degree $N_1 + N_2$, which splits into $N_1 + N_2 + 1$ equations corresponding to coefficients of the polynomial [7, 8].

In the special case $N_1 = N_2 = 1$, where U, V each have one simple pole the compatibility condition (12) reduces to a degree 2 polynomial in λ from which three equations are obtained. For instance, assuming $\lambda_1 = -\mu_1 = 1$ and under appropriate assumptions on U_j, V_j, (11) can be reduced to a Lax formulation of sine-Gordon equivalent to the one given in Example 2. A complete analysis appears in [9, 13]. Note that just as in the previous examples, (11) naturally defines the covariant differentiation operations given by $\partial_\zeta - \frac{A}{1-\lambda}$ and $\partial_\eta - \frac{B}{1+\lambda}$.

Remark 1. Under the coordinate transformation $\zeta = \frac{1}{2}(t + x), \eta = \frac{1}{2}(t - x)$, the Lagrangian density of the action (9) differs by a null Lagrangian from $\frac{1}{2}\mathrm{Tr}\,(\tilde{B}^2 - \tilde{A}^2)$, where $\tilde{A} = g_t g^{-1}$ and $\tilde{B} = g_x g^{-1}$. Noting that $\tilde{A}, \tilde{B}$ must lie in the Lie algebra $\mathfrak{g}$ of the Lie group G, the transformed action may be viewed as an inner product of the matrix-valued one-form $W = \tilde{A}dt + \tilde{B}dx$ with itself, relative to the Minkowski metric on $\mathbb{R}^2$. As we shall see, this one form can be interpreted as the pull-back of the Maurer-Cartan form $w = -dgg^{-1}$ under the mapping $g(x, t)$.

3 Gravitational Equations

Motivated by the last example, we turn our attention to the Einstein vacuum equations with vanishing cosmological constant

$$\mathbf{R}_{\mu\nu} = 0, \qquad \mu, \nu = 0, \ldots 3. \tag{14}$$

This quasilinear system is satisfied by the metric tensor of a four-dimensional Lorentian manifold $(\mathcal{M}, \mathbf{g})$. Here, $\mathbf{R}_{\mu\nu}$ denotes the Ricci curvature tensor of $\mathbf{g}$. Complexity of these field equations is reduced considerably using symmetry reductions, corresponding to existence of Killing fields for the metric $\mathbf{g}$. In this context, existence of timelike Killing fields correspond to stationary metrics, and existence of spacelike Killing fields correspond to spherically symmetric or axisymmetric metrics (depending on whether the action generated is an $SO(3)$ or $SO(2)$ rotation, respectively). Under such symmetries, the simplest nontrivial asymptotically flat solutions[6] are Schwarzschild spacetimes, which are static and spherically symmetric, and the Kerr spacetimes, which are stationary and axisymmetric [14, 15].

[6]Asymptotically flat solutions represent vacuum outside an isolated body.

We shall describe two approaches for the analysis of a two-Killing field reduction of (14), namely the Zakharov-Belinski and the Ernst formulations, and establish a concrete connection between them by way of harmonic maps.

3.1 Integrable Formulation of the Einstein Vacuum Equations

First, we briefly outline the techniques in [16, 17] used to establish integrability of the stationary, axisymmetric Einstein vacuum equations. Assuming the existence of two commuting Killing fields, the spacetime metric $\mathbf{g}$ can be expressed in block-diagonal form

$$\mathbf{g}_{\mu\nu}dx^\mu dx^\nu = f(\rho, z)(d\rho^2 \pm dz^2) + \tilde{\mathbf{g}}_{ab}(y, z)dx^a dx^b \quad a, b = 1, 2, \tag{15}$$

with $x^0 = \rho$ and $x^3 = z$ [18]. Here the $+$ sign is to be used when both Killing fields are spacelike (and hence $\tilde{\mathbf{g}}$ is Lorentzian), while the $-$ sign is for the case of one timelike and one spacelike Killing field (in which case the quotient metric $\tilde{\mathbf{g}}$ is Riemannian). In the former case, let $\zeta = \rho + z, \eta = \rho - z$ be null coordinates on the quotient manifold, and in the latter, set $\zeta = \rho + iz, \eta = \bar{\zeta}$. The equations for $\tilde{\mathbf{g}}$ are given by

$$A_\eta - B_\zeta = 0 \tag{16}$$

$$A_\eta + B_\zeta + \frac{1}{\alpha}[A, B] - \frac{\alpha_\eta}{\alpha}A + \frac{\alpha_\zeta}{\alpha}B = 0,$$

for $\det \tilde{\mathbf{g}} = \alpha^2$, $A = -\alpha\tilde{\mathbf{g}}_\zeta\tilde{\mathbf{g}}^{-1}$, $B = \alpha\tilde{\mathbf{g}}_\eta\tilde{\mathbf{g}}^{-1}$. Compare with the chiral field model (10). The task is to find a linear system for U, V analogous to (11) for which the above PDE system appears as a compatibility condition; note that once $\tilde{\mathbf{g}}$ is determined, f can be found by quadrature (see, e.g. [19]).

Just as in (4) for the sine-Gordon equation and (11) for the chiral field, assume U and V have simple poles in λ. The key idea used in establishing integrability in this case is a generalization of ∂_j to operators D_j which include differentiation in the *spectral parameter* λ, namely

$$D_1\psi = U\psi := \frac{A}{\lambda - \alpha}\psi, \quad D_2\psi = V\psi := \frac{B}{\lambda + \alpha}\psi, \tag{17}$$

where $D_j = \partial_j - p_j\partial_\lambda$, and $p_j = p_j(\zeta, \eta, \lambda)$ is a rational function of λ. We shall interpret these generalized derivatives geometrically in Sect. 4. Note that here $\alpha = \alpha(\rho, z)$, thus in contrast to the chiral field model, the poles in (17) are not fixed, but *moving*.

With integrability established, the authors of [16, 17] implement a vesture or dressing method to generate new solutions to the Einstein vacuum equations.

Formally, given a solution ψ_0 of the Lax system having initial data $\mathbf{g}_0$, dressing refers to the procedure of finding a matrix χ for which $\psi = \chi\psi_0$ also solves the system. It is from this "dressed" generating matrix ψ that new solutions $\tilde{\mathbf{g}}$ (and hence $\mathbf{g}$) are constructed. Determining the dressing matrix χ reduces to solving a finite system of algebraic equations. In the next section we shall describe a generalization of this technique in the context of dressing harmonic maps.

3.2 Ernst Formulation of the Vacuum Equations

Prior to the works of [16, 17], another approach to the two-Killing field reduction of EVE had been investigated in [20, 21], giving what is now known as the Ernst Equation for a complex potential function $\varepsilon = \varepsilon(\rho, z)$. We briefly review this now.

Let the two commuting Killing fields for the metric $\mathbf{g}$ be denoted by $\mathbf{K} = \frac{\partial}{\partial t}$ and $\tilde{\mathbf{K}} = \frac{\partial}{\partial \varphi}$. Let us assume that $\mathbf{K}$ is timelike and $\tilde{\mathbf{K}}$ spacelike. Set

$$X := -\mathbf{g}(\mathbf{K}, \mathbf{K}) > 0,\ \tilde{X} = \mathbf{g}(\tilde{\mathbf{K}}, \tilde{\mathbf{K}}),\ W := \mathbf{g}(\mathbf{K}, \tilde{\mathbf{K}}),\ \Omega := \frac{W}{X}, \tag{18}$$

and define

$$\rho := \sqrt{W^2 + X\tilde{X}}, \qquad \mathbf{b} := i_{\mathbf{K}} * d\mathbf{K}. \tag{19}$$

Thus, ρ is the element of area of the cylindrical group orbits and $\mathbf{b}$ is the *twist form* of $\mathbf{K}$. The line element of $\mathbf{g}$ is then

$$\begin{aligned} ds^2_{\mathbf{g}} &= -Xdt^2 + 2Wdtd\varphi + \tilde{X}d\varphi^2 + \tilde{\mathbf{g}}_{ab}dy^a dy^b \\ &= -X(dt + \Omega d\varphi)^2 + \frac{1}{X}(\rho^2 d\varphi^2 + ds^2_{\mathbf{m}}), \end{aligned} \tag{20}$$

where $\mathbf{m}$ is a metric on the two-dimensional Riemannian quotient manifold conformal to $\tilde{\mathbf{g}}$.

Suppose now that the metric $\mathbf{g}$ is a solution of Einstein vacuum equations. It then follows that (i) the function ρ is harmonic $\Delta_{\mathbf{m}}\rho = 0$, and (ii) $\mathbf{b}$ is closed, i.e. $d\mathbf{b} = 0$. Thus (assuming the domain is simply connected) $\mathbf{b} = dY$. The function Y is called the *twist potential* for the Killing field $\mathbf{K}$ (see e.g. [19]). Choosing a conjugate harmonic function z for ρ, and using (ρ, z) as *isothermal* coordinates on the quotient brings the line element to the form

$$ds^2_{\mathbf{g}} = -X(dt + \Omega d\varphi)^2 + \frac{1}{X}(\rho^2 d\varphi^2 + e^{2u}(d\rho^2 + dz^2)). \tag{21}$$

This is known as the Lewis-Papapetrou form of the metric. There are now only three unknown metric coefficients left: X, Ω and u. Next, by virtue of $\mathbf{K}$ and $\tilde{\mathbf{K}}$ being

Killing fields (and the Einstein vacuum equations being satisfied), it is possible to show that the quantities X and Y satisfy a coupled system of elliptic PDEs, which can be combined into a single equation by introducing a complex-valued potential $\varepsilon = X + iY$, called the *Ernst potential*:

$$(\varepsilon + \bar{\varepsilon})\Delta\varepsilon + 2\nabla\varepsilon \cdot \nabla\varepsilon = 0. \tag{22}$$

Once the potential ε is found, the remaining metric coefficients Ω and u can be found using quadratures [19].

Equation (22) is derivable from a variational principle, namely, it is the Euler-Lagrange equation of the following action

$$\int_{\mathbb{R}^2} \frac{1}{(\Re\varepsilon)^2}|\nabla\varepsilon|^2 d\mathbf{x} = \int\int \frac{1}{4X^2}\left(|\nabla X|^2 + |\nabla Y|^2\right)\rho\, d\rho dz \tag{23}$$

We will see below that this is also the action for a harmonic map into the hyperbolic plane.

3.3 *Harmonic Maps and Gravitation*

Recall that a harmonic map $f : (\mathcal{M}, \mathbf{g}) \to (\mathcal{N}, \mathbf{h})$ is a critical point of the Dirichlet Energy

$$\mathcal{E}[f, \mathcal{D}] = \int_{\mathcal{D}} \frac{1}{2}\mathrm{tr}_{\mathbf{g}} f^*\mathbf{h}. \tag{24}$$

This definition generalizes the notion of a geodesic to higher-dimensional domains: letting $(\mathcal{M}, \mathbf{g}) = (\mathbb{R}^1, id)$, the Euler-Lagrange equations for the harmonic map action reduce precisely to the equations of an energy-minimizing geodesic on $\mathcal{N}$. Harmonic maps also generalize harmonic functions to nonlinear targets: letting $(\mathcal{N}, \mathbf{h}) = (\mathbb{R}^1, id)$, the Euler-Lagrange equation for (24) corresponds to that of a harmonic function on the base manifold $\mathcal{M}$. With this notation in place, we can cast some of our prior examples from the previous sections in the context of harmonic maps.[7]

Example 3, revisited. To express the chiral field appearing in Example 3 as a *wave map*, the hyperbolic analog of a harmonic map, let $g = f(\eta, \zeta)$ with $f : \mathbb{R}^{1,1} \to G$ a mapping from the $1+1$-dimensional Minkowski space into the group G, and (η, ζ) null coordinates on the domain. The Maurer-Cartan form of G is $w = -dgg^{-1}$, a Lie-algebra-valued 1-form. Let W denote its pull-back under f. Thus $W = \theta^a X_a$ where $\{X_a\}$ is a basis for the Lie algebra $\mathfrak{g}$ of the group G, and θ^a are 1-forms on

[7] The sine-Gordon equation can also be put into this framework, see [22].

the domain: $\theta^a = \theta^a_\mu dx^\mu$. On the other hand, if G is semi-simple, the Killing-Cartan quadratic form $\kappa_{ab} := \frac{1}{2}\mathrm{tr}\,(X_a X_b)$ is non-degenerate, and it thus endows G with a pseudo-Riemannian[8] metric $\mathbf{h} = \frac{1}{2}\mathrm{tr}\,(w^2)$. With $\mathbf{g}$ denoting the Minkowski metric, we thus have

$$\frac{1}{2}\mathrm{tr}_{\,\mathbf{g}} f^*\mathbf{h} = \frac{1}{2}\mathbf{g}^{\mu\nu}\kappa_{ab}\theta^a_\mu\theta^b_\nu = \frac{1}{2}\mathbf{g}^{\mu\nu}\mathrm{tr}\,(X_a X_b)\theta^a_\mu\theta^b_\nu = \frac{1}{2}\mathrm{tr}\,(W^2), \tag{25}$$

which agrees with (9). More precisely, $W = (-g_\xi g^{-1}, -g_\eta g^{-1})$, so that $\mathrm{tr}\,(W^2) = \mathrm{tr}\,(g_\xi g_\eta^{-1})$, the chiral field Lagrangian, as before.

We now move on to the reduction of vacuum Einstein equations under the assumption of existence of two commuting Killing fields.

Example 4. Let $(\mathcal{M}, \mathbf{g})$ denote a solution of Einstein's vacuum equations possessing two commuting Killing fields. We may then express the line element of $\mathbf{g}$ in the Lewis-Papapetrou form 21. Let $(X, Y) \in \mathbb{R}^+ \times \mathbb{R}$ be coordinates on the real hyperbolic plane $\mathcal{N} = \mathbb{H}_\mathbb{R}$ with its standard metric $\mathbf{h}$,

$$ds^2_{\mathbf{h}} = \frac{dX^2 + dY^2}{X^2}. \tag{26}$$

Suppose $f : (\mathcal{M}, \mathbf{g}) \to (\mathcal{N}, \mathbf{h})$ is a harmonic map that is invariant under the action of the two given Killing fields, so that f is well-defined on the quotient manifold, $f = (X(\rho, z), Y(\rho, z))$. The harmonic map Lagrangian density is therefore

$$\begin{aligned}\mathrm{tr}_{\,\mathbf{g}} f^*\mathbf{h}\sqrt{-\det\mathbf{g}} &= \frac{1}{4X^2}\left\{g^{\rho\rho}[(\partial_\rho X)^2 + (\partial_\rho Y)^2] + g^{zz}[(\partial_z X)^2 + (\partial_z Y)^2]\right\}\frac{\rho e^{2u}}{X} \\ &= \frac{1}{4X^2}\left(|\nabla X|^2 + |\nabla Y|^2\right)\rho \end{aligned}\tag{27}$$

which is the same as (23), the Lagrangian obtained for the Ernst equation. Indeed, since the action is independent of all other features of the spacetime metric, we can pretend that the domain is simply $\mathbb{R}^3$ in cylindrical coordinates (ρ, z, φ), and that we have an axisymmetric (i.e. φ-independent) harmonic map from $\mathbb{R}^3$ into the hyperbolic plane.

Finally, observing that the real hyperbolic plane may be realized as a symmetric space $\mathbb{H}_\mathbb{R} \cong SL(2,\mathbb{R})/SO(2)$, we may use the Cartan embedding of a symmetric space G/K into its mother Lie group G, to represent (X, Y) as a mapping into $SL(2,\mathbb{R})$, namely

$$(X, Y) \to \frac{1}{X}\begin{pmatrix} X^2 + Y^2 & Y \\ Y & 1 \end{pmatrix}. \tag{28}$$

[8]The metric is Riemannian if G is a compact group.

The fact that the target of the map is now embedded in a linear space[9] is the key to the success of the vesture method, described in the next section.

Example 5. The Einstein-Maxwell system, equations governing the interaction of the spacetime metric $\mathbf{g}$ with an electromagnetic field $\mathbf{F}$, given by

$$\mathbf{R}_{\mu\nu} - \frac{1}{2}\mathbf{g}_{\mu\nu} R = \kappa \mathbf{T}_{\mu\nu}, \quad \mathbf{T}_{\mu\nu} := \mathbf{F}_{\mu}^{\lambda} * \mathbf{F}_{\nu\lambda} - \mathbf{g}_{\mu\nu}\mathbf{F}_{\alpha\beta}\mathbf{F}^{\alpha\beta}, \quad d\mathbf{F} = 0, \quad d * \mathbf{F} = 0, \tag{29}$$

also represent the equations of an axially symmetric harmonic map, assuming existence of two commuting Killing fields which also leave the field $\mathbf{F}$ invariant. In this case, the mapping is given by $f : \mathcal{M} \to \mathcal{N}$, where $\mathcal{M}$ is $\mathbb{R}^2_+ \times \mathbb{S}^1$ as in the previous example, and $\mathcal{N}$ is the *complex* hyperbolic plane realized as $\mathbb{H}_{\mathbb{C}} = G/K$ with $G = SU(2,1)$, and $K = S(U(2) \times U(1))$ [21,23,24], and the corresponding Ernst formulation now involves a *pair* of complex potentials (ε, Φ) [20,21,25].

Appearances of harmonic maps in physics have been extensively documented, as has the integrability of harmonic maps into particular targets; see [26–28], for instance. The primary observation we now make is the following: the chiral field, Einstein vacuum and Einstein-Maxwell equations (under a two-Killing-field reduction) are integrable and admit a dressing mechanism simply because *harmonic maps into symmetric spaces are integrable and admit a generalized dressing mechanism.*

4 Integrability and Vesture of Axially Symmetric Harmonic Maps

Motivated by previous work, we prove a pair of theorems in [29] which unify the integrability and solution-generating mechanism (vesture or dressing technique) of special cases considered in the literature for the Einstein vacuum, Einstein-Maxwell and Chiral Field models; the results further establish integrability and vesture for a broader class of geometric field theories. We shall outline the main points surrounding their proof, after which the second half of this chapter will focus on control of the solution-generating mechanism.

---

[9]Note that by virtue of the Cartan embedding, a symmetric space can be embedded into a Lie group, for which one may choose a matrix representation, thus making addition and scalar multiplication of elements possible.

4.1 Integrability of Axially Symmetric Harmonic Maps

Theorem 1 (Integrability). *Let G be a real semisimple Lie group and K a maximal compact subgroup. Then any axially symmetric harmonic map from $\mathbb{R}^3$ into the Riemannian symmetric space G/K satisfies an integrable system of equations.*

The proof of this statement involves two main steps. Let $(\mathcal{M}, \mathbf{g}) = (\mathbb{R}^3, \mathbf{g})$ where the metric is given in cylindrical coordinates $(\rho, z, \varphi) \in \mathbb{R}^2_+ \times \mathbb{S}^1$. Suppose $f : \mathcal{M} \to G/K$ is an axially symmetric harmonic map and let q be a parametrization of the symmetric space via the Cartan embedding $G/K \to G$, so that $q = f(\rho, z) \in G$.

Step a. Rewrite the equations of a harmonic map as a Hodge System. Using the pull-back of the Maurer-Cartan form $W = dqq^{-1} \in \bigwedge^1(\mathcal{M})$, the harmonic map equations are

$$dW + W \wedge W = 0, \qquad d(\rho * W) = 0, \tag{30}$$

where the domain is the right-half plane in $\mathbb{R}^2$ with coordinates $\mathbf{x} = (\rho, z)$.

Step b. Exhibit a Lax Pair whose compatibility condition is the Hodge System. In our case, we generalize the linear operators D_j appearing in (17) by defining

$$D\Psi = -\Omega\Psi, \qquad \Psi|_{\lambda=0} = q, \tag{31}$$

$$D_\mu := \partial_\mu - \omega_\mu \frac{\partial}{\partial\lambda}, \quad \omega_\mu := \frac{1}{\partial\varpi/\partial\lambda}\partial_\mu\varpi, \quad \Omega_\mu := aW_\mu + b\rho(*W)_\mu, \quad \mu = 1, 2,$$

where $\lambda \in \bar{\mathbb{C}}$, $\mathbf{x} = (\rho, z)$, and $\varpi(\lambda, \mathbf{x})$, $a(\lambda, \mathbf{x})$, $b(\lambda, \mathbf{x})$ are three specific $\bar{\mathbb{C}}$-valued functions on $\bar{\mathbb{C}} \times \mathbb{R}^2_+$ defined to be

$$\varpi(\lambda, \mathbf{x}) = \frac{\rho^2}{2\lambda} + z - \frac{\lambda}{2}, \quad a(\lambda, \mathbf{x}) = \frac{\rho^2}{\lambda^2 + \rho^2}, \quad b(\lambda, \mathbf{x}) = \frac{\lambda}{\lambda^2 + \rho^2}. \tag{32}$$

The unknown is $\Psi : \mathbb{C} \times \mathbb{R}^2_+ \to \mathbb{C}^{n\times n}$, solving the Lax System (31). Here n is the dimension of a regular representation of the Lie group $G \subset GL(n, \mathbb{R})$. Note that this treatment subsumes approaches appearing in [16, 17, 30, 31].

Remark 2. We remark that just as in Examples 2, 3, the operator D defines a connection on the domain of Ψ, a Riemann surface bundle $\mathcal{B} - \bigcup_{\mathbf{x}\in\mathbb{R}^2_+} \mathcal{R}_\mathbf{x}$, where $\mathcal{R}_\mathbf{x}$ is the two-sheeted Riemann surface defined by

$$\mathcal{R}_\mathbf{x} = \{(\lambda, \varpi) \in \mathbb{C}^2 \mid \lambda^2 - 2\lambda(z - \varpi) - \rho^2 = 0\}. \tag{33}$$

The mapping $T_\mathbf{x} : \bar{\mathbb{C}} \to \bar{\mathbb{C}}$, $T_\mathbf{x}(\lambda) = -\rho^2/\lambda$ is a deck transformation on the universal cover of $\mathcal{R}_\mathbf{x}$ and $T_\mathbf{x}^2 = id$ for all $\mathbf{x} \in \mathbb{R}^2_+$. We prove in [29] that the compatibility condition of the D_μ is equivalent to a zero curvature condition $d\Omega + \Omega \wedge \Omega = 0$, for the 1-form Ω on the domain (cf. [32], p. 54).

4.2 Vesture of Axially Symmetric Harmonic Maps

Theorem 2 (Vesture). *If the Lie group G is such that the two involutions τ and σ defining the symmetric space G/K can be given by conjugation with the same element, then the vesture method can be used to construct new harmonic maps starting from any given harmonic map.*

As stated earlier, vesture refers to the procedure of finding a dressing matrix $\chi(\mathbf{x}, \lambda)$ such that if Ψ_0 is a solution of (31) and

$$\Psi = \chi \Psi_0, \tag{34}$$

then Ψ also solves the Lax system (31). Using the notation set up in Theorem 1, we prove this result in two main steps as well.

Step c. Construct a dressing matrix χ possessing the appropriate symmetries. An invertible mapping $\chi : \overline{\mathbb{C}} \times \mathbb{R}^2_+ \to H$ is a *dressing matrix* for q_0, an axially symmetric harmonic map of $\mathbb{R}^3$ into G/K, if it satisfies

$$\begin{aligned} \chi(\infty) &= I \\ \tau(\chi(\overline{\lambda})) &= \chi(\lambda) \\ \chi(\lambda) &= \chi(0) q_0 \sigma(\chi(T_{\mathbf{x}}(\lambda)))\sigma(q_0). \end{aligned} \tag{35}$$

Here, τ, σ are commuting involutions characterizing the symmetric space G/K and H is a complex group containing G [33]; the second and third equations in (35) are referred to as the G-reality and involutive symmetry conditions, movitated by the study of wave maps into compact symmetric spaces in [34, 35]. The symmetry conditions in (35) ensure that the resulting axially symmetric harmonic map $q(\mathbf{x}) = \chi(0, \mathbf{x}) q_0(\mathbf{x})$, found by setting $\lambda = 0$ in the generating matrix Ψ, takes its values in the appropriate target space, G/K. We show that there always exists an element of the equivalence class of Ψ which satisfies such imposed symmetries [29].

Step d. Reduce the problem of finding the dressing matrix to an algebraic problem. Motivated by simplifications appearing in [16, 17, 31], we assume χ is a rational function of λ and unknown matrices $R_k = R_k(\mathbf{x})$ of the form

$$\chi(\mathbf{x}, \lambda) = I + \sum_{k=1}^{2N} \frac{R_k(\mathbf{x})}{\lambda - \lambda_k(\mathbf{x})}. \tag{36}$$

The moving poles, $\lambda_k(\mathbf{x})$, are specified in the next section. Under further assumptions on the involutions τ, σ we prove solvability of the linear system in R_k.

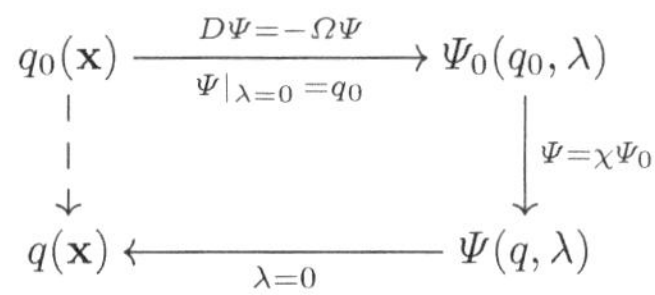

Fig. 2 ISM for harmonic maps

We summarize Theorems 1 and 2 in Fig. 2. Steps $a.$ and $b.$ comprise the integrability component (top arrow) and steps $c.$ and $d.$ comprise the vesture component (rightmost and bottom arrows) of the results. Compare with Fig. 1. In our case, the dressing procedure results in axisymmetric harmonic maps q possessing ring singularities [29].

The linear system resulting from this reduction unfortunately still contains too many unknown (and seemingly free) parameters. Thus, the question posed at the beginning of this chapter is restated as: Can we control the output of this mechanism to generate desired harmonic maps from a given initial seed map q_0, and if so, how?

4.3 *The Dressing Algorithm*

In order to control the number of parameters appearing in the dressing mechanism, we will need to outline the solution-generating algorithm afforded by the proof of Theorem 2. We do so now.

Let $\{\varpi_k\}_{k=1}^{N}$ be N distinct non-real complex numbers in the upper half plane. For every $\mathbf{x} = (\rho, z) \in \mathbb{R}^2_+$, each ϖ_k corresponds to a pair of points on the Riemann surface $\mathcal{R}_x$ defined in (33). Number these $2N$ prescribed poles $\lambda_k(\mathbf{x})$ in such a way that they are related by the deck transformation on $\mathcal{R}_\mathbf{x}$:

$$\lambda_{N+k} = T_\mathbf{x}(\lambda_k) = \frac{-\rho^2}{\lambda_k}, \qquad 1 \le k \le N. \tag{37}$$

Suppose that the involutions τ, σ characterizing the symmetric space target G/K can be given by conjugation with respect to the same element Γ. In conjunction with the G-reality and involutive symmetry conditions in (35), this assumption allows the algebraic reduction to be carried out efficiently.[10]

The overdetermined system in R_k is further simplified using a rank-one assumption. Define $M_k := R_k(\mathbf{x})\Psi_{0k}(\mathbf{x})$, where $\Psi_{0k}(\mathbf{x}) = \lim_{\lambda\to\lambda_k} \Psi_0(\lambda, \mathbf{x})$ and assume there exist non-zero vector functions $\mathbf{u}_k, \mathbf{v}_k$ for $k = 1, \dots, 2N$ such that

$$M_k = \mathbf{u}_k \mathbf{v}_k^*, \tag{38}$$

[10] In particular the pseudo-unitary groups $SU(p, q)$ satisfy this assumption. Moreover, groups possessing Dynkin diagrams with no symmetries may also qualify. Modification of this procedure to address other families of involutions will be taken up in a future work.

and furthermore that the vectors $\mathbf{v}_k$ satisfy

$$\mathbf{v}_k = \Gamma \mathbf{v}_{N+k} \qquad k = 1, \ldots, N. \tag{39}$$

This rank-one ansatz is consistent with the prior conditions imposed on the R_k system by symmetry, and the relation on the vectors $\mathbf{v}_k$ imposes a relation on the vectors $\mathbf{u}_k$ which does not overdetermine the problem.

Although the R_k system is equivalent to a nonlinear system for the M_k, it is shown in [29] that $\mathbf{v}_k$ can be taken to be constant vectors, and that the rank one assumption reduces the M_k system to a *linear* system for the unknown vector functions $\mathbf{u}_k$, given by

$$\sum_j a_{kj}\mathbf{u}_j^* = \mathbf{b}_k^*, \qquad a_{kj} := \frac{1}{\lambda_k - \bar{\lambda}_j}\mathbf{v}_k^* S_{kj}\mathbf{v}_j, \qquad \mathbf{b}_k^* := -\mathbf{v}_k^* \Psi_{0k}^{-1}\Gamma, \tag{40}$$

where we have set

$$S_{kj} := \Psi_{0k}^{-1}\Gamma(\Psi_{0j}^*)^{-1} = \Gamma\Psi_{0k}^*\Gamma\Psi_{0j}\Gamma. \tag{41}$$

Thus, a complete set of unknowns for the problem are the vectors $\{\mathbf{u}_k\}_{k=1}^{2N}$, given an arbitrary set of constant vectors $\{\mathbf{v}_k\}_{k=1}^{N}$ in $\mathbb{C}^n$. Equation (40) can be written more compactly in matrix form as

$$AU^* = B^*. \tag{42}$$

Here $U = U(\mathbf{x})$ is the $n \times 2N$ matrix whose columns are the vector functions $\mathbf{u}_k$, $A = A(\mathbf{x}) = (a_{kj})$ is a $2N \times 2N$ matrix function, and $B = B(\mathbf{x})$ is the $n \times 2N$ matrix whose columns are the vector functions $\mathbf{b}_k$ defined above.

If the matrix A can be shown to be invertible (at least in a neighborhood in $\mathbb{R}_+^2$), then the above system has a unique solution $U^* = A^{-1}B^*$ in that neighborhood. From there, one can then calculate the matrices R_k and the dressing matrix χ, and setting $\lambda = 0$, the new solution q is found to be

$$q(\mathbf{x}) = q_0(\mathbf{x}) - \sum_{k=1}^{2N} \frac{1}{\lambda_k(\mathbf{x})}\mathbf{u}_k(\mathbf{x})\mathbf{v}_k^*\Psi_{0k}^{-1}(\mathbf{x})q_0(\mathbf{x}). \tag{43}$$

In general, A is *not* invertible everywhere in the domain $\mathbb{R}_+^2$, and indeed the zero set of $\det A$ has a geometric significance for the dressed harmonic map $q(\mathbf{x})$. However, we can prove that under appropriate conditions on the arbitrary vectors $\mathbf{v}_k$, the matrix $A(\mathbf{x})$ is, in general, invertible *for large* $|\mathbf{x}|$ (i.e., in a neighborhood of infinity). Once a $q(\mathbf{x})$ has been found, its domain of definition may be maximally extended by analytic continuation.

Remark 3. Observe that (40) and hence (43) satisfies scaling invariance: if we rescale the constant vectors $\mathbf{v}_k$ by (nonzero) complex constants $t_k \in \mathbb{C}$, $k = 1, \ldots N$, then the expressions for a_{kj} also rescale by a factor of $t_k^* t_j$. This suggests that by rescaling each $\mathbf{u}_j$ by $\frac{1}{t_j^*}$, one preserves the equality in (43). Thus, we may view $\mathbf{u}_j$ not as vectors in $\mathbb{C}^{2N}$, rather as elements of $(\mathbb{CP})^{2N}$. This also means that the actual number of free real parameters in our dressing procedure is nN, and not $2nN$ as it initially appeared to be (from N arbitrary constant vectors in $\mathbb{C}^n$). We shall invoke the scale invariance of the system explicitly when it is convenient to do so in our subsequent calculations.

5 N-Solitonic Solutions of Einstein's Equations

The solution-generating algorithm outlined in the previous section is realized concretely for the case of the pseudo-unitary groups, $G = SU(p, q)$ in [29]. Define involutions σ, τ on $H = SL(p+q, \mathbb{C})$ by

$$\tau(g) = \Gamma g^{-*} \Gamma, \quad \sigma(g) = \Gamma g \Gamma, \quad \Gamma = \begin{bmatrix} I_{p\times p} & 0 \\ 0 & -I_{q\times q} \end{bmatrix}. \tag{44}$$

Then $G = \{g \in SL(p+q, \mathbb{C}) | g^* \Gamma g = \Gamma\}$, $K = \{g \in G \mid \sigma(g) = g\}$, and using the Cartan embedding, the symmetric space G/K is the complex Grassmann manifold

$$\mathcal{G}_{p,q} := G/K = SU(p,q)/S(U(p) \times U(q)) = \{q \in G \mid q\Gamma q\Gamma = I\}. \tag{45}$$

In the special case where $p = q = 1$, one obtains $G/K = \mathbb{H}_{\mathbb{R}}$, the target in the Ernst reduction of Einstein vacuum equations, appearing in Example 4, and for the case $p = 1, q = 2$, one obtains $G/K = \mathbb{H}_{\mathbb{C}}$, the target in the Ernst reduction of Einstein Maxwell equations, appearing in Example 5.

5.1 *Kerr Spacetimes from 1-Soliton Dressed Minkowski Metrics*

We now specialize to the case $p = q = 1$ to demonstrate the dressing algorithm explicitly, showing that dressing Minkowski spacetime with one soliton results in a family of (naked) Kerr spacetimes.

1. (Choose initial seed) Recall from Example 4, in the Ernst formulation of the vacuum Einstein equations, $\varepsilon = X + iY$, where as a harmonic map,

$q = \frac{1}{X}\begin{pmatrix} X^2+Y^2 & Y \\ Y & 1 \end{pmatrix}$. Thus, the Minkowski spacetime metric can be expressed as the harmonic map into G/K given by

$$q_0(\mathbf{x}) = I_{2\times 2} : \mathbb{R}^2_+ \longrightarrow SU(1,1)/S(U(1)\times U(1)), \tag{46}$$

where $\mathbf{x} = (\rho, z)$. We solve the Lax system (31) by inspection and find $\Psi_0 = I_{2\times 2}$.

2. (Dress q_0) Fix $\varpi = is$, $s > 0$ and let λ_1, λ_2 be the two roots of $p(\mathbf{x},\lambda) := \lambda^2 - 2(z - is)\lambda - \rho^2$ so that

$$\chi(\mathbf{x},\lambda) = I + \frac{R_1(\mathbf{x})}{\lambda - \overline{\lambda_1}} + \frac{R_2(\mathbf{x})}{\lambda - \overline{\lambda_2}}. \tag{47}$$

3. (Linear Algebra) Set $\mathbf{v}_1 = [\alpha\ \ \delta]^t$ for $\alpha, \delta \in \mathbb{C}$, so that by (39), $\mathbf{v}_2 = \Gamma \mathbf{v}_1 = [\alpha\ \ -\delta]^t$. Rewrite R_k, as in (40), in terms of α, δ and vectors $\mathbf{u}_k$:

$$\begin{bmatrix} \frac{1}{\lambda_1 - \overline{\lambda_1}}(|\alpha|^2 - |\delta|^2) & \frac{1}{\lambda_1 - \overline{\lambda_2}}(|\alpha|^2 + |\delta|^2) \\ \frac{1}{\lambda_2 - \overline{\lambda_1}}(|\alpha|^2 + |\delta|^2) & \frac{1}{\lambda_2 - \overline{\lambda_2}}(|\alpha|^2 - |\delta|^2) \end{bmatrix} \begin{bmatrix} \mathbf{u}_1^* \\ \mathbf{u}_2^* \end{bmatrix} = \begin{bmatrix} -\overline{\alpha} & \overline{\delta} \\ -\overline{\alpha} & -\overline{\delta} \end{bmatrix} =: \begin{bmatrix} -\mathbf{v}_1^* \\ -\mathbf{v}_2^* \end{bmatrix}. \tag{48}$$

Solving for $\mathbf{u}_1, \mathbf{u}_2$ above, the new harmonic map q in (43), expressed in terms of the complex constants α, δ and $\lambda_j = \lambda_j(\mathbf{x})$, is given by

$$q(\mathbf{x}) = \chi(\mathbf{x}, 0) q_0(\mathbf{x}) = I_{2\times 2} - \frac{1}{\lambda_1}\mathbf{u}_1(\mathbf{x})\mathbf{v}_1^* - \frac{1}{\lambda_2}\mathbf{u}_2(\mathbf{x})\mathbf{v}_2^*. \tag{49}$$

4. (Change coordinates) We now change from Weyl coordinates $\mathbf{x} = (\rho, z)$ to Boyer-Lindquist coordinates (r, θ), defined as

$$\rho = \sqrt{(r-m)^2 + s^2}\sin\theta, \qquad z = (r-m)\cos\theta, \tag{50}$$

for real parameters m, s. Then the harmonic map is re-expressed as

$$q = \begin{pmatrix} 1 + \frac{8|\alpha|^2|\delta|^2 s^2}{F} & \frac{1}{F}[-4s\alpha\overline{\delta}(i\mathcal{A}(r-m) - \mathcal{B}s\cos\theta)] \\ \frac{1}{F}[4s\overline{\alpha}\delta(i\mathcal{A}(r-m) + \mathcal{B}s\cos\theta)] & 1 + \frac{8|\alpha|^2|\delta|^2 s^2}{F} \end{pmatrix} I_{2\times 2},$$

$$\mathcal{A} = |\alpha|^2 - |\delta|^2, \quad \mathcal{B} = |\alpha|^2 + |\delta|^2, \quad F = \mathcal{A}^2((r-m)^2 + s^2) - \mathcal{B}^2 s^2 \sin^2\theta. \tag{51}$$

At this point, we would like to systematically deduce whether particular choices of constants α, δ lead to recognizable solutions, such as Kerr spacetimes. We use this formulation of the harmonic map to develop our control theory approach.

5.2 Control Theory Meets GR: An Asymptotic Calculation

In the solution-generating algorithm, it is necessary to invert the matrix A to solve the linear system $AU^* = B^*$ appearing in (40) and (48) for the unknown vectors $\mathbf{u}_k$. However, for each soliton added into the system, that is, for each ϖ, the matrices U, A, and B grow in considerably size (order N^2). The plan will be to replace (r, θ) with coordinates at infinity (R, Θ) and write down an asymptotic expansion for A in powers of $\frac{1}{R}$, for which truncations of the matrix will be more readily inverted. This is made possible by that fact that since the roots λ_1, λ_2 can be expressed in terms of the elliptical coordinates

$$\lambda_1 = (r + m + is)(\cos\theta - 1), \quad \lambda_2 = (r + m - is)(\cos\theta + 1), \tag{52}$$

we have $|\lambda_j| \to \infty$ as $|\mathbf{x}| \to \infty$ (since $r_j \to \infty$ as well) for all j.

First observe that in the coordinates (R, Θ), the matrix A can be expressed as $A = D + \frac{1}{R}N$, where D is a diagonal matrix and N is a nilpotent skew-hermitian matrix, $N^* = -N$. Consequently,

$$\begin{aligned} A &= D + \frac{1}{R}N = D(I + \frac{1}{R}D^{-1}N) \\ \Rightarrow A^{-1} &= (I + \frac{1}{R}D^{-1}N)^{-1}D^{-1} = D^{-1} - \frac{1}{R}D^{-1}ND^{-1} \\ &+ \frac{1}{R^2}D^{-1}ND^{-1}ND^{-1} - \cdots \end{aligned} \tag{53}$$

This, in turn, allows us to deduce constraint equations on the constants α, δ by explicitly inverting A in (39) using truncations of this expansion. In particular

$$U = -BD^{-1} + \frac{1}{R}BD^{-1}ND^{-1} - \frac{1}{R^2}BD^{-1}(ND^{-1})^2 + \frac{1}{R^3}BD^{-1}(ND^{-1})^3 - \cdots \tag{54}$$

can be calculated to various desired orders of R.

In our example, we shall truncate the expansion of U at the $\frac{1}{R^2}$ term and, recalling the definitions of $\mathcal{A}, \mathcal{B}$ from (51), we use

$$B = \begin{bmatrix} -\alpha & -\alpha \\ \delta & -\delta \end{bmatrix}, \quad D = \frac{\mathcal{A}}{2is}\begin{bmatrix} \frac{1}{\cos\theta - 1} & 0 \\ 0 & \frac{-1}{\cos\theta + 1} \end{bmatrix}, \quad \frac{1}{R}N = \frac{\mathcal{B}}{2}\begin{bmatrix} 0 & \frac{-1}{R-m+is} \\ \frac{1}{R-m-is} & 0 \end{bmatrix}. \tag{55}$$

Calculating U explicitly to obtain the column vectors $\mathbf{u}_j$ and combining them with the constant row vectors $\mathbf{v}_j$, we compute each component $\frac{1}{\lambda_j}\mathbf{u}_j\mathbf{v}_j^*$ of the new (truncated) mapping $q^{(R)}$. Note that each $\mathbf{u}_j\mathbf{v}_j^*$ contains a prefactor of $\frac{1}{\lambda_j} \sim \frac{1}{R}$, resulting in the following components for $q^{(R)} \doteq I_{2\times 2} - \frac{1}{\lambda_1}\mathbf{u}_1\mathbf{v}_1^* - \frac{1}{\lambda_2}\mathbf{u}_2\mathbf{v}_2^*$ in (49), or equivalently (51):

$$
\begin{aligned}
q_{11}^{(R)} &= 1 + \frac{8s^2|\alpha|^2|\delta|^2}{\mathcal{A}^2[(R-m)^2+s^2]} - \frac{4\mathcal{B}^2 s^4 \sin^2\theta|\alpha|^2}{\mathcal{A}^3[(R-m)^2+s^2]^2} \\
q_{12}^{(R)} &= \frac{4s\alpha\overline{\delta}}{\mathcal{A}^3[(R-m)^2+s^2]}\{\mathcal{A}\mathcal{B}s\cos\theta[(R-m)^2+s^2] \\
&\qquad -i(R-m)\left(\mathcal{A}^2[(R-m)^2+s^2]+\mathcal{B}^2s^2\sin^2\theta\right)\} \\
q_{21}^{(R)} &= \frac{4s\overline{\alpha}\delta}{\mathcal{A}^3[(R-m)^2+s^2]}\{\mathcal{A}\mathcal{B}s\cos\theta[(R-m)^2+s^2] \\
&\qquad +i(R-m)\left(\mathcal{A}^2[(R-m)^2+s^2]+\mathcal{B}^2s^2\sin^2\theta\right)\} \\
q_{22}^{(R)} &= 1 + \frac{8s^2|\alpha|^2|\delta|^2}{\mathcal{A}^2[(R-m)^2+s^2]} + \frac{4\mathcal{B}^2 s^4 \sin^2\theta|\delta|^2}{\mathcal{A}^3[(R-m)^2+s^2]^2}.
\end{aligned}
\tag{56}
$$

We apply the Cayley Transform $q' = QqQ^*$, with $Q = \frac{1}{\sqrt{2}}\begin{bmatrix} 1 & i \\ i & 1 \end{bmatrix}$, to adjust from the target space $G/K = SU(1,1)/S(U(1)\times U(1))$ to the real hyperbolic plane $\mathbb{H}_{\mathbb{R}} = SL(2,\mathbb{R})/SO(2)$, in order to compare the resulting mapping with that of the Ernst formulation. Recall that in this setting, (26) and (28) dictate that $X = \frac{1}{q'_{22}}$, so we need only expand the q'_{22} entry in R in order to devise constraints on the constants α, δ. For the truncated mapping $q_{22}^{(R)}$,

$$
X^{(R)} \sim 1 + \frac{-4sn_1}{\mathcal{A}R} + \left(\frac{-4sn_1 m}{\mathcal{A}} + \frac{16sn_1^2 - 8s|\alpha|^2|\delta|^2 + 4s^2 n_2 \mathcal{B}\cos\Theta}{\mathcal{A}^2}\right)\frac{1}{R^2},
\tag{57}
$$

where we have set $\alpha\overline{\delta} = n_1 + in_2$. By the definitions of $\mathcal{A}, \mathcal{B}$ in (51), this implies $\mathcal{B}^2 - \mathcal{A}^2 = 4(n_1^2 + n_2^2)$.

The specification of the ADM mass as M, and the condition of vanishing dipole moment for the metric in the coordinate system (R, Θ) impose natural restrictions on the coefficients of the expansion in $\frac{1}{R}$ and $\frac{1}{R^2}$, namely

$$
\frac{-4sn_1}{\mathcal{A}} = -2M, \quad \frac{-4sn_1 m}{\mathcal{A}} + \frac{16sn_1^2 - 8s|\alpha|^2|\delta|^2 + 4s^2 n_2 \mathcal{B}\cos\Theta}{\mathcal{A}^2} = 0.
\tag{58}
$$

Subjecting the second equation in (58) to the first reduces the conditions to

$$
\frac{-4sn_1}{\mathcal{A}} = -2M, \quad \frac{4s^2 n_2(\mathcal{B}\cos\Theta - 2n_2)}{\mathcal{A}^2} = 2M^2 - 2Mm.
\tag{59}
$$

Here, a, m, s are free coordinate parameters, with $m, s > 0$. Now, since these equations must hold for all Θ, we have

$$\frac{4s^2 n_2 \mathcal{B}\cos\Theta}{\mathcal{A}^2} = 0, \quad \frac{8s^2 n_2^2}{\mathcal{A}^2} = 2M^2 - 2Mm, \tag{60}$$

whereby $n_2 = 0$ and thus $M = m$. Returning to the relation between $\mathcal{A}, \mathcal{B}$ and n_1, n_2, it is easy to verify that

$$\mathcal{B}^2 - \mathcal{A}^2 = 4n_1^2 = 4\left(\frac{2m\mathcal{A}}{4s}\right)^2 \quad \Rightarrow \quad \mathcal{B}^2 = \mathcal{A}^2\frac{m^2+s^2}{s^2}, \tag{61}$$

yielding a real scaling invariance in the choice of complex constants α and δ (see Remark 3). Making the choice $\mathcal{A} = s$ gives $\mathcal{B} = a := \sqrt{m^2+s^2}$, and we have automatically that $n_1 = \pm\frac{m}{2}$ (and $n_2 = 0$ is imposed already), recovering the (naked) Kerr family, as

$$X = \frac{r^2 - 2mr + a^2\cos^2\theta}{r^2 + a^2\cos^2\theta}, \quad Y = \frac{2ma\cos\theta}{r^2 + a^2\cos\theta}, \quad a = \sqrt{m^2+s^2}. \tag{62}$$

We note that the quantity F appearing in (51) is the determinant of the linear system (48), and that with the above choices for $\mathcal{A}$ and $\mathcal{B}$, it vanishes precisely when $r = 0$ and $\theta = \pi/2$, i.e. on the ring singularity of the Kerr metric. In inverting the coefficient matrix of (48) we have assumed $F \neq 0$, which is assured for r large enough, provided $\mathcal{A} \neq 0$ (i.e., the extremal Kerr is also excluded from our setup). On the other hand, the analyticity in (r, θ) of the resulting expressions for (X, Y) ensures that by taking the analytic continuation of the solution one can recover the maximal extension of the (naked) Kerr spacetime we have hereby obtained.

5.3 Einstein-Maxwell Equations

Since the Einstein-Maxwell equations also have a harmonic maps formulation, as in Example 5, the vesture method can equally well be used to construct new solutions, comparing with [31, 36].

For the $N = 1$ case, dressing the Minkowski seed $q_0(\mathbf{x}) = I_{3\times3}$ with one soliton, the roots of $\lambda_{1,2}$ of $p(\mathbf{x}, \lambda)$ are exactly as in the previous example. Thus, set $\mathbf{v}_1 = [\alpha, \beta, \gamma]^t$ for $\alpha, \beta, \gamma \in \mathbb{C}$, so that $\mathbf{v}_2 = \Gamma\mathbf{v}_1 = [\alpha, \beta, -\gamma]^t$, for $\Gamma = \mathrm{diag}(1, 1, -1)$. The resulting linear system in (40) becomes

$$\begin{bmatrix} \frac{\mathcal{A}}{\lambda_1 - \overline{\lambda_1}} & \frac{\mathcal{B}}{\lambda_1 - \overline{\lambda_2}} \\ \frac{\mathcal{B}}{\lambda_2 - \overline{\lambda_1}} & \frac{\mathcal{A}}{\lambda_2 - \overline{\lambda_2}} \end{bmatrix} \begin{bmatrix} \mathbf{u}_1^* \\ \mathbf{u}_2^* \end{bmatrix} = \begin{bmatrix} -\overline{\alpha} & -\overline{\beta} & \overline{\gamma} \\ -\overline{\alpha} & -\overline{\beta} & -\overline{\gamma} \end{bmatrix}, \tag{63}$$

where $\mathcal{A} := |\alpha|^2 + |\beta|^2 - |\gamma|^2$, and $\mathcal{B} := |\alpha|^2 + |\beta|^2 + |\gamma|^2$. Isolating each of the vectors $\mathbf{u}_1, \mathbf{u}_2$, the harmonic map is given by

$$q(\mathbf{x}) = I_{3\times 3} - \frac{1}{\lambda_1}\mathbf{u}_1\mathbf{v}_1^* - \frac{1}{\lambda_2}\mathbf{u}_2\mathbf{v}_2^* \tag{64}$$

$$= \begin{pmatrix} 1 + \frac{8|\alpha|^2|\gamma|^2s^2}{F} & \frac{1}{F}[8\alpha\bar{\beta}|\gamma|^2s^2] & \frac{1}{F}[-4s\alpha\bar{\gamma}(i\mathcal{A}(r-m) - \mathcal{B}s\cos\theta)] \\ q_{2,1} & 1 + \frac{8|\beta|^2|\gamma|^2s^2}{F} & \frac{1}{F}[-4s\beta\bar{\gamma}(i\mathcal{A}(r-m) - \mathcal{B}s\cos\theta)] \\ q_{3,1} & q_{3,2} & 1 + \frac{8(|\alpha|^2+|\beta|^2)|\gamma|^2s^2}{F} \end{pmatrix}$$

Here, $F := \mathcal{A}^2((r-m)^2 + s^2) - \mathcal{B}^2s^2\sin^2\theta$ and $q_{ji} = -\overline{q_{ij}}$.

Just as before, it is possible to transform to Boyer-Lindquist coordinates and consider an expansion of the matrix A in powers of $\frac{1}{R}$. We will spare the reader these details, since an identical application of the technique used in the Einstein vacuum case applies. The above six-parameter family (three complex-parameter family α, δ, γ) of harmonic maps into $\mathcal{G}_{2,1}$ contains as a special case, the three real-parameter family of Kerr-Newman metrics in Boyer-Lindquist coordinates (r, θ). To that end, we set $\mathcal{A} = s$, whereby $\mathcal{B} = -a := \sqrt{m^2 + s^2 - e^2}$, and obtain a system of equations for the constants $\alpha\bar{\gamma} = n_1 + in_2$ and $\beta\bar{\gamma} = n_3 + in_4$ in which one finds $n_2 = n_4 = 0$. Matching the notation used in the previous example, we choose $n_1 = \frac{m}{2}$ and $n_3 = -\frac{e}{2}$ to then obtain exactly the Ernst potentials for the Kerr-Newman spacetimes (see Eq. (21.26), p. 326 in [37]),

$$\Phi = \frac{e}{r - ia\cos\theta} \qquad \varepsilon = 1 - \frac{2m}{r - ia\cos\theta}, \tag{65}$$

for real parameters e, a, m.

6 Outlook, Conclusions and Future Directions

In this section we discuss concrete ways to interpret, generalize and extend the integrability results as well as the solution-generating mechanism described in this chapter to other physical theories of interest.

As we have seen, in each of the Einstein Vacuum and Einstein-Maxwell cases, the Kerr and Kerr-Newman families, respectively, can be recovered by dressing a Minkowski initial seed metric with one soliton. This program has been completed for the broader class of complex Grassmann manifolds $\mathcal{G}_{p,q}$. The groups $SU(p,q)$ appear as universal covers of the conformal groups $SO(p,q)$, and the $\mathcal{G}_{p,q}$-nonlinear sigma model described in [29] has found applications in other settings (e.g., [27, 38, 39]), and in particular in the study of higher-dimensional gravity.

Current work-in-progress includes calculating 2-soliton solutions to the Einstein Vacuum and Einstein Maxwell Equations. Such calculations are not straightforward, since the resulting solution and hence its singular structure is unknown. For instance,

Table 1 Nonlinear σ-models/harmonic maps $f : \mathcal{M} \longrightarrow G/K$

Gravitational theory	G/K	n-soliton dressing
Einstein vacuum $D = 3+1, k = 2$ ($m = 2$ SUGRA)	$\frac{SU(1,1)}{S(U(1)\times U(1))}$	Kerr family
Einstein-Maxwell $D = 3+1, k = 2$ ($m = 2$ SUGRA)	$\frac{SU(2,1)}{S(U(2)\times U(1))}$	Kerr-Newman
Einstein gravity $D = n+4, k = n$	$\frac{SL(n+2)}{SO(n+2)}$	$\frac{SU(p,q)}{S(U(p)\times U(q))}$, ($n = p+q$ [29])
$D = 3+1, k = 2$ ($m = 4$ SUGRA)	$\frac{SO(8,2)}{SO(8)\times SO(2)}$	⋆
$D = 5, k = 2$ or 3 ($m = 1$ SUGRA)	$\frac{G_{2(2)}}{SL(2,\mathbb{R})\times SL(2,\mathbb{R})}$	Myers-Perry, Cvetič-Youm ([43] partial)
$D = 3+1, k = 2$ (m = 8 SUGRA)	$E_{8(+8)}/SO^*(16)$ ([38] partial)	$E_{7(+7)}/SU(8)$ ⋆

in the 1-soliton cases, an asymptotically flat spacetime with one singular ring must presumably belong to the Kerr family of spacetimes, by virtue of uniqueness results.[11] On the other hand, in dressing Minkowski with two poles, the location of the second pole gives rise to additional free parameters, and it is no longer clear whether any additional singularities (other than a second ring singularity) have been introduced into the spacetime, such as for example a singular "strut" on the axis of symmetry between the centers of the two rings.

We observe that four-dimensional black holes arising from Kaluza-Klein Theories naturally set the stage for work in integrability of harmonic maps. In [38], several supergravity (SUGRA) models are formulated precisely as harmonic maps into symmetric spaces. Table 1, therefore comprises a collection of other potential gravitational theories which may be better understood by way of the integrability apparatus established here and in [29]; furthermore, calculation of new solutions using the dressing apparatus may now be within reach. Entries marked with ⋆ are currently under investigation and results will be reported in a future work.

Notation for the nonlinear σ-models $f : \mathcal{M} \to G/K$ refer to harmonic maps where D=dim $\mathcal{M}$, k denotes the number of Killing fields, and m denotes the ratio of the number of supersymmetries of the minimal model; supergravity (SUGRA) models closely related to the discussed harmonic maps are indicated in parentheses.[12]

[11]Note however, that all of the known uniqueness results for the Kerr metric are in the black hole regime, and cover only the outside of the event horizon (cf. [40] and references therein.) The same holds for "double-Kerr" non-existence results (e.g. [41, 42]).

[12]The first two columns of Table 1 are excerpted directly from [38], except the fifth row (see [43]).

Although integrability of the two classical equations, Einstein vacuum and Einstein-Maxwell equations, has been studied (e.g., [16, 17, 30, 31, 36]), explicit construction of solutions from this unified approach does not yet appear in the literature. In the case of minimal five-dimensional supergravity, it would be a significant milestone to recover the Myers-Perry or Cvetič-Youm solutions [43] systematically from this technique, identifying the seed metric as well as the pole structure imposed. More generally, explicating black-hole solutions in d-dimensional vacuum gravity for $d > 4$ has been of recent interest both in (minimal) supergravity and in string theory [43, 44]. In this context, stationary solutions possessing $d - 3$ rotational Killing fields have been extensively studied; imposing an additional timelike Killing field results in effectively two-dimensional theories, to which integrability techniques may apply. This and other applications will be pursued elsewhere.

Acknowledgements SB gratefully acknowledges support from M. Kiessling and the NSF through grant DMS-0807705, during which the first part of this project was conceived; this material is also based upon work supported by the NSF under Grant #0932078000, while SB was in residence at the Mathematical Science Research Institute in Berkeley, California, during the 2013 Autumn semester. STZ thanks the Institute for Advanced Study for their hospitality and the stimulating environment provided during Spring 2011 while the authors were working on this project.

References

1. P.D. Lax, Comm. Pure Appl. Math. **21**, 467 (1968)
2. P.G. Drazin, R.S. Johnson, *Solitons: An Introduction* (Cambridge University Press, Cambridge, 1989)
3. C.S. Gardner, J.M. Greene, M.D. Kruskal, R.M. Miura, Comm. Pure Appl. Math. **27**, 97 (1974)
4. M.J. Ablowitz, P.A. Clarkson, *Solitons, Nonlinear Evolution Equations and Inverse Scattering* (Cambridge University Press, Cambridge, 1991)
5. M.J. Ablowitz, D.J. Kaup, A.C. Newell, H. Segur, Stud. Appl. Math. **53**(4), 249 (1974)
6. M.J. Ablowitz, D.J. Kaup, A.C. Newell, H. Segur, Phys. Rev. Lett. **30**, 1262 (1973)
7. V.E. Zaharov, A.B. Šabat, Funktsional. Anal. i Prilozhen. **8**(3), 43 (1974)
8. V.E. Zaharov, A.B. Šabat, Funktsional. Anal. i Prilozhen. **13**(3), 13 (1979)
9. V.E. Zaharov, L.A. Tahtadžjan, L.D. Faddeev, Dokl. Akad. Nauk SSSR **219**, 1334 (1974)
10. V.E. Zakharov, A.V. Mikhaĭlov, Sov. Phys. JETP **47**, 1017 (1978)
11. K. Uhlenbeck, J. Differential Geom. **30**(1), 1 (1989)
12. F. Gürsey, Il Nuovo Cimento **16**(2), 230 (1960)
13. L.D. Faddeev, L.A. Takhtajan, *Hamiltonian Methods in the Theory of Solitons*, English edn. (Springer, Berlin, 2007)
14. K. Schwarzschild, Gen. Relat. Gravit. **35**(5), 951 (2003) [Translated from the original German article (Sitzungsber. Königl. Preussich. Akad. Wiss. Berlin Phys. Math. Kl. 1916, 189–196)]
15. R.P. Kerr, Phys. Rev. Lett. **11**, 237 (1963)
16. V.A. Belinskiĭ, V.E. Zakharov, Sov. Phys. JETP **48**(6), 985 (1979)
17. V.A. Belinskiĭ, V.E. Zakharov, Sov. Phys. JETP **77**(1), 3 (1979)
18. A. Papapetrou, Ann. Inst. H. Poincaré Phys. Théor. **IV**(2), 83 (1966)
19. G. Weinstein, Trans. Am. Math. Soc. **343**(2), 899 (1994)
20. F.J. Ernst, Phys. Rev. **167**, 1175 (1968)
21. F.J. Ernst, Phys. Rev. **168**, 1415 (1968)

22. J. Shatah, W. Strauss, Physica D **99**(2–3), 113 (1996)
23. B. Carter, Gen. Relat. Gravit. **41**, 2873 (2009)
24. P.O. Mazur, Acta Phys. Polon. B **14**(4), 219 (1983)
25. S. Chandrasekhar, Selected Papers: The Mathematical Theory of Black Holes and of Colliding Plane Waves, vol. 6 (University of Chicago Press, Chicago, 1991)
26. C.W. Misner, Phys. Rev. **D18**, 4510 (1978)
27. N. Sanchez, Phys. Rev. D **26**, 2589 (1982)
28. J.C. Wood, in *Harmonic Maps and Integrable Systems* (Vieweg, Braunschweig, 1994), pp. 29–55
29. S. Beheshti, S. Tahvildar-Zadeh, *Integrability and Vesture for Harmonic Maps into Symmetric Spaces*. Submitted, arXiv:1209.1383
30. G.A. Alekseev, Dokl. Akad. Nauk SSSR **256**(4), 827 (1981)
31. A. Eriş, M. Gürses, A. Karasu, J. Math. Phys. **25**(5), 1489 (1984)
32. M.A. Guest, *From Quantum Cohomology to Integrable Systems* (Oxford University Press, Oxford, 2008)
33. S. Helgason, *Differential Geometry, Lie Groups, and Symmetric Spaces* (American Mathematical Society, Providence, 2001)
34. C.L. Terng, K. Uhlenbeck, Comm. Anal. Geom. **12**(1–2), 345 (2004)
35. C.L. Terng, in *Fourth International Congress of Chinese Mathematicians* (American Mathematical Society, Providence, 2010), pp. 367–381
36. G.A. Alekseev, Pis'ma Zh. Eksp. Teor. Fiz. **32**(4), 301 (1980)
37. H. Stephani, D. Kramer, M. MacCallum, C. Hoenselaers, E. Herlt, *Exact Solutions of Einstein's Field Equations*, 2nd edn. (Cambridge University Press, Cambridge, 2003)
38. P. Breitenlohner, D. Maison, G. Gibbons, Comm. Math. Phys. **120**(2), 295 (1988)
39. A.A. Pomeransky, Phys. Rev. D **73**(4), 044004 (2006)
40. P.T. Chrusciel, J.L. Costa, M. Heusler, Living Rev. Relat. **15**(7) (2012)
41. G. Neugebauer, J. Hennig, Gen. Relat. Gravit. **41**(9), 2113 (2009)
42. P.T. Chruściel, M. Eckstein, L. Nguyen, S.J. Szybka, Classical Quant. Grav. **28**, 245017 (2011)
43. P. Figueras, E. Jamsin, J.V. Rocha, A. Virmani, Classical Quant. Grav. **27**, 135011 (2010)
44. R. Emparan, H. Reall, Living Rev. Relat. **11**(6), 1 (2008)

A Planar Skyrme-Like Model

Ramón J. Cova

Abstract After an introduction of solitons with emphasis in elementary particle theory, we present a non-linear $O(3)$ model in $(2+1)$ dimensions modified by the addition of both a potential-like and a Skyrme-like term. We study some basic scattering properties of the model via numerical simulation using a general field configuration. The skyrmion-scattering is found to be quasi-elastic, the skyrmions' energy density profiles remaining unscathed after collisions. In low-energy processes the skyrmions exhibit back-scattering, while at larger energies they scatter at right angles. These results confirm those obtained in previous investigations, in which a similar problem was studied for a different choice of the potential-like term.

Keywords CP^1 model • Scattering • Sigma model • Skyrme model • Skyrmion • Soliton • Topological charge

1 Introduction

Elementary particles are described by fields which obey relativistic wave equations. These fields have basic properties that follow from a postulated invariance of the wave equations under certain groups of transformations, the symmetry groups. Examples of these symmetries are the independence of the laws of physics from the origin of time and from the position and orientation of laboratories in ordinary space. Invariance of the wave equations under the groups of time and space translations and rotations leads to conservation of energy and momentum and angular momentum.

R.J. Cova (✉)
School of Mathematics and Statistics, Carleton University, Ottawa, Canada K2E 7E8

The University of Zulia, Department of Physics, Maracaibo, Venezuela
e-mail: rcova@math.carleton.ca

J. Cuevas-Maraver et al. (eds.), *The sine-Gordon Model and its Applications*, Nonlinear Systems and Complexity 10, DOI 10.1007/978-3-319-06722-3_10,

The most important invariance principle in quantum field theory (QFT) is the principle of relativity, which imposes that the equations of motion be invariant under the proper orthochronous Poincare group. Wave fields belong to representations of this group and their space-time geometrical nature is thus determined: the fields can only be spinors or tensors under the Poincare group.

In addition to their Poincare geometrical nature there is a rich variety of fields which may also be spinors or tensors with respect to certain symmetry groups related to internal quantum numbers like isospin, flavour, colour, etc. These internal degrees of freedom result from the invariance of the wave equations under gauge transformations which act on the (gauge) field regarded as an entity in an internal space, such as the group $SU(2)$ of phase operator transformations of isospinors. Whereas the Poincare invariance of the equations of field theory is universal, invariance under internal symmetry groups (gauge groups) is obeyed by specific models, e.g., quantum chromodynamics (QCD). Internal symmetries can also be approximate, as in flavourdynamics.

Gauge fields play a fundamental role in unification models where the nuclear interaction is analogous to quantum electrodynamics (QED) but with consideration to the specific characteristics of a non commutative gauge group. Such a prosperous idea was introduced by Yang and Mills (YM) in 1954 [1]. The utility of gauge fields lies chiefly in their ability to express underlying relations among forces that superficially appear to be distinct. An example of an unification model is the standard model (SM) whose gauge group is $SU(3)\times SU(2)\times U(1)$. Roughly, the SM is based on six leptons and six accompanying quarks in three separate generation levels of energy. It amalgamates strong, electro-weak and classical gravitational interactions. The forces responsible for these interactions are transmitted by another set of particles, the gauge bosons (a modern version of Yukawa's bosons).

Of utmost importance for the formulation of unification models is the Brout-Englert-Higgs mechanism [2–4] of mass generation for the weak bosons by spontaneous symmetry breaking. This phenomenon occurs when the basic states of a given system do not enjoy a symmetry of the associated Lagrangian. Experimental confirmation of the Higgs mechanism has been recently found (see [5] and references therein).

For a long time the only means for obtaining numerical quantities in field theory was through the perturbation method, which gives excellent results in QED but it is less useful in low-energy QCD due a to a rising coupling constant. However, the development of YM theories has given rise to a promising alternative area of research: classical non-linear equations that admit non-dissipative non-singular solutions of finite total energy; lump-like structures that propagate without diffusion. These entities, generically referred to as *solitons* owing to their 'solitary wave' and particle-like behaviour allow to model complicated phenomena which lie beyond the scope of linear description. The properties of solitons in the quantum version of these schemes are not analysed through a standard perturbative approach but via an expansion which is a generalisation of the familiar JWKB method in field theory. In this expansion the quantum solitons receive their leading contributions from their classical counterparts, around which the fluctuations are quantised [6].

The existence of solitons is quite special since in most equations dispersion effects lead to the break-up or collapse of the energy lumps. Solitons only emerge from equations where diffusion and non-linear effects exactly balance each other, permitting the travelling solitonic lumps to conserve their initial integrity as if the medium were linear and dispersionless. Furthermore, soliton solutions possess the linear-like feature whereby two solutions can be combined to produce a third one. Not unnaturally, the localised nature of soliton-bearing systems, resembling extended objects with finite energy such as might be the classical limit of hadrons, make them candidates to describe elementary particles. It is noteworthy that solitons appear in most theories with spontaneous symmetry breaking.

The theory of solitons brings together many branches of mathematics, e.g, topology, differential geometry, group theory and complex analysis. It has many applications in physics, including nuclear and particle theory, condensed matter, hydrodynamics, optics, etc, and in biophysics solitons arise in the description of DNA and protein dynamics [7]. Soliton theory illustrates the interaction between physics and mathematics, of how one is drawn from one subject to understand the other [8]. It brings new insight to research in technology affecting related concepts of noise figure degradation and information entropy increase. In communication theory, for example, the possibility of transmitting optical signals over virtually unlimited lengths using soliton-pulse propagation augurs well, and it certainly checks previous engineering knowledge. Much research has been done, and continues today, on system topologies, management and communication protocols for solitonic fiber-optic networks [9].

The first recorded observation of solitonic behaviour was made by an English shipbuilding engineer on the Edinburgh-Glasgow canal in 1834, although the matter was officially reported a decade later [10]. In 1895 Korteweg and de Vries [11] derived an equation for waves in shallow water, taking into account dispersion but ignoring the dissipation of energy. The KdV equation $u_t + 6uu_x + u_{xxx} = 0$ has the travelling wave solution $u = 2k^2\{\cosh[k(x - 4k^2t - x_0)]\}^{-2}$, with the amplitude $2k^2$ equal to half the speed of the wave.

For a long time solitons were treated as an unimportant piece of exotica which one encountered in two-dimensional problems of non-linear waves. One supposed that when two such waves collided they fell apart; there was no basis for considering soliton solutions to be sufficiently general. But things took a turn in 1965 when Zabusky and Kruskal [12], studying a periodic unidimensional anharmonic lattice whose continuum limit is the KdV equation, discovered via numerical computations that solitons were not destroyed under collisions. The theoretical basis for such properties was found 3 years later [13] when it was shown that equations of the KdV type have an infinite number of conserved quantities in a class of multi-soliton solutions, described by a Schrodinger equation. In 1972 it was brought to light yet another class of non-linear Schrodinger equations possessing similar properties [14].

Starring topic of the present book, the sine-Gordon model is another most important solitonic system; it may be regarded as the limit of a mechanical system of coupled pendula as the spacing of the rods tends to zero. The sine-Gordon system was introduced by the French engineer Edmond Bour in 1862 whilst studying surfaces of constant negative curvature. Rediscovered by Frenkel and Kontorova in 1939 in the context of crystal dislocations, the sine-Gordon theory began to attract greater interest since the 1970s due to its soliton solutions. The classical scheme consists of a single scalar field $\phi(x,t)$ in $(1+1)$-dimensions (1 space, 1 time) whose Lagrangian density and field equation have the form (α is a constant):

$$\mathcal{L} = \alpha[(\partial_\mu\phi)(\partial^\mu\phi) + (\cos\phi - 1)] \;\; \rightarrow \;\; \phi_{tt} - \phi_{xx} + \sin\phi = 0. \tag{1}$$

The applications of the sine-Gordon abound [15, 16]: nonlinear field theory, solid state physics, nonlinear optics, fluid dynamics, DNA dynamics and elementary particles (the sine-Gordon model is a nonlinear version of the well-known Klein–Gordon linear model). Considerable research has been dedicated lately to the supersymmetric extension of the sine-Gordon model [17] which has been shown to be equivalent to the supersymmetric CP^1 sigma model [18, 19]. The standard, classical CP^1 or $O(3)$ model is the substance of the present chapter.

The sine-Gordon equation and KdV-like equations may be solved by powerful methods like (a) the inverse scattering transform [20] which can be considered as the non-linear analogue of the Fourier transform; (b) the Backlund transformations [21] whereby multi-solitons are generated via a recursive method and (c) the Hirota method [22] which relies on reducing the given non-linear equation to a bilinear system through a change of variables. Now, as we have already mentioned, such *integrable* systems are constrained by the presence of infinitely many conserved quantities and they have a dynamics where, for instance, collisions occur always elastically where the solitons essentially undergo no more than a phase shift. These integrable models are of enormous interest in soliton theory itself and in several areas of science but in particle physics where phenomena like particle annihilation must be accounted for, integrable models have a somewhat limited scope.

The majority of integrable equations are encountered in unidimensional systems, the panorama becoming more difficult as we augment the dimensionality of space. In $(2+1)$ dimensions only few integrable systems are known, and none of them enjoy the requisite relativistic invariance. Examples are the Davey-Stewardson [23] and the Kadomtsev–Petviashvilli [24] equations, which are basically simple generalisations of familiar unidimensional equations. The Kadomtsev-Petviashvill equation, for example, is a two-dimensional version of the KdV: $\frac{d}{dx}(u_t + 6uu_x + u_{xxx}) + 3u_{yyy} = 0$. Therefore, when the space dimensionality is greater than one, explicit soliton solutions of relativistically invariant *non-integrable models* are mostly restricted to static lumps and Lorentz transformations thereof; their dynamics is studied using numerical simulations and approximation techniques. Possessors of a non-trivial scattering behaviour whereby a soliton and an anti-soliton can collide and annihilate each other producing radiation, non-integrable solitons seem more suitable than integrable solitons as far as particle physics is concerned.

The properties of higher dimensional solitons are closely related to the topological aspects of the gauge fields, instances of more abstract mathematical structures known as connections in fiber bundles. Through topology one can usually explain the complexity of these models even without knowing the analytic expression of their solutions. This is superlatively useful for, as already outlined, analytical soliton solutions are certainly not in oversupply. One of the principal results in non-integrable systems is the appearance of conserved quantities having no dynamic but purely topological origin. Solitonic entities bearing these so-called topological charges are frequently referred to as *topological solitons* [25]. These charges invest the solitons with a stability that prevents them from decaying into the basic states and relies on the global, topological aspects of the system which do not depend critically on the self-interaction of the system and so the circumstances that facilitate the appearance of topological solitons in nature (where they are widely observed) is by large and far less restrictive than in the case of integrable solitons, which owe their stability to the local, delicate details of the governing evolution equations. It is worthy of note that although the sine-Gordon model is fully integrable, it does feature a rich topological structure. Famous examples of topological solitons are the 't Hooft-Polyakov monopole (tHP) [26, 27], a structure that dons a fresh look upon the familiar Dirac magnetic monopole and that appears in a whole class of models of weak and electromagnetic interactions, and the skyrmions of nuclear physics.

Another important aspect in soliton theory is the study of instantons: localised finite-action classical solutions of the Euclidean version of the field equations of any given model. Classically, instantons in D Euclidean dimensions are essentially the same as *static* solitons of the same model in $(D+1)$ Minkowskian dimensions. This is basically because static solutions involve only the spatial coordinates, the Euclidean sub-space of Minkowskian space-time. In spite of their classical similitude, the impact of solitons and instantons on the corresponding quantum theory is very different: solitons lead to extended-particle states and instantons lead to tunnelling effects that can affect the structure of the vacuum state. The contribution of classical instantons to quantum effects is proportional to $\exp(-S)$ (where S is the Euclidean action), a quantity that cannot be substantial unless S is finite. This and other important elements of Minkowskian QFT can be conveniently explored through instantons. Such methods have been applied in, for example, the problems of a double-well potential [28] and a periodic potential [29].

Returning to the topological charge, one can interpret it in a natural way: we can imagine the soliton as a subatomic particle carrying the topological charge as one of its constants of motion. Among the most successful models which have made use of this appealing idea is the $(3+1)$-dimensional *Skyrme model* in hadron physics [30]. It deals with an effective theory of pions and how to derive baryons and their interactions within such a theory. Its soliton solutions, the skyrmions, are thought of as classical protons and neutrons with the conserved topological charge being the baryon number. This model, where just two free parameters are required, leads to qualitative results which are in good qualitative agreement with experimental results of nuclear physics [31]. We remark that solitons in the Skyrme model appear directly, by construction. In contrast, solitons within the framework of grand unified theories come about as an offshoot: they emerge in the form of domain walls, cosmic strings and monopoles through the Kibble mechanism.

In the beginning the Skyrme model received relatively short shrift, partly because of the advent of QCD: the QFT-notion of a free particle described by a quantised mode of a linear equation with its interactions being described through perturbation theory would not give way. Nevertheless, the Skyrme model acquired popularity towards the end of the 1970s due to speculations that it might serve as a link between QCD and the familiar old theory (valid in the low-energy regime) where the inter-baryonic forces occur via the exchange of Yukawa π mesons. In this low-energy regime QCD encounters the difficulty of having no small parameter to describe the dynamics of quarks and gluons. But in the limit as the number of colours N tends to infinity, QCD reduces to a theory of effective mesons with interactions of order $1/N$. Amazingly, it transpires that in such a limit baryons may be regarded as solitons of an effective meson theory without any further reference to their quark content [19]. Hence the re-newed interest in the Skyrme model. In 1983 the model was boosted further, when it was shown that its Lagrangian, supplemented with a Wess–Zumino term, reproduces the quantum numbers of baryons in QCD [32]. Remarkably, this latter result comes from simply eliminating a certain discrete symmetry of the skyrmion Lagrangian which is not a symmetry of QCD. A scheme that connects the Skyrme model with quarks is the so-called hybrid model [33]. It pictures the space in two parts: a very small volume with only quarks and gluons interacting perturbatively as dictated by QCD, and the rest, large part of the whole containing pions and skyrmions.

The Skyrme system is just one example of a large family of non-linear models known as *chiral* or *sigma models* introduced in the 1960s to describe β-decay and strong interactions where topology played no role [34]. The baryons were not lumps of energy but were non-linear quantum fields themselves. The calculations were made utilising the techniques of current algebra. Soliton-like structures occur abundantly in the context of sigma models. In the plane for instance they bear several properties in common with the more involved $(3 + 1)$-dimensional YM schemes. Amongst these properties we have conformal invariance, spontaneous symmetry breaking, asymptotic freedom and existence of soliton solutions. One of the ideas behind the study of sigma models is to obtain information about the quantum field theory of gauge systems in three-space, starting from classical solutions of the corresponding, less cumbersome, equations of low-dimensional analogues. In this vein, Skyrme used the sine-Gordon system as a toy model before launching his full $(3+1)$ skyrmion model. Extensive treatments of the skyrmion model may be found in [35, 36].

Sigma models are interesting by themselves; they represent a rich industry of research in pure mathematics where they are known as harmonic maps [37]. Besides, there is an ample spectrum of applications of sigma models in nature that motivates one's dedication to the subject. For example, the *non-linear O(3) model* has been used in solid state physics as the continuum limit of the Heisenberg ferromagnet, and also appears to provide mechanisms for high-temperature superconductivity [38] and the quantum Hall effect [39].

In the present chapter we study aspects of the classical theory of two particular sigma CP^1 models in $(2+1)$ dimensions: the non-linear $O(3)$ model and a modified version of it which is a low-dimensional analogue of the Skyrme model in three spatial dimensions. Specifically, we are concerned with the stability and scattering properties of these solitons. As these CP^1 models are non-integrable, their evolution in time must be done via numerical simulations, and so a significant proportion of our work is numerical. The foresaid two models have links not only with the $(3 + 1)$ dimensional skyrmion theory but also with several other important theories in particle physics, like the tHP monopoles and the vortices of the abelian Higgs model. Our class of low-dimensional, planar Skyrme models is similar to these extended entities in that parameters exist in the Lagrangian which determine the size of the extended structures. Moreover, these parameters possess a limit where the interaction forces between the particles vanish: this is the so-called Bogomolny-Prasad-Sommerfield (BPS) limit for the tHP monopoles and the $O(3)$ limit for the planar skyrmion model. Also, the CP^1 models of our concern have scattering properties akin to those observed in the foresaid $(3 + 1)$ dimensional systems.

2 The Non-linear $O(3)$ Sigma Model in $(2 + 1)$-D

One of the simplest Lorentz-invariant models in $(2 + 1)$ dimensions is the $O(3)$ model. It involves three real scalar fields $\boldsymbol{\phi}(x^\mu) \equiv \{\phi_a(x^\mu),\ a = 1, 2, 3\}$ with the constraint that $\forall\ x^\mu \equiv (x^0, x^1, x^2) = (t, x, y)$ [speed of light set equal to unity] the fields lie on the unit sphere $S_2^{(\Phi)}$:

$$\boldsymbol{\phi}.\boldsymbol{\phi} = 1. \tag{2}$$

Subject to this constraint, the Lagrangian density reads

$$\begin{aligned}\mathcal{L}_{O(3)} &= \frac{1}{4}\sum_{a=1}^{3}\sum_{\mu=0}^{2} \partial_\mu\phi_a\partial^\mu\phi_a \\ &= \frac{1}{4}(\partial_\mu\boldsymbol{\phi}).(\partial^\mu\boldsymbol{\phi}). \end{aligned} \tag{3}$$

The model is invariant under global $O(3)$ rotations in internal space. The dynamics of the $O(3)$ fields is governed by

$$\partial^\mu\partial_\mu\boldsymbol{\phi} - (\boldsymbol{\phi}.\partial_\mu\partial^\mu\,\boldsymbol{\phi})\boldsymbol{\phi} = \mathbf{0}, \tag{4}$$

which for the static case reduces to

$$\nabla^2\boldsymbol{\phi} - (\boldsymbol{\phi}.\nabla^2\boldsymbol{\phi})\boldsymbol{\phi} = \mathbf{0}. \tag{5}$$

Were it not for the constraint imposed on $\boldsymbol{\phi}$, the second term in the left-hand-side of the above equations would not be present, and the static non-singular solutions would be trivial. The condition (2) enriches the system, leading to finite-energy non-singular solutions: solitons. Furthermore, the interaction of the system is of a pure geometrical nature; it is defined by Eq. (2) which determines the curvature of the internal space Φ. This is a particularity of the sigma models.

It is straightforward to see that the kinetic and potential energies are given by

$$K = \frac{1}{4}\int (\partial_t \boldsymbol{\phi}).(\partial_t \boldsymbol{\phi})\, dxdy, \tag{6}$$

$$\begin{aligned} V &= \frac{1}{4}\int (\partial_i \boldsymbol{\phi}).(\partial_i \boldsymbol{\phi})\, dxdy \qquad [i = 1, 2] \\ &= \frac{1}{4}\int (\partial_r \boldsymbol{\phi}, \frac{1}{r}\partial_\theta \boldsymbol{\phi}).(\partial_r \boldsymbol{\phi}, \frac{1}{r}\partial_\theta \boldsymbol{\phi})\, r dr d\theta. \end{aligned} \tag{7}$$

The problem is completely specified by giving the boundary conditions. We take

$$\lim_{r\to\infty} \boldsymbol{\phi}(r, \theta) = \boldsymbol{\phi}^{(0)}, \quad \forall\, t, \tag{8}$$

where the unit vector $\boldsymbol{\phi}^{(0)}$ is independent of the polar angle θ. This condition ensures a finite potential energy. In effect, its finiteness demands

$$\lim_{r\to\infty} r|\nabla \boldsymbol{\phi}| = \lim_{r\to\infty} \sqrt{(r\partial_r \boldsymbol{\phi})^2 + (\partial_\theta \boldsymbol{\phi})^2} \to 0, \tag{9}$$

which implies (8).

It is interesting to note that the classical vacua [V=0] ought to be represented by $\boldsymbol{\phi}^{(0)}$ for all (x, y). Since such unit vector can point in any direction, we have a continuous family of zero-energy solutions connected by $O(3)$ rotations in internal space (this is an example of spontaneous symmetry breaking).

The boundary condition (8) defines a one-point compactification of $\Re_2$ allowing us to consider $\boldsymbol{\phi}$ on the extended plane $\Re_2 \cup \{\infty\}$, topologically equivalent to $S_2^{(x)}$ (the superscript indicating that the sphere refers to compactified plane). Consequently, the field configurations we want are maps $S_2^{(x)} \mapsto S_2^{(\Phi)}$ which can be labelled by an integral topological index (the topological charge) using the formula [40]

$$Q = (constant) \int_{\mathcal{N}} (\phi^* w), \tag{10}$$

where $\phi^* w$ is a suitable volume-form on $\mathcal{N}$. The mapping $\phi^* : \mathcal{M} \mapsto \mathcal{N}$ is the pull-back map [41] induced by $\phi : \mathcal{N} \mapsto \mathcal{M}$. The constant in (10) normalises Q to an integer. In our case, we pull back the differential form

$$\begin{aligned} w &= \boldsymbol{\phi}.\boldsymbol{dS}^{(\boldsymbol{\Phi})} \\ &= (\phi_1, \phi_2, \phi_3).(d\phi_2 \wedge d\phi_3, d\phi_3 \wedge d\phi_1, d\phi_1 \wedge d\phi_2) \end{aligned} \tag{11}$$

from the internal sphere $S_2^{(\phi)}$ to the 'physical' sphere $S_2^{(x)}$. Using coordinates (x, y) in the latter, expansion of w yields

$$\phi^* w = \begin{vmatrix} \phi_1 & \partial_x\phi_1 & \partial_y\phi_1 \\ \phi_2 & \partial_x\phi_2 & \partial_y\phi_2 \\ \phi_3 & \partial_x\phi_3 & \partial_y\phi_3 \end{vmatrix} dx \wedge dy. \tag{12}$$

Relaxing the wedge notation and setting the constant in (10) equal to $1/4\pi$ we get

$$Q = \frac{1}{4\pi} \int_{S_2^{(x)}} \boldsymbol{\phi}.(\partial_x \boldsymbol{\phi} \times \partial_y \boldsymbol{\phi})\, dx\, dy, \tag{13}$$

quantity sometimes also called the winding number because it gives the number of times that $\boldsymbol{\phi}$ ranges over the internal sphere as (x, y) ranges over the compactified plane once. We can convince ourselves of that by observing that (11) is nothing but the element of area of the unit sphere $S_2^{(\Phi)}$: expanding w in terms of local space polar coordinates (ϑ, φ) in internal space and parametrising $\boldsymbol{\phi} = (\sin\vartheta \cos\varphi,\ \sin\vartheta \sin\varphi,\ \cos\vartheta)$, we find $w = \sin\vartheta\, d\vartheta d\varphi$. The topological charge now stems from

$$\int_{S_2^{(x)}} \phi^* w = Q \int_{S_2^{(\Phi)}} w,$$

hence Eq. (10). Note as well that Q may be considered as the zero component of the topological current $k^\mu = \epsilon^{\mu\nu\xi}\epsilon^{abc}\phi_a \partial_\nu \phi_b \partial_\xi \phi_c$, where ϵ^{jkl} is the Levi-Civita pseudo-tensor.

3 A CP^1 Formulation of the $O(3)$ Sigma Model

Let us display the $O(3)$ model in the CP^1 formulation where the soliton solutions adopt a simpler form. This formulation involves just one independent complex field, W, related to the fields $\boldsymbol{\phi}$ via the stereographic projection [42]

$$W = \frac{\phi_1 + i\phi_2}{1 - \phi_3}. \tag{14}$$

Introducing complex coordinates $z = x + iy$ and $\bar{z} = x - iy$ on the extended plane and using the handy notation $\partial_z W = W_z$, $\partial_z(\partial_{\bar{z}} W) = W_{z\bar{z}}$, &c, the field equation (5) becomes (an overline denotes complex conjugation)

$$W_{z\bar{z}} - \frac{2\bar{W} W_z W_{\bar{z}}}{|W|^2 + 1} = 0. \tag{15}$$

In terms of W the potential energy and the topological index read

$$V = 2\int_{S_2^{(x)}} \frac{|W_z|^2 + |W_{\bar{z}}|^2}{(|W|^2+1)^2}\, dxdy, \tag{16}$$

$$Q = \frac{1}{\pi}\int_{S_2^{(x)}} \frac{|W_z|^2 - |W_{\bar{z}}|^2}{(|W|^2+1)^2}\, dxdy. \tag{17}$$

Clearly

$$V = \begin{cases} 2\pi Q + 4\int_{S_2^{(x)}} \dfrac{|W_{\bar{z}}|^2}{|W|^2+1}\, dxdy \\ \\ 2\pi(-Q) + 4\int_{S_2^{(x)}} \dfrac{|W_z|^2}{|W|^2+1}\, dxdy \end{cases} \rightarrow V \geq 2\pi|Q|. \tag{18}$$

The static solitons or instanton solutions correspond to the equality in (18): solutions with $Q > 0$ (solitons) and solutions with $Q < 0$ (anti-solitons) obey, respectively,

$$W_{\bar{z}} = 0, \qquad W_z = 0, \tag{19}$$

the Cauchy-Riemann conditions for W being an analytic function of z or $\bar{z}$ [in terms of $\boldsymbol{\phi}$ equations (19) read $\partial_i\boldsymbol{\phi} \pm \epsilon_{ij}\boldsymbol{\phi}\times\partial_j\boldsymbol{\phi} = \mathbf{0}$].

Therefore, the most general static solutions of the planar $O(3)$ model may be written in the form

$$W(z) = \lambda\prod_{j=1}^{\kappa}\frac{z-a_j}{z-b_j}, \qquad W(\bar{z}) = \lambda\prod_{j=1}^{\kappa}\frac{\bar{z}-a_j}{\bar{z}-b_j}, \tag{20}$$

where λ is a free parameter. The degree κ of the polynomials is numerically equal to $|Q|$. For a degree-one soliton, the potential energy density is

$$\varepsilon = 2\frac{|\lambda(a-b)|^2}{\{|z-b|^2 + |\lambda|^2|z-a|^2\}^2}, \tag{21}$$

which possesses a bell-like shape whose maximum value

$$\varepsilon_{max} = 8\frac{(|\lambda|^2+1)^2}{|\lambda(a-b)|^2} \tag{22}$$

is positioned at

$$z_{max} = \frac{a|\lambda|^2 + b}{|\lambda|^2+1}. \tag{23}$$

In any given topological sector the potential energy is minimised when one of the Cauchy conditions is satisfied. A solution of (19) automatically solves the original second order equation (15) but the converse need not be true. However, all the static finite-energy solutions of (15) are exhausted by Eq. (19) [43, 44]. This is a special asset of the CP^1 model on the sphere which is absent in its generalisation CP^n. The latter possesses static solutions $W(z, \bar{z})$ which are non-meromorphic and correspond to saddle points of the energy [45]. Furthermore, in the CP^1 model itself with *periodic* boundary conditions [46] there are solutions to (15) which disobey Eqs. (19).

The simplest one-soliton solution has the form

$$W = \lambda\,(z - a),$$

configuration that has been numerically studied in [47]. When viewed as an evolving structure in $(2+1)$ dimensions this soliton is unstable under any small perturbation, either explicit (e.g., by setting the soliton into motion) or implicit (as inevitably introduced by the discretisation procedure). This behaviour has been seen not only in the full simulation of the model but also in the approximation known as the collective coordinate approach [48]. The said instability, which eventually collapses the numerical procedure by infinitely shrinking or expanding the soliton, is associated with the conformal invariance of the $O(3)$ Lagrangian in two dimensions: the solitons can change their size at the expense of no energy at all [49].

With regards to collisions, two-soliton solutions of the form

$$W = \lambda(z-a)(z-b), \quad \lambda\frac{(z-a)(z-b)}{z-c},$$

scatter at right angles with the initial direction of motion in the centre of mass frame, the instability taking over a short while afterwards [49]. Nonetheless, it has been demonstrated [50] that the instability of the discretised $O(3)$ model can be cured by the addition of two extra terms to the Lagragian. The first one resembles the term introduced by Skyrme in his nuclear model in four dimensional space-time, and the second one is an additional potential-like term. The solitonic configurations of this modified model give stable extended structures which repel each other and scatter at 90° when sent towards each other with sufficient speed. In the present chapter we follow [50] and study a version of such modified, Skyrme-like model, corresponding to a more general choice of the potential term in the Lagrangian.

4 A Skyrme-Like Model in the Plane

Our modified model corresponds to a Lagrangian density

$$\mathcal{L}_{sky} = \mathcal{L}_{O(3)} - \frac{1}{4}\theta_1[(\partial^{\mu}\boldsymbol{\phi}.\partial_{\mu}\boldsymbol{\phi})^2 - (\partial^{\mu}\boldsymbol{\phi}.\partial^{\nu}\boldsymbol{\phi})(\partial_{\mu}\boldsymbol{\phi}.\partial_{\nu}\boldsymbol{\phi})]$$

$$-\frac{1}{4}\theta_2[-2Re(\lambda)\phi_1 - 2Im(\lambda)\phi_2 + (1-|\lambda|^2)\phi_3$$
$$+(1+|\lambda|^2)]^4, \qquad \theta_1, \theta_2 \in \Re^+, \tag{24}$$

where $\mathcal{L}_{O(3)}$ is given by Eq. (3). Both the kinetic and potential energies can be read-off from (24):

$$K_{sky} = \frac{1}{4}\int dxdy\{(\partial_t\boldsymbol{\phi}.\partial_t\boldsymbol{\phi})[1 + 2\theta_1(\partial_x\boldsymbol{\phi}.\partial_x\boldsymbol{\phi} + \partial_y\boldsymbol{\phi}.\partial_y\boldsymbol{\phi})]$$
$$-2\theta_1[(\partial_t\boldsymbol{\phi}.\partial_x\boldsymbol{\phi})^2 + (\partial_t\boldsymbol{\phi}.\partial_y\boldsymbol{\phi})^2]\}, \tag{25}$$

$$V_{sky} = \frac{1}{4}\int dxdy\{\,\partial_x\boldsymbol{\phi}.\partial_x\boldsymbol{\phi} + \partial_y\boldsymbol{\phi}.\partial_y\boldsymbol{\phi}$$
$$+2\theta_1[(\partial_x\boldsymbol{\phi}.\partial_x\boldsymbol{\phi})(\partial_y\boldsymbol{\phi}.\partial_y\boldsymbol{\phi}) - (\partial_x\boldsymbol{\phi}.\partial_y\boldsymbol{\phi})^2]$$
$$+\theta_2[-2Re(\lambda)\phi_1 - 2Im(\lambda)\phi_2 + (1-|\lambda|^2)\phi_3$$
$$+(1+|\lambda|^2)]^4\}. \tag{26}$$

It can be verified that the potentials corresponding to the terms θ_1 and θ_2 scale like

$$\begin{aligned} V_{\theta_1}[\boldsymbol{\phi}(\gamma x, \gamma y)] &= \gamma^2 V_{\theta_1}[\boldsymbol{\phi}(x, y)], \\ V_{\theta_2}[\boldsymbol{\phi}(\gamma x, \gamma y)] &= \gamma^{-2} V_{\theta_2}[\boldsymbol{\phi}(x, y)], \end{aligned} \tag{27}$$

so the expression (26) scales in the following manner:

$$V_{sky}[\boldsymbol{\phi}(\gamma \boldsymbol{x})] = V[\boldsymbol{\phi}(\boldsymbol{x})] + \gamma^2 V_{\theta_1}[\boldsymbol{\phi}(\boldsymbol{x})] + \gamma^{-2} V_{\theta_2}[\boldsymbol{\phi}(\boldsymbol{x})], \tag{28}$$

where the scale-free V is the pure $O(3)$ potential (7). The θ-terms break the conformal invariance and their combined effect stabilises the solitons. If the size of the solitons is appropriately chosen, it is now energetically unfavourable for them to change it. Also note that the skyrmion Lagrangian is no longer $O(3)$ invariant, but it still respects the requirement of relativistic invariance. The θ_1 term is the Skyrme-like term and it prevents the solitons from shrinking, whereas the θ_2 term resembles a potential that prevents their expansion. Judicious choices of the θ_2 term, which unlike the Skyrme term is nonunique [51], opens up the possibility of writing different interesting versions of the skyrmion model, a realisation of which is our proposition (24). The form of the θ_2 term in our Skyrme Lagrangian density is precisely the one that will make the field $W(z)$ of degree one in (20) a solution of the equation of motion (see below); we are interested in a variant of the model in which the most general expression for a single-soliton be a solution. So, a method for obtaining a suitable potential term in the planar skyrmion system is to ask oneself what the potential should be in order to have the desired soliton configuration as a solution of the equation of motion. Upon inserting the field in question into the governing static equation, one can work out the required θ_2 expression as well as the relation between the θs and the parameter λ, which determines the size of the extended structures.

In the W-formulation the Lagrangian density (24) acquires the form

$$\begin{aligned}\mathcal{L}_{sky} &= \frac{|W_t|^2 - |W_x|^2 - |W_y|^2}{(|W|^2+1)^2} \\ &\quad + \frac{2\theta_1}{(|W|^2+1)^4}[(\bar{W}_t W_x - W_t \bar{W}_x)^2 + (\bar{W}_t W_y - W_t \bar{W}_y)^2 \\ &\quad -(\bar{W}_x W_y - W_x \bar{W}_y)^2] - 4\theta_2 \frac{|W-\lambda|^8}{(|W|^2+1)^4}, \end{aligned} \tag{29}$$

the ensuing field equation being

$$\begin{aligned}0 &= W_{tt} - W_{xx} - W_{yy} - \frac{2\bar{W}}{|W|^2+1}[(W_t)^2 - (W_x)^2 - (W_y)^2] \\ &\quad + \frac{4\theta_1}{(|W|^2+1)^2}\{2\bar{W}_{tx}W_t W_x + 2\bar{W}_{ty}W_t W_y - 2\bar{W}_{xy}W_x W_y \\ &\quad -\bar{W}_{tt}[(W_x)^2 + (W_y)^2] + \bar{W}_{xx}[(W_y)^2 - (W_t)^2] + \bar{W}_{yy}[(W_x)^2 - (W_t)^2] \\ &\quad +W_{tt}(|W_x|^2 + |W_y|^2) + W_{xx}(|W_t|^2 - |W_y|^2) + W_{yy}(|W_t|^2 - |W_x|^2) \\ &\quad -W_{tx}(\bar{W}_t W_x + W_t \bar{W}_x) - W_{ty}(\bar{W}_t W_y + W_t \bar{W}_y) + W_{xy}(\bar{W}_x W_y + \bar{W}_y W_x) \\ &\quad +\frac{2W}{|W|^2+1}[(\bar{W}_t W_x - W_t \bar{W}_x)^2 + (\bar{W}_t W_y - W_t \bar{W}_y)^2 - (\bar{W}_x W_y - W_x \bar{W}_y)^2]\} \\ &\quad +4\theta_2 \frac{|W-\lambda|^6}{(|W|^2+1)^3}[\bar{\lambda}W^2 + (1-|\lambda|^2)W - \lambda]. \end{aligned} \tag{30}$$

For the static case we may write more conveniently

$$\begin{aligned}0 &= W_{z\bar{z}} - \frac{2\bar{W}W_z W_{\bar{z}}}{|W|^2+1} \\ &\quad + \frac{4\theta_1}{(|W|^2+1)^2}[2\bar{W}_{z\bar{z}}W_z W_{\bar{z}} - \bar{W}_{zz}(W_z)^2 - \bar{W}_{\bar{z}\bar{z}}(W_{\bar{z}})^2 + W_{zz}\bar{W}_z W_{\bar{z}} \\ &\quad + W_{\bar{z}\bar{z}}\bar{W}_{\bar{z}}W_z - W_{z\bar{z}}(|W_z|^2 + |W_{\bar{z}}|^2) + \frac{2W}{|W|^2+1}(|W_z|^2 - |W_{\bar{z}}|^2)^2] \\ &\quad + 4\theta_2 \frac{|W-\lambda|^6}{(|W|^2+1)^3}[\bar{\lambda}W^2 + (1-|\lambda|^2)W - \lambda]. \end{aligned} \tag{31}$$

Ignoring anti-solitonic terms like $W_{\bar{z}}$ in (31), some re-arrangement produces

$$0 = \{ 8\theta_1 |W_z|^4 - 4\theta_2 |\mathcal{V}|^2 + [-4\theta_1 (W_z)^2 \bar{W}_{zz} + \theta_2 \mathcal{V} \frac{d\bar{\mathcal{V}}}{\bar{W}}] \bar{W} \} W$$

$$+[-4\theta_1 (W_z)^2 \bar{W}_{zz} + \theta_2 \mathcal{V} \frac{d\bar{\mathcal{V}}}{d\bar{W}}], \qquad \mathcal{V} \equiv (W - \lambda)^4, \tag{32}$$

from which it is directly checked that the configuration

$$W = \lambda \frac{z - a}{z - b}, \tag{33}$$

where

$$\lambda = \frac{\sqrt[4]{2\theta_1/\theta_2}}{a - b}, \qquad a, b \in \mathcal{C}, \tag{34}$$

satisfies the field equation. The field (33) is the general solution for a single CP^1 soliton but now with λ fixed by (34). A soliton with its size thus fixed we will call a 'skyrmion' [an anti-skyrmion is derived by complex conjugation of (33)]. As touched on in the introduction, theories like the tHP monopoles also have a parameter in the Lagrangian which determines the size of the corresponding solitons.

The skyrmion's potential or static energy density E is found by inserting (33) into

$$E = 2\frac{|W_z|^2 + |W_{\bar{z}}|^2}{(|W|^2 + 1)^2} + 8\theta_1 \frac{(|W_z|^2 - |W_{\bar{z}}|^2)^2}{(|W|^2 + 1)^4}$$

$$+4\theta_2 \frac{|W - \lambda|^8}{(|W|^2 + 1)^4}, \tag{35}$$

whose maximum value is calculated via

$$\begin{aligned} E_{max} &= \varepsilon_{max}[1 + \frac{\theta_1}{2}\varepsilon_{max} + \frac{\theta_2 |\lambda(a - b)|^4}{4}\varepsilon_{max}], \\ \varepsilon_{max} &= 8\frac{(|\lambda|^2 + 1)^2}{|\lambda(a - b)|^2}. \end{aligned} \tag{36}$$

Through relation (34) it is possible to cast (36) into the simpler

$$E_{max} = \varepsilon_{max}(1 + \theta_1 \varepsilon_{max}). \tag{37}$$

The position of E_{max} is still determined by formula (23). Observe also that in the $O(3)$ limit where the θs go to zero we have $E_{max} = \varepsilon_{max}$, Eq. (22).

One can retrieve the formulae obtained in reference [50] by setting $\mathcal{V} = 1$ [instead of $\mathcal{V} = (W - \lambda)^4$] in Eq. (32); we get

$$E_{max} = 2|\lambda|^2 + 8\theta_1|\lambda|^4 + 4\theta_2, \qquad \lambda = \sqrt[4]{\frac{\theta_2}{2\theta_1}},$$

whose location is found at $z_{max} = a$.

In order to study processes involving two skyrmions we are going to consider fields of the appearance

$$W = \lambda \frac{z-a}{z-b} \frac{z+c}{z+d}. \tag{38}$$

They do not satisfy identically the field equation (31) and hence describe two solitons in an approximate manner. For objects of the simpler form $\lambda(z-a)(z-b)$ it is possible to find a relationship similar to (34) such that the model actually have static two-skyrmion solutions [52]. Inspection of Eq. (32) reveals that such an expression for a configuration (38) is quite involved to obtain. For instance, the term

$$-4\theta_1(W_z)^2\bar{W}_{zz} + \theta_2\mathcal{V}\frac{d\bar{\mathcal{V}}}{d\bar{W}}$$

in (32) gives, with the help of (38) [take $a = c$, $b = d$ real for simplicity],

$$-16\theta_1[\frac{b}{\lambda(a^2-b^2)}]^3\sqrt{\mathcal{V}}\frac{d\bar{\mathcal{V}}}{\bar{W}} + \theta_2\mathcal{V}\frac{d\bar{\mathcal{V}}}{\bar{W}}, \qquad \mathcal{V} \equiv (W-\lambda)^3(W-\lambda\frac{a^2}{b^2}),$$

where it is uncertain how to define a relation like (34) so that a two-skyrmion solution is produced.

5 Numerical Procedure for the Evolution of the CP^1 Lumps

The fields W are treated as the basic initial configurations, their analytic values being used at each lattice site in the discrete approximation of the model. After giving some initial speed to the W fields, we pass on to the ϕ-formulation by means of Eq. (14). Then $\boldsymbol{\phi}$ is numerically evolved according to the full Eq. (30) written in terms of $\boldsymbol{\phi}$.

It is quite common for ϕ_3 to have values near to 1, in which case W in (14) becomes too large for numerical comfort. So in our simulations we have preferred, instead of (14), the projection

$$W' = \frac{1-\phi_3}{\phi_1 + i\phi_2}, \tag{39}$$

clearly a more numerically tractable selection. The dynamics of the system is unchanged because the Lagrangian density is unchanged, ie, $\mathcal{L}(W') = \mathcal{L}(W)$.

Our simulations employ the fourth-order Runge-Kutta method and approximate the spatial derivatives by finite-differences. The Laplacian is evaluated using the standard nine-point formula. We use double-precision arithmetics on a 200×200 ($n_x = n_y = 200$) lattice with spatial and time steps $\delta x = \delta y = 0.02$ and δt=0.005.

Unavoidable numerical truncation errors introduced at various stages of the calculations gradually shift the fields away from the unit sphere (2), thereby building-up numerical inaccuracies in the evolution equations. So we rescale

$$\boldsymbol{\phi} \rightarrow \frac{\boldsymbol{\phi}}{\sqrt{\boldsymbol{\phi}.\boldsymbol{\phi}}}$$

every few iterations. The error associated with this procedure is of the order of the accuracy of our calculations. Each time, just before the rescaling operation, we evaluate the quantity

$$\mu \equiv \boldsymbol{\phi}.\boldsymbol{\phi} - 1$$

at each lattice point. Treating the maximum of the absolute value of μ as a measure of the numerical error, we found that $|\mu|_{max} \approx 10^{-8}$. This magnitude is useful as a guide to determine how reliable a given numerical result is. Usage of an unsound numerical procedure like, say, taking $\delta x < \delta t$ in the Runge-Kutta evolution, shows itself as a rapid growth of $\max|\mu|$; such increase also occurs when the solitons become exceedingly thin.

We also include along the boundary a narrow strip to absorb the various radiation waves, reducing their effect on the skyrmions via the reflections from the boundary. The absorption is implemented by setting

$$\partial_t \boldsymbol{\phi} \rightarrow \chi \partial_t \boldsymbol{\phi},$$

where the damping function χ has the form

$$\chi(j) = \begin{cases} 1, & j \in [0, j_1], \\ 1 - \dfrac{j - j_1}{j_2 - j_1}(1 - \chi_0), & j \in [j_1 + 1, j_2 - 1], \\ \chi_0, \quad j \in [j_2, n_x] & (\chi_0 = 0.95), \end{cases}$$

where the absorbing band is small, no more than about 10 % of mesh-points. As time elapses, the absorption of radiation manifests itself through a small decrease of the total energy, which gradually stabilises as the radiation waves are gradually absorbed. The damping device is specially useful when studying soliton stability but it is dispensable when considering collision processes.

For the parameters we have chosen the values:

$$\begin{aligned} a &= c = 0.75, \qquad b = d = 0.05, \\ \theta_1 &= 0.015006250, \qquad \theta_2 = 0.1250. \end{aligned} \tag{40}$$

The global $U(1)$ symmetry of (33) has been used to choose λ real. From (34) and (40) it follows that λ=1, a calculational-friendly number as can be inferred from (26).

Noteworthy is the fact that the finite-difference expressions for the derivatives of fields like λz and λz^2, used in previous works, are exact. This is no longer true for our choices (33) and (38), whose numerical versions are in this sense more perturbed. It turns out that this factor has no telling effect in the qualitative behaviour of the system.

6 Results of the Numerical Simulation

6.1 The Case of Static Skyrmions

Firstly consider the situation where the skyrmions have zero initial velocity. Our single-skyrmion field is

$$W = \frac{z - 0.75}{z - 0.05}, \tag{41}$$

a typical picture of which energy density is displayed in the upper half of Fig. 1.

Our simulations show that the energy density corresponding to this soliton evolves only very slightly and does not change its shape. At the initial time the amplitude of the energy density has the numerical value of $\approx$ 128.47, which quickly re-adjusts and stabilises itself around 129.5. This last number agrees very well with the analytical result as calculated from Eq. (36) [or (37)] and the data provided in (40):

$$E_{max} \approx 65.3 + 31.99 + 31.99 = 129.3. \tag{42}$$

In the lower half of the Fig. 1 a graph showing the stable evolution of $Emax$ is exhibited.

In Fig. 2 we show some pictures of the radiation waves emitted by the soliton-hump. They propagate out to the boundary at the speed of light, leaving the central region of the lattice essentially free of kinetic energy. The smallness of the kinetic energy indicates that our soliton is almost perfectly static. This is in fact numerically observed: at the initial time the lump of energy is situated at z_{max}=(0.40,0) and by

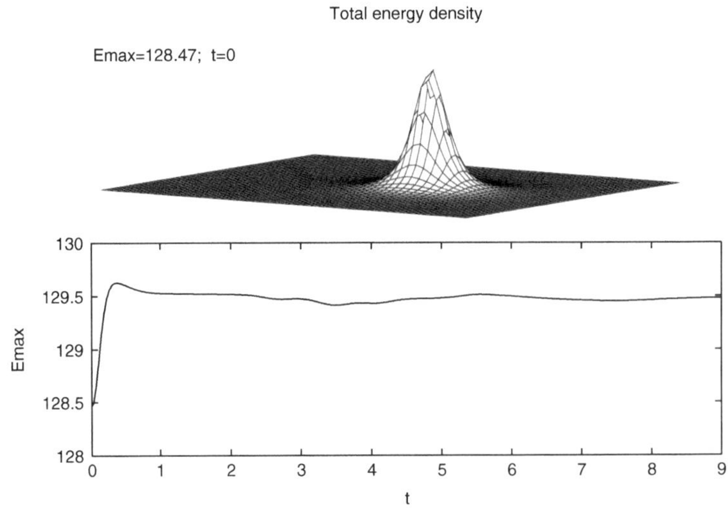

Fig. 1 Total energy density for one skyrmion at the initial time and the evolution of its height

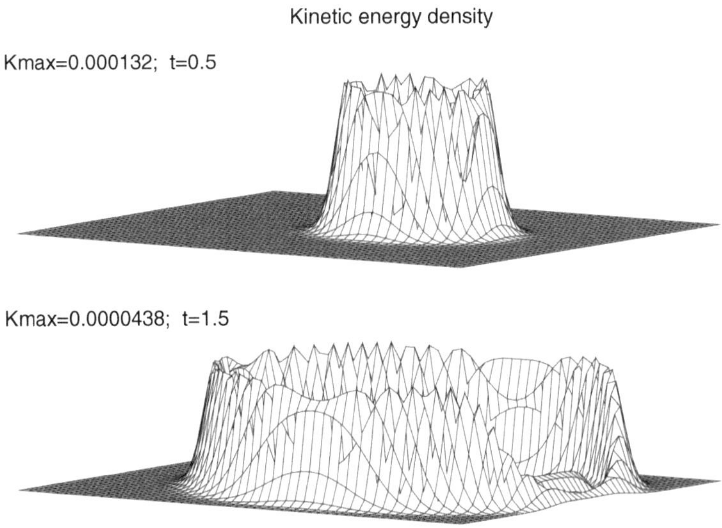

Fig. 2 The skyrmion-lump shakes off some kinetic energy waves that spread out to the boundary at the speed of light

t=10 it has slowly shifted to (0.4013,0). Note that the theoretical value of z_{max}, as per formula (23), is precisely (0.40,0).

Although small enough as to cause no preoccupation in the problem under study, the said kinetic energy is considerably larger than the one emitted by the solitons of the model with $\mathcal{V} = 1$ in (32). This is not surprising since our choice of potential leads to a more perturbed, complicated discrete representation of the analytic soliton fields than in the $\mathcal{V} = 1$ case.

A cursory glance at (24) shows that by simply equating $\theta_1 = \theta_2 = 0$ we recover the unmodified $O(3)$ model:

$$\lim_{\theta_1, \theta_2 \to 0} \mathcal{L}_{sky} \to \mathcal{L}_{O(3)}, \tag{43}$$

result that resembles the BPS limit in the tHP monopole theory.

Our simulations for this limiting case show that (41) represents a static $O(3)$ solution which behaves stably on the mesh, corroborating the results found in [47]. It possesses an oscillating quasi-periodic energy density whose amplitude decreases as time goes by, eventually becoming quite small (recall there is an absorbing set-up operating along a small band near the edges of the grid). However, the source of such stability resides on the scale set by the finite net, where the kinetic waves are reflected in such a way as to stabilise the hump. In fact, upon effectively moving the boundaries to infinity, the discrete $O(3)$ system is unstable [47]. This kind of lattice-stabilisation is clearly of no applicability within a general field theoretical framework. By contrast, the skyrmion scheme has an intrinsic scale that renders a stable soliton regardless, being therefore alluring from a general perspective. A depiction of the said $O(3)$ lattice-stability is given in Fig. 3.

Without the absorption arrangement the kinetic waves, rather than petering out as shown in the previous pictures, keep travelling from the soliton-lump out to the boundary and inward back again. The energy density in this case exhibits a periodic behaviour, with the amplitude of its oscillations being about constant in time; the question of stability under these circumstances meets with difficulties. Hence the importance of the absorbing band that gradually extinguishes the radiation, as seen in Fig. 3.

We now shift our attention to the approximate two-soliton configuration

$$W = \frac{z - 0.75}{z - 0.05} \frac{z + 0.75}{z + 0.05}, \tag{44}$$

which gives two skyrmion-lumps of equal size initially well separated from each other (so that they have a minimal overlapping) but still far away from the borders of the mesh (thus avoiding reflections from the boundaries as much as possible). Since the above field is not an exact solution of the model it should undergo some evolution even for the static case $v = 0$.

The energy density corresponding to the *ansatz* (44) at $t = 0$ is illustrated in the upper part of Fig. 4. We can see that the amplitude starts at a value somewhat bigger than twice the value for a single soliton. As soon as the evolution commences the skyrmions shake off some radiation and alter their size by getting broader. In so doing, they slowly move away from each other, uncovering the presence of a repulsive force between them; this is apparent from Fig. 4. During this process the peak E_{max} decreases and undergoes damped oscillations around the analytical

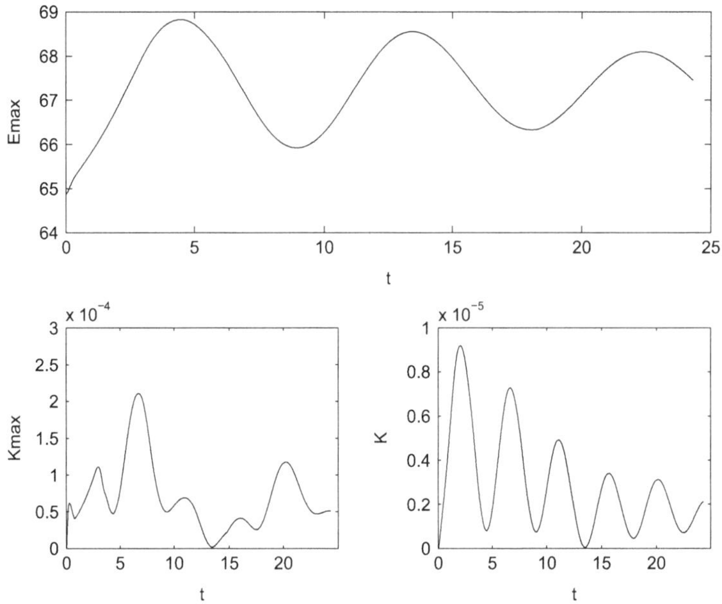

Fig. 3 As the kinetic waves fade away with time we are left with a stable pure $O(3)$ lump on the lattice

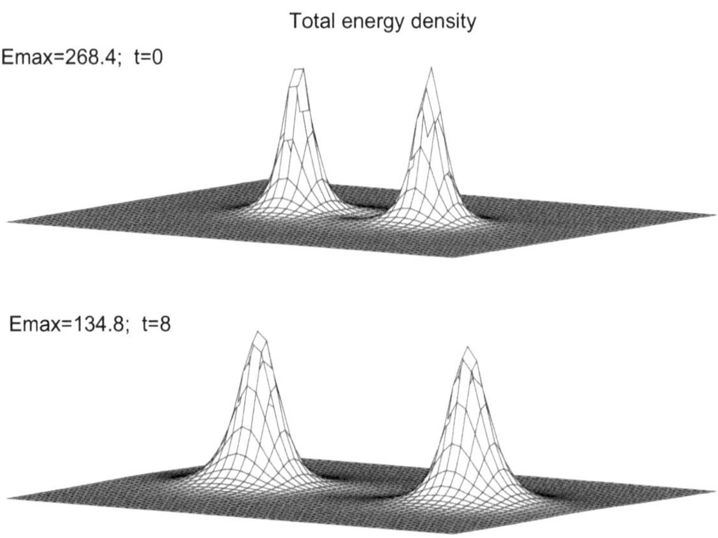

Fig. 4 A repulsive force exists between the two skyrmions

result (42); by $t \approx 8$ the oscillations are quite small and the energy stabilises near the theoretical value. A feel for this course of events can be developed through Fig. 5, where the graph of the maximum of the energy density is plotted versus

Fig. 5 Development of the energy density for two skyrmions which start from rest

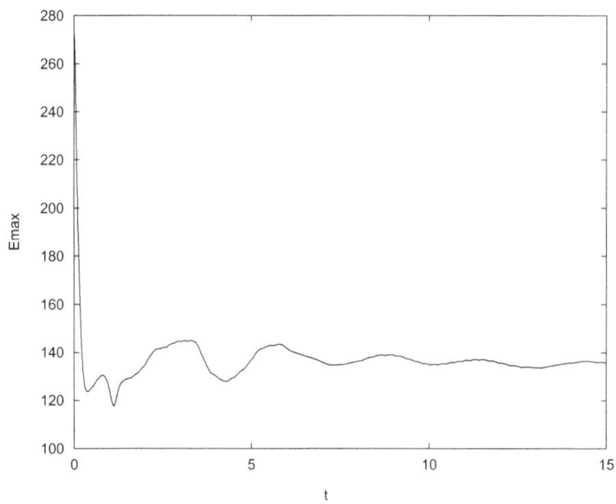

time. It is re-assuring that this kind of weak repulsion, also characteristic of the unidimensional kink model [25] has been observed using a collective coordinate method as well [53].

By taking a value of the parameter λ which does not satisfy (34) we can introduce explicit perturbations into the system, other than the implicit ones brought about by the discretisation procedure. We have checked that our skyrmions are indeed stable under all such disturbances. If the initial λ is much greater (smaller) than the value dictated by (34), unity in our case, the initial amplitude of the solitons will be manifestly above (below) the analytic value (42). But the pattern is always the same: maintaining its shape unscathed, the skyrmion-hump corrects its height to around 129.5, emitting kinetic waves in so doing.

In the limit (43) we have verified that the repulsion between the lumps disappears, and these remain motionless in their initial positions [as in the upper half of Fig. 4] throughout the simulation if started off from rest. However, the solitons are now unstable and their energy density increases non-stoppingly meaning that their breadths go down to the order of the lattice spacing, eventually collapsing the numerics. A prototype picture after this time would reveal that the soliton-lumps become not so much thin as blown asunder.

In the upper half of Fig. 6 we show the said rise of the peak of the energy density corresponding to (44) in the $O(3)$ limit. It is pedagogical to compare the unstable behaviour of these two static solitons with the stable performance of the $O(3)$ single-soliton shown in Fig. 3. One may naturally ask himself why the two lumps, having been given no initial velocity, behave unstably. The answer relies on the fact that the discrete representation of the two-solitons (44) brings forth an extra perturbation which spells unfavourably in terms of stability, the boundary-reflected radiation being now incapable of balancing things out. This result is very important, for if the kinetic waves reflected from the boundary stabilised the two $O(3)$ solitons as in the single-lump case, then it would be necessary to study the behaviour of the

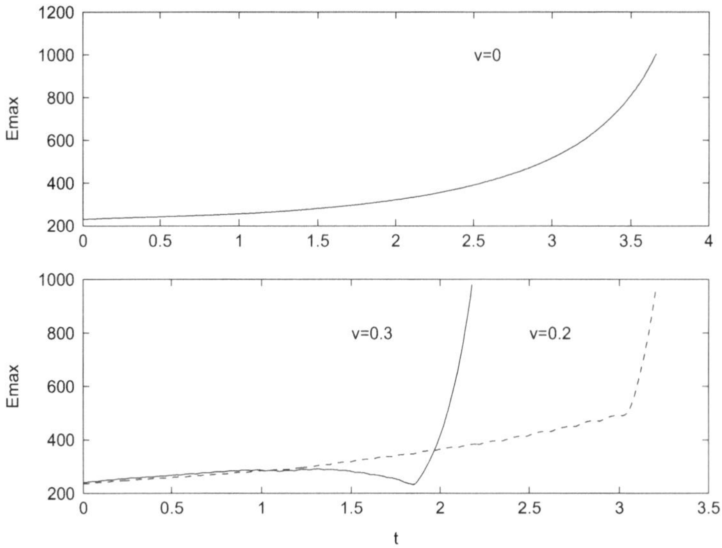

Fig. 6 *Above*: Total energy density corresponding to two pure $O(3)$ solitons with no initial speed. *Below*: The same $O(3)$ lumps with different non-zero initial speeds (both curves correspond to head-on collisions that lead to scattering at right angles). Compare with the stable case shown in diagram 3.11

skyrmion model in an infinite lattice to make sure that its extended objects are stable regardless. But given the actual circumstances, such study would be only academic.

6.2 *The Case of Skyrmion Scattering*

We now study the scattering behaviour of our two-skyrmion field (44) for different initial velocities. Let us consider head-on collisions.

There is always an initial burst of radiation as the skyrmions re-arrange themselves towards their stable size. At small speeds the two entities approach each other, but the repulsive force between them results in the reversal of their motion. This situation can be viewed in Fig. 7, where we present some pictures of the total energy density for skyrmions sent towards each other along the horizontal axis with a relative initial speed of 0.2; the corresponding contour plots are shown in Fig. 8.

A qualitatively similar behaviour is observed for speeds up to approximately 0.3. For $v \approx 0.3$ and higher the skyrmions acquire enough kinetic energy to overcome their mutual repulsion; during their collision they form a complicated ringish state (at which stage they attain a minimum height and hence maximum width) and re-emerge at 90° to the original direction of motion in the centre-of-mass frame. The emerging skyrmions are initially shrinking but, after they have

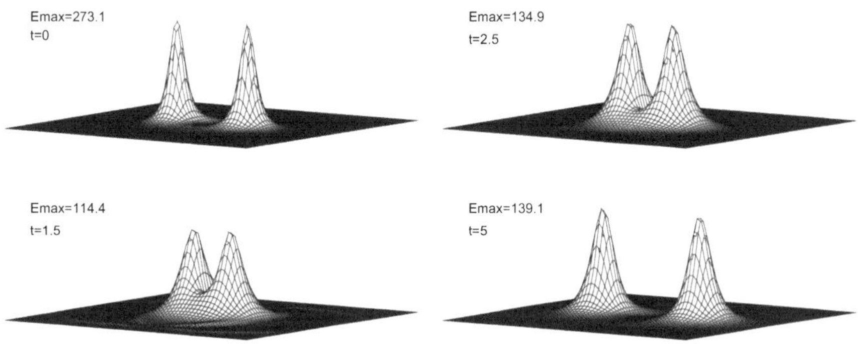

Fig. 7 Collision featuring two skyrmion-lumps with relative initial velocity $\boldsymbol{v} = (0.2, 0.0)$. As the figures show, a speed of 0.2 is not big enough to overcome the repulsion between the lumps and back-to-back scattering occurs

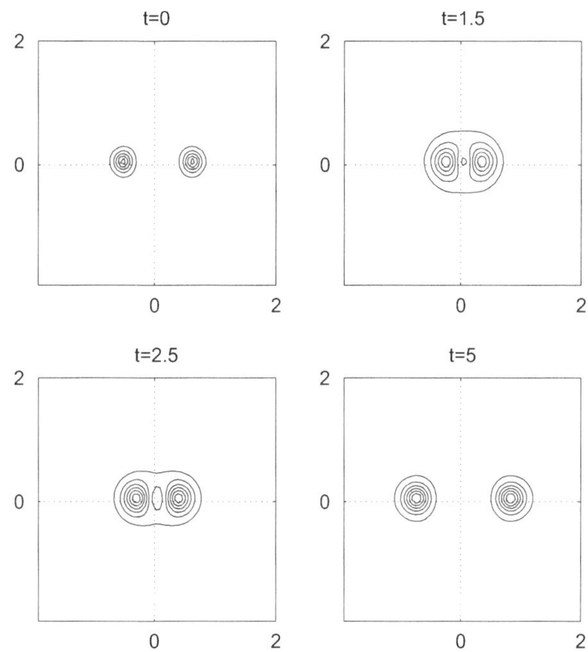

Fig. 8 Contour plots for Fig. 7

travelled some distance, they expand once more. The final state is achieved after some oscillations of the energy density. In Fig. 9 we present a quartet of pictures for this 90°-scattering; the corresponding contour plots may be seen in Fig. 10.

The existence of a critical velocity above (below) which the lumps scatter at right angles (180°) is a major difference between the pure $O(3)$ model and its modified

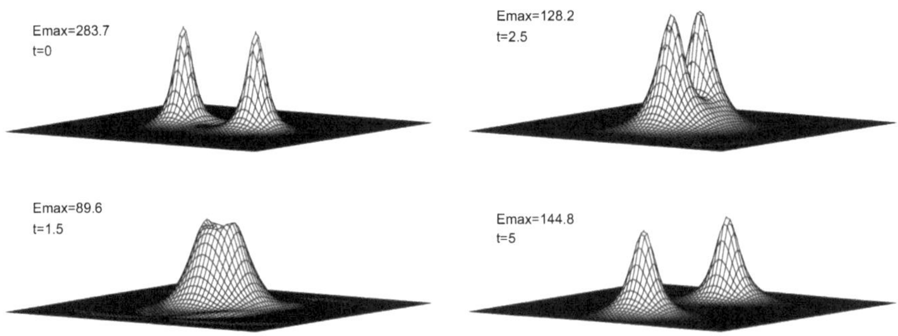

Fig. 9 Two skyrmion-humps in collision course; their relative initial velocity is $\boldsymbol{v} = (0.3, 0.0)$. As the figures show, above the critical value, the speed leads to 90° skyrmion scattering

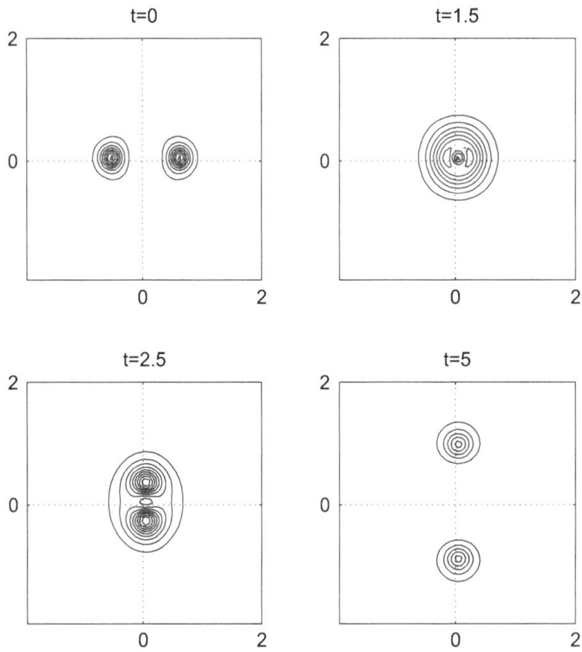

Fig. 10 Contour plots for Fig. 9

Skyrme version. Going to the limit (43) we have been able to confirm that this critical velocity ceases to exist and 90°-scattering occurs as long as $v > 0$: events unfold very much like the skyrmion situation of Fig. 10 but now the system is no longer stable. The soliton-humps continue to grow thin, increasing in height, to eventually break down the numerical procedure as an aftermath. In the bottom half of Fig. 6 these episodes are illustrated with typical $E_{max}(t)$ plots. They are to be

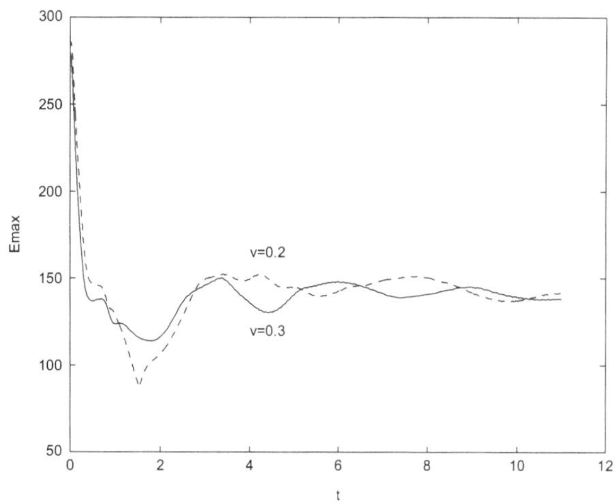

Fig. 11 Amplitude of the total energy density corresponding to skyrmion scattering. A speed of 0.2 (0.3) leads to back-to-back (right angles) scattering. In both cases the lowest E_{max} occurs at impact time, when the two lumps coalesce

compared with their skyrmion counterpart plots of Fig. 11, which correspond to both the back-to-back (180°) and 90° scattering cases already discussed. It is noteworthy that this CP^1-like scattering behaviour and the existence of a critical velocity are exhibited by some other important soliton models, e.g., the ϕ^4 kink [54], the vortex model [55] and the tHP monopoles [56].

The fact that in our model the radiation emitted by the soliton is relatively large (as compared to other versions of the model) is immaterial for the modified $O(3)$ model because the lumps are stable anyway and can be studied for as long as necessary. But in the unstable, pure $O(3)$ scheme, such larger perturbation shows itself in a quicker collapse of the numerics, making the analysis of scattering processes more difficult to follow. For instance, the shrinking of the $O(3)$ solitons exemplified in Fig. 6 takes place significantly faster than fields of the aspect

$$(z-a)(z-b), \qquad \frac{(z-a)(z-b)}{z-c},$$

studied in reference [49]. In any case, in all the situations considered the collisions are quasi-elastic (in the $O(3)$ case the quasi-elastic process is as long as the numerics runs smoothly).

For collisions with small but non-zero impact parameter the results are not at variance with prognostication: the skyrmion-lumps scatter either bouncing back or at nearly right-angles to the initial direction of motion, depending on the velocity. A 90° event is pictured in Fig. 12. In general, the larger the impact parameter, the smaller the scattering angle, and the more the lumps conserve their identity during the process.

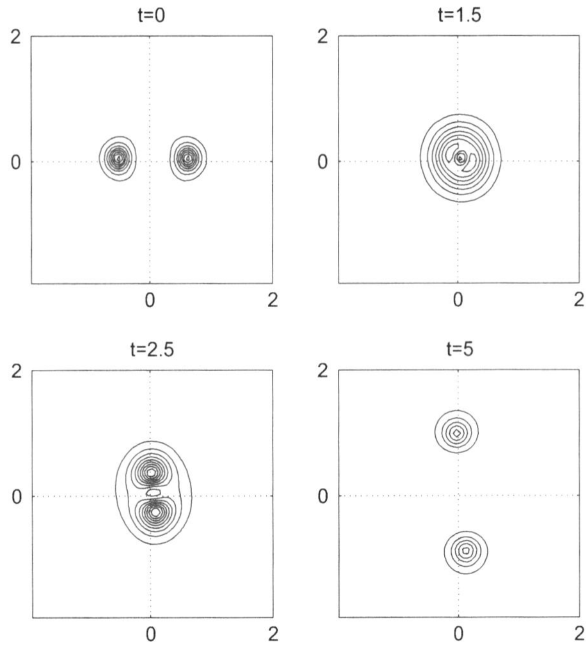

Fig. 12 Non-zero impact parameter collision between skyrmions. The initial velocity is $\boldsymbol{v} = (0.3, 0.01)$

7 Conclusions

The scattering properties of a general two-soliton field configuration of our version of the Skyrme model in $(2 + 1)$-D confirm the results previously found in other variants of the model where, unlike the case studied in this paper, the finite-difference expressions for the derivatives of the fields enjoyed the numerically convenient feature of being exact (corresponding to simpler form of the field solution). This factor, however, did not affect the qualitative scattering properties of the model under study.

Although our solutions resemble two instantons in an approximate manner -the model is not integrable-, they exhibit a clear soliton-like behaviour. All radiation effects are small and the skyrmions' total energy density profiles are preserved during the scattering process. They may get distorted, but always recover when the distance between the skyrmions becomes large enough. Their interaction is of a repulsive nature, at least at large distances, the interaction being more difficult to asses when the skyrmions are close together. For head-on collisions there is a velocity below which the skyrmions bounce back, and above which they scatter at ninety degrees. There is a resemblance both with the properties of kinks in the ϕ^4 model, which has a critical velocity, and with monopole scattering at 90°.

Acknowledgements The author is greatly indebted to Dr. David Amundsen and Dr. Patrick Farrell for their hospitality at the School of Mathematics and Statistics, Carleton University. He is also very grateful to Dr. Amundsen for helpful research discussions.

References

1. C.N. Yang, R.L. Mills, Phys. Rev. **96**, 191 (1954)
2. F. Englert, R. Brout, Phys. Rev. Lett. **13**, 321 (1964)
3. P.W. Higgs, Phys. Rev. **13**, 508 (1964)
4. G.S. Guralnik, C.R. Hagen, T.W.B. Kibble, Phys. Rev. Lett. **13**, 585 (1964)
5. J.L. Miller, Phys. Today **65**(12), 9 (2012)
6. R. Rajaraman, *Solitons and Instantons* (North-Holland, New York, 1987)
7. L.V. Yakushevich, Q. Rev. Biophys. **20**, 201 (1993)
8. D. Iagolnitzer (ed.), Proceedings of the International Congress of Mathematical Physics, vol. 691 (International Press, Paris, 1994)
9. E. Desurvire, Phys. Today **41**(1), 20 (1994)
10. J.S. Russell, Report of the Fourteenth Meeting of the British Association for the Advancement of Science, York, 1844, pp. 311–390
11. D.J. Korteweg, G. de Vries, Philos. Mag. Ser. 5 **39**, 422 (1895)
12. N.J. Zabusky, M.D. Kruskal, Phys. Rev. Lett. **15**, 240 (1965)
13. R.M. Miura, C.S. Gardner, M.D. Kruskal, J. Math. Phys. **9**, 1204 (1968)
14. V.E. Zakharov, A. Shabat, Sov. Phys. JETP **34**, 62 (1972)
15. A. Barone, F. Esposito, C.J. Magee, A.C. Scott, Riv. Nuovo Cimento **1**, 227 (1971)
16. A.C. Scott, F.Y.F. Chu, D.W. MacLaughlin, Proc. IEEE **61**, 1443 (1973)
17. A.M. Grundland, A.J. Hariton, L. Šnobi, J. Phys. A: Math. Theor. **42**, 335203 (2009)
18. J. Hruby, Czech. Jour. Phys. B **32**, 1100 (1982)
19. E. Witten, Phys. Rev. D **16**, 2991 (1977)
20. C.S. Gardner, J.M. Greene, M.D. Kruskal, R.M. Miura, Phys. Rev. Lett. **19**, 1095 (1967)
21. G.L. Lamb, Phys. Lett. A **25**, 181 (1967)
22. R. Hirota, Phys. Rev. Lett. **27**, 1192 (1971)
23. A. Davey, K. Stewardson, Proc. Roy. Soc. Lond. A **338**, 101 (1974)
24. S. Novikov, *Theory of Solitons* (Consultants Bureau, New York, 1984)
25. N. Manton, P. Sutcliffe, *Topological Solitons* (Cambridge University Press, Cambridge, 2004)
26. G.'t-Hooft, Nucl. Phys. B **79**, 276 (1974)
27. A.M. Polyakov, JETP Lett. **20**, 194 (1974)
28. A.M. Polyakov, Nucl. Phys. B **120**, 429 (1977)
29. S. Coleman, in *The Whys of Subnuclear Physics*, ed. by A. Zichichi (Plenum Press, New York, 1979)
30. T.H.R. Skyrme, Nucl. Phys. **31**, 556 (1962)
31. G. Holzwarth, B. Schwesinger, Rep. Prog. Phys. **49**, 825 (1986)
32. E. Witten, Nucl. Phys. B **160**, 57 (1979)
33. A. Chodos, C.B. Thor, Phys. Rev. D **12**, 2733 (1975)
34. M. Gell-Mann, M. Levy, Nuovo Cimento **16**, 705 (1960)
35. E. Brown Gerald (ed.), *Selected Papers, with Commentary, of Tony Hilton Royle Skyrme* (World Scientific, River Edge, 1994)
36. V.G. Makhankov, *The Skyrme Model* (Springer, New York, 1993)
37. J. Eells, J.C. Wood, Topology **15**, 263 (1976)
38. I. Dzyaloshinskii, A. Polyakov, P. Wiegmann, Phys. Lett. A **127**, 112 (1988)
39. A.G. Green, I.I. Kogan, A.M. Tsvelik, Phys. Rev. B **53**, 6981 (1996)
40. N. Steenrod, *The Topology of Fiber Bundles* (Princeton University Press, Princeton, 1951)
41. M. Nakahara, *Geometry, Topology and Physics* (IOP publishing, Bristol/Philadelphia, 1990)

42. L.V. Ahlfors, *Complex Analysis* (McGraw Hill, New York, 1953)
43. A.A. Belavin, A.M. Poliakov, JETP Lett. **22**, 245 (1975)
44. G. Woo, J. Math. Phys. **18**, 1264 (1977)
45. A.M. Din, W.J. Zakrzewski, Nucl. Phys. B **174**, 397 (1980)
46. R.J. Cova, W.J. Zakrzewski, Nonlineartity **10**, 1305 (1997)
47. R.A. Leese, M. Peyrard, W.J. Zakrzewski, Nonlinearity **3**, 387 (1990)
48. R.A. Leese, M. Peyrard, W.J. Zakrzewski, Nucl. Phys. B **344**, 33 (1990)
49. W.J. Zakrzewski, Nonlinearity **4**, 429 (1991)
50. R.A. Leese, M. Peyrard, W.J. Zakrzewski, Nonlinearity **3**, 773 (1990)
51. J.A. Azcárraga, M.S. Rashid, W.J. Zakrzewski, J. Math. Phys. **32**, 1921 (1991)
52. B. Piette, W.J. Zakrzewski, Nucl. Phys. B **393**, 65 (1993)
53. P.M. Sutcliffe, Nonlinearity **4**, 1109 (1991)
54. D.K. Campbell, M. Peyrard, Physica D **18**, 47 (1986)
55. E.P.S. Shellar, P.J. Ruback, Phys. Lett. B **209**, 262 (1988)
56. M.F. Atiyah, N.J. Hitchin, *The Geometry and Dynamics of Magnetic Monopoles* (Princenton University Press, Princenton, 1988)

Index

J. Cuevas-Maraver et al. (eds.), *The sine-Gordon Model and its Applications*,
Nonlinear Systems and Complexity 10, DOI 10.1007/978-3-319-06722-3,

MIX
Papier aus verantwortungsvollen Quellen
Paper from responsible sources
FSC® C105338

If you have any concerns about our products,
you can contact us on
ProductSafety@springernature.com

In case Publisher is established outside the EU,
the EU authorized representative is:
Springer Nature Customer Service Center GmbH
Europaplatz 3, 69115 Heidelberg, Germany

Printed by Libri Plureos GmbH
in Hamburg, Germany